Archaeology and Environment in the Libyan Sahara

The excavations in the Tadrart Acacus, 1978 – 1983

edited by

Barbara E. Barich

with contributions by

Massimo Baistrocchi, Giorgio Belluomini, Isabella Caneva, Angela E. Close, Achilles Gautier, Luigia Manfra, Bruno Marcolongo, Alberto M. Palmieri, Erhard Schulz and Sandra Y. Vons-Comis

Cambridge Monographs in African Archaeology 23

BAR International Series 368

1987

B.A.R.

122 Banbury Rd, Oxford, OX2 7BP, UK

BAR -S368, 1987: 'Archaeology and Environment in the Libyan Sahara: The excavations in the Tadrart Acacus, 1978-1983'

ISBN 9780860544746 paperback
ISBN 9781407346434 e-book
DOI https://doi.org/10.30861/9780860544746
A catalogue record for this book is available from the British Library

This book is available at www.barpublishing.com

CAMBRIDGE MONOGRAPHS IN AFRICAN ARCHAEOLOGY

General Editor

John Alexander, M.A., Ph.D., F.S.A.

St. John's College,
Cambridge CB2 1TP, England

Manuscripts should be submitted to
Dr. Alexander in the first instance.

Titles already published:

Northern Africa

No. 4. 'Later Pleistocene Cultural Adaptations in Sudanese Nubia' by Yousif Mukhtar el Amin. BAR-S114. Out of print.

No. 6. 'The Neolithic Period in the Sudan, c. 6000 - 2500 B.C.' by Abbas S. Mohammed-Ali. BAR-S139. Out of print.

No. 8. 'Meroitic Settlement in the Central Sudan: an analysis of sites in the Nile Valley and the Western Butana' by Khidir Abdelkarim Ahmed. BAR-S197. £14.00 post free.

No. 14. 'The Archaeology of Central Darfur (Sudan) in the 1st Millennium A.D.' by Ibrahim Musa Mohammed. BAR-S285. £19.00 post free.

No. 16. 'L'art rupestre préhistorique des massifs centraux sahariens' by Alfred Muzzolini. BAR-S318. £21.00 post free.

No. 18. 'The Royal Crowns of Kush' by L. Török. BAR-S338 £12.00 post free.

No. 20. 'Socio-Economic Differentiation in the Neolithic Sudan' by Randi Haaland. BAR-S350. £18.00 post free.

Western Africa

No. 1. 'The Niger Delta: aspects of its Prehistoric Economy and Culture' by Nwanna Nzewunwa. BAR-S75. Out of print.

No. 2. 'Prehistoric Investigations in the Region of Jenne, Mali' by Susan Keech McIntosh and Roderick J. McIntosh. BAR-S89. Out of print.

No. 7. 'History and Ethnoarchaeology in Eastern Nigeria: a Study of Igbo-Igala relations with special reference to the Anambra Valley' by Philip Adigwe Oguagha and Alex Ikechukwu Okpoko. BAR-S195. £15.00 post free.

No. 11. 'Archaeology and History in Southern Nigeria: the ancient linear earthworks of Benin and Ishan' by P.J. Darling. BAR-S215. £28.00 post free.

Eastern Africa

No. 3. 'Off-Site Archaeology and Human Adaptation in Eastern Africa: an analysis of Regional Artefact Density in the Amboseli, Southern Kenya' by Robert Foley. BAR-S97. Out of print.

No. 13. 'Fisher-Hunters and Neolithic Pastoralists in East Turkana, Kenya' by John Webster Barthelme. BAR-S254. £18.00 post free.

No. 17. 'Spheroids and Battered Stones in the African Early and Middle Stone Age' by Pamela R. Willoughby. BAR-S321. £16.00 post free.

Southern Africa

No. 5. 'Settlement Patterns in the Iron Age of Zululand: an Ecological Interpretation' by Martin Hall. BAR-S119. Out of print.

No. 9. 'Economy and Technology in the Late Stone Age of Southern Natal' by Charles Cable. BAR-S201. £13.00 post free.

No. 10. 'Frontiers: Southern African Archaeology Today' ed. M. Hall, G. Avery, D.M. Avery, M.L. Wilson and A.J.B. Humphreys. BAR-S207. £18.00 post free.

No. 12. 'The Latest Stone Age of Southernmost Africa' by Janette Deacon. BAR-S213. £20.00 post free.

No. 15. 'Stable Carbon Isotopes and Prehistoric Diets in the South-Western Cape Province, South Africa' by Judith Sealy. BAR-S293. £10.00 post free.

No. 19. 'The Later Stone Age of the Drakensberg Range and its foothills' by H. Opperman. BAR-S339. £15.00 post free.

No. 21. 'Later Stone Age Settlement Patterns in the Sandveld of the South-Western Cape Province, South Africa' by Anthony Manhire. BAR-S351. £11.00 post free.

TABLE OF CONTENTS

Please note that additional material is available to download from www.barpublishing.com/additional-downloads.html.
The original foldouts have been reduced in size to match the A4 format of this book, the images are therefore not as clear as the original foldouts. Please refer to the original foldouts via the download for the original content.

ACKNOWLEDGEMENTS

Besides thanking the scholars who were directly involved in this publication, I would like to express my sincere thanks to all those who, year by year in the research team, contributed to the common task. I am grateful for the cooperation of the archaeologists A. Zarattini and B. Pecci; the geologists F. Bonadonna, A.M. Palmieri and C. Arias; the topographers R. Caciagli and C. Amici. I acknowledge with particular emphasis the help of Piergiovanni Checchi who took part in all the campaigns, seeing to the reproduction of the rock art evidence and the organization of the base camp. I wish to express my thanks to Dr. Giorgio Lattanzi for the histograms of the ceramics from Uan Muhuggiag, their computing and for advice on the presentation of the archaeological tables. Cesare Placidi, a member of the 1972 Mission, has drawn all the lithic and ceramic materials. The photographic reproductions of all the artifacts were by Filiberto Scarpelli. I am also grateful to Dr. Gianfranco Catania (Tripoli ICE Office) for his assistance in the organization phase and during the field work.

I wish also to express my thanks to Dr. Graeme Barker, Director of the British School at Rome, and to Mrs. Barker, who helped me revising some parts of the English translation (Chapters 1, 5, 6, 15). For the same reason I am deeply indebted to Dr. John Alexander who polished the whole text of this Volume.

The success of the undertaking was made possible through the continuous support of the Libyan Directorate for Antiquities and the equipment they placed at our disposal. My sincere thanks goes to the General Directors, Mr. Mohamed Nemri, Mr. Salahaddin Hassan and Mr. Abdallah Sheibub. I also acknowledge with gratitude their officers who have directly taken part in the field work of the joint Mission; Mr. Abd-el-Gader Sherif, Mr. Mahdi Abbderahman of the Sebha Department, Mr. Giuma Anag of the Tripoli Department, Mr. Abdulatif Gerbi and Mr. Mohamed Breik of the Derna and Cirene Departments. Last but not least, I am deeply grateful to Salem Wadawi - a mechanic of the Tripoli Department and the guide and interpreter of the joint Mission - as well as to Mohamed Carnefuda of the Sebha Department - a peerless guide and expert in the Tadrart Acacus area.

It goes without saying that I wish to express special thanks to Fabrizio Mori who was responsible for making this important archaeological area known, and for beginning an exciting although tough research project.

B.E.B.

DEDICATION

This book is dedicated to Salvatore M. Puglisi, Director of the Rome University Mission in the Libyan Sahara, who upheld the need for the paleoethnological research in that area to be an integrated reconstruction, covering all the different cultural phenomena. At the same time, he indicated a wider prospect to be investigated, in which that specific area would be included within a more general Saharan context and related to the Upper Nile region.

The standards set by S.M. Puglisi have been my constant and inescapable reference framework in working in the Tadrart Acacus, initially under his direction and, later on, independently. The results attained confirm the correctness of his approach and the clearness with which the whole scheme was conceived.

Barbara E. Barich

CONTRIBUTORS

Barbara E. Barich, Editor
Dipartimento di Scienze Storiche, Archeologiche e Antropologiche dell'Antichità, University of Rome "La Sapienza", Rome, Italy.

- Massimo Baistrocchi, Ministero degli Affari Esteri, Rome, Italy.

- Giorgio Belluomini, Centro di Studio per la Geocronologia e Geochimica delle Formazioni Recenti, C.N.R., Dipartimento di Scienze della Terra, University of Rome "La Sapienza", Rome, Italy.

- Isabella Caneva, Dipartimento di Scienze Storiche, Archeologiche e Antropologiche dell'Antichità, University of Rome "La Sapienza", Rome, Italy.

- Angela E. Close, Department of Anthropology, Southern Methodist University, Dallas, Texas, U.S.A.

- Achilles Gautier, Laboratorium voor Paleontologie, Rijksuniversiteit Gent, Gent, Belgium.

- Luigia Manfra, Dipartimento di Scienze della Terra, University of Rome "La Sapienza", Rome, Italy.

- Bruno Marcolongo, Istituto di Geologia Applicata, C.N.R., Padova, Italy.

- Alberto M. Palmieri, Istituto per le Tecnologie Applicate ai Beni Culturali, C.N.R., Rome, Italy.

- Erhard Schulz, Geographisches Institut der Universität Würzburg, Würzburg, Germany.

- Sandra Y. Vons-Comis, Rijksdienst voor het Oudheidkunding Bodemonderzoek, Amersfoort, Netherlands.

PART I: Wadi Auis and the Northern Acacus

CHAPTER 1

METHODOLOGY AND HISTORY OF THE RESEARCH IN THE TADRART ACACUS AREA

by Barbara E. BARICH*

1. INTRODUCTION

The Tadrart Acacus Mountains are located close to the Algerian border, in the Fezzan province in the extreme South-West of Libya (1). The nearest oases - our reference point during field work - are Serdeles (El Auenat) and Ghat. The geomorphologic and lithologic features of this massif are analysed in detail by B. Marcolongo in a subsequent chapter of this Volume (Chapter 11). On the whole, its height is below 1000 m and thus even the shelters on the highest terraces could be explored. It is an extensive geographic area (<u>ca</u> 11000 sq.Km) and, owing to its wealth of evidence, Italian research teams have been actively involved in its exploration for approximately thirty years (2) (fig. 1.1).

The investigation of the prehistory of the Tadrart Acacus massif started in the Sixties, on F. Mori's initiative, and in a series of research campaigns, he succeeded in bringing to light and illustrating an impressive complex of rock art having characteristics that were quite similar to the Tassili art in Algeria (Mori 1965). The core of the research was represented by the Wadi Teshuinat, which allowed us to observe a long succession of these works of art related to human occupation deposits. Stratigraphic excavations were carried out and the results are reported in the aforementioned volume. They supply the first evidence of human occupation in the Tadrart Acacus massif in a period that, based on the C14 datings of the major deposits, should range from the 7th to the 6th millennia B.P. (3).

At that time, however, the research effort was aimed at classifying and understanding the rock art. Mori offered an overall classification from the point of view of its stylistic aspects (Big Wild Fauna, Round-Heads, Pastoral, Horse and Bi-triangular styles) which, later on - contrary to his own intentions - became used as chronological phases.

---

* Dipartimento di Scienze Storiche, Archeologiche e Antropologiche dell'Antichità, University of Rome "La Sapienza", Via Palestro 63, Rome, Italy.

1) The area has a rectangular shape, included within the system of coordinates: 24°00'N - 10°20'E and 26°00'N - 11°20'E.

2) Since the beginning, the research has been carried out through contributions by the Italian Ministry of Foreign Affairs and the National Research Council. At present, the Scientific Commission of the University of Rome is also allocating funds to this purpose.

3) This issue will be dealt with, once again, in Part II, Chapter 6.

In the early Seventies, the state of knowledge of this area of the Central Sahara showed a huge gap between the understanding of the art and the understanding of human settlement. Scholars felt the need to develop stratigraphical research to be able to compare the results in either sector. Starting from 1970, thanks to the initiative of the Institute of Paleoethnology of the University of Rome (Director: Professor Salvatore M. Puglisi) the research scheme developed according to the criteria S.M. Puglisi himself had set out: "... the paleoethnological research in Libya must point in the direction of an overall reconstruction of the cultural framework through the use of data relative to the environment and archaeological documents resulting from systematic excavations, and drawing from the analysis of rock art those ethnological contents that different "material" evidence fails to provide in full ..." (Puglisi 1981: 112).

This statement clearly draws its inspiration from the new acquisitions of the prehistoric archaeology of the 1960's. Indeed our studies have adopted methods of cultural and social anthropology, sharing their concept of culture and cultural process (Binford 1965, 1968, 1977). As a result, the paleoethnological reconstruction turns into an integrated reconstruction that takes into account all the evidence left by prehistoric groups; archaeological deposits and pictorial representations.

The Tadrart Acacus offered the possibility of an overall reconstruction of this type. Following the guidelines given by S.M. Puglisi, the pictorial evidence was viewed as one of the kind of data to be explained in the light of all the evidence and, in such a way, it was given a prominence that differed from direct enjoyment of it as a merely aesthetic fact. This involved the adoption of a multidisciplinary approach which synthesised data from the natural sciences in the reconstruction of the habitat (paleobotany, archaeozoology, geochemistry, geomorphology, paleoecology) and subsequently interpreted them in terms of ethno-anthropological models.

At the same time and, once again, due to Prof. Puglisi, the Mission was re-organised on the basis of Libyan-Italian cooperation at both the operational and technical-scientific levels. Its name, Libyan-Italian Joint Mission for Saharan Researches, presupposed a new kind of prehistoric research in Libya. On the one hand, this involved the interaction between the Italian initiative and the Libyan support, and on the other hand, the Italian commitment increasingly to involve the young Libyan scholars in the study of the prehistoric past of their country, through field training and critical assessment.

The field work was carried out in two complementary sectors: stratigraphic excavation and the study of rock art (4).

The excavations were first organised in the northern area of the Tadrart Acacus, under the direction of Puglisi and with my collaboration. The selected region was the Wadi Auis on the eastern side of the massif, the moderately high piedmont section sloping towards the eastern valley system. Due to the scarcity of its pictorial art, this sector of the Acacus had been only marginally involved in previous research. Its different topography

---

4) The scholars in charge of the two sectors are B.E. Barich and F. Mori respectively.

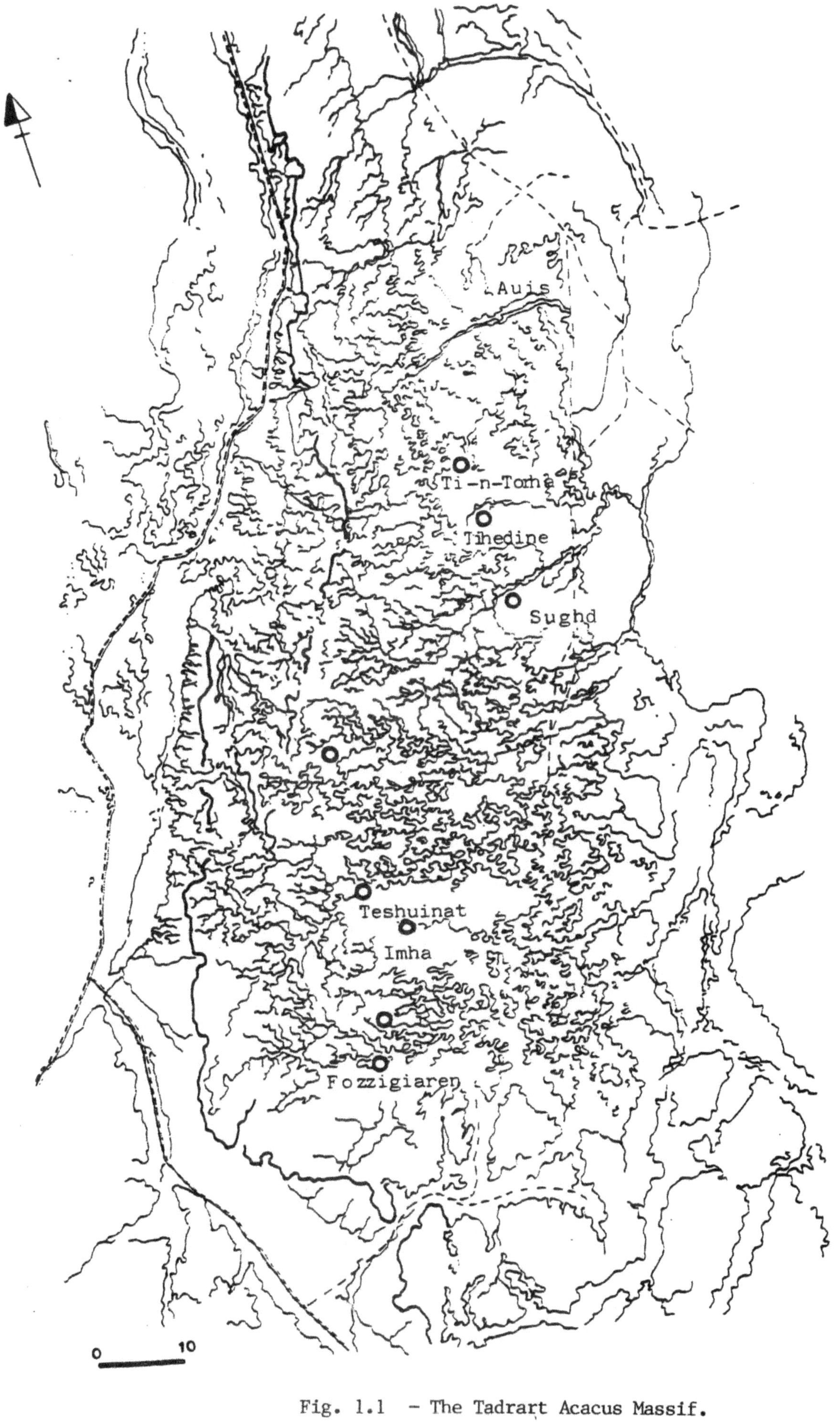

Fig. 1.1 - The Tadrart Acacus Massif.

compared to the higher Teshuinat (that will be dealt with in more detail by B. Marcolongo in Chapter 11), suggested straightaway a difference in the type of human inhabitation. This was later confirmed by the discovery of a form of occupation which appeared more archaic and with a different orientation in its economy from other areas.

The C14 dates currently available take the human occupation of the Acacus back to the final phase of the Pleistocene, that is to the renewed occupation of the region after the arid post-Aterian period. Indeed this evidence answers the question whether or not the Tadrart Acacus art should be considered in a Pleistocene context (Mori 1971).

The first research concentration was in the Wadi Ti-n-Torha, where the systematic investigation of two areas (Ti-n-Torha East and Ti-n-Torha North), allowed us to observe a wadi occupation sequence covering a broad chronological span. The most ancient facies (Torha East) dates from the 10th-8th millennia B.P. and involved a semi-permanent occupation by hunter-gatherers of Epipaleolithic tradition who already possessed ceramics. After some time (7th-6th millennia B.P.) these were replaced by pastoral groups, the upper levels (II-I) of Torha North (Barich 1974).

In 1978, an earlier phase of the facies that had already come to light at Ti-n-Torha East, was found devoid of ceramics, at the site known as Two Caves, a few hundred meters away in the wadi running parallel to the Ti-n-Torha (fig. 1.2 a,b). The 'Two Caves' deposit is extensively dealt with in the next chapter and is the reference point for discussing the evidence in this northern area.

2. FIELD METHODS: OBJECTS AND RESULTS

The field work of the Joint Paleoethnological Mission aimed at attaining the maximum amount of stratigraphic information by establishing, within the area under consideration, a reference sequence allowing the various scattered findings to be referred to it. In fact, it met the need to identify a diachronic sequence that ascribed a meaning and a specific placing in a sequence to the known archaeological record.

On this level, the achievements were significant even in a broader perspective: the Saharan sphere as a whole. The Saharan environment was indeed considered, and depicted, as a unitary system, and the processes that had affected it were interpreted as present throughout (Puglisi 1976).

We succeeded in outlining, within our territory, a development of an economic and social pattern from the system of exploitation of the hunter-gatherers of the early Holocene to the advent of an economy and a society of herdsmen. The opportunity of tracing the changes that took place, through different settlement phases, led us to accept the mechanisms of autocthonous development, thus rejecting the traditional models relying on diffusion (Barich 1980 b).

There was evidence that the pastoral organisation found in Levels II and I at Ti-n-Torha North (5) had its formative phase in the inner as well as

5) Also found at Wadi Athal, the first deposit excavated under the direction of Puglisi (Barich, Mori 1970).

a

b

Fig. 1.2 - Tadrart Acacus. a: a panoramic view of the Wadi Ti-n-Torha; b: Ti-n-Torha Two Caves.

the higher area of the massif. In particular, at the site of Uan Muhuggiag, investigated in 1959 and 1960 by the Mori team, the evidence of unequivocally domestic cattle dated to 5952 ± 120 B.P. In 1982, the excavation of the Uan Muhuggiag shelter was resumed for the purpose of checking this and the analytical study is reported in the second part of this Volume. The occupation deposit, that was perfectly preserved, indicated a stratigraphy of closely succeeding habitation layers, symptomatic of the temporary nature of the occupation by pastoral nomadic groups; moreover, the pastoral nature of the settlement was emphasised by the archaeozoological study of A. Gautier (Chapter 12).

It should be pointed out that the results we achieved, especially in respect of economic development, are in keeping with the results obtained by other missions working in the Sahara, as far as the eastern, Nubian and Sudanese, limits (Wendorf, Schild 1980). A comparison of all the results allows us to define an autocthonous process in North Africa that has its roots in the Late-Paleolithic period. There is therefore, a distinct sphere of cultural development here, with reference to both food and pottery production (Barich 1984 a, 1984 b) independent of the long known Neolithic development in the Near East.

3. THE ROLE OF REMOTE SURVEYING IN THE SPATIAL ANALYSIS

Since 1978, the stratigraphic investigations (that allowed a vertical reading and, therefore, an understanding of the cultural dynamics of the region) began to be accompanied by systematic surveys aimed at evaluating the use of the environment by prehistoric groups. These involved a spatial analysis, the object of which was to survey the horizontal attributes of the ecosystem (reciprocal site dislocation, distance from water sources, distance from raw material supplying sources, paths followed by the groups for hunting needs, alternative settlements of nomadic groups) (Clarke ed. 1977; Dunnel, Dancey 1983).

This type of study had the advantage of allowing a synthetic reading of the way in which the environment was used and it was an approach extensively used for a number of years by research teams working in the Sahara. Generally, the incentive for developing spatial analyses has been the difficulty in obtaining dependable stratigraphies. We were fortunate to have the opportunity of using both types, that is the horizontal and the vertical analyses, even though the latter was of greater importance since it was useful in shedding light on distant but comparable situations.

It recently became possible to plan large scale surveys thanks to the availability of remote sensing or satellite-surveys that made up for the lack of aerial photographic cover in our survey territory; without it that type of work could not be adequately performed. In Chapter 11 B. Marcolongo details the possibilities opened to our research by the use of remote sensing, showing how it developed as a decisive factor in land evaluation and palaeoenvironmental reconstruction. For our purposes it was also necessary, for our study of the relationships between archaeological settlements, physiographic conditions and contemporary resources, to be preceded by an introductory research phase based on this technique. The relationships between the distribution/typology of the sites and physio-environmental parameters were worked on the basis of satellite (Landsat 3) multispectral images. The cartographic and physiographic-geological information was subsequently checked in the field by comparing it with the

a

b

Fig. 1.3 - Tadrart Acacus. a: the Baraccat area in the north-western side of the Acacus; b: a rock shelter in the Wadi Tihedine.

environmental conditions and the archaeological record.

This type of activity occupied the Mission in 1983 and was continued through subsequent survey phases. The first phase concerned the western slope of the massif, along the Serdeles-Ghat route (fig. 1.3a). This checked the lithological and hydrogeological structure that had been ascertained by satellite analysis. It allowed us to verify that this slope lacked the geomorphological conditions suitable for permanent occupation and it pointed to the presence of water zones, especially around the Tahala area and the Wadi Tasbät in the Tanezzuft which might have been exploited by the population on the eastern slope since they could be reached through the 'Tulla and Tasbät passes (corresponding to Wadis Auis and Ti-n-Torha on the opposite slope). This was checked through a second phase of exploration in the Auis/Ti-n-Torha area. The archaeological typology of the settlements there (shelters with drystone sandstone walling at a certain height from the wadi bottom) was augmented by the following characteristics: southern exposure; special altimetry (about 5-10 m from the wadi bottom); lithology; directly usable water resources.

A third exploration phase has concentrated on the Wadis Tihedine and Sughd, where cartographic study had pointed to the same geomorphological and environmental conditions as the Wadi Ti-n-Torha (6).

The Wadi Sughd complex comprised three shelters, sited at a height of approximately 10 m above the wadi bottom, always S or SE facing. While the shelters contained no visible deposits, a large quantity of lithic tools, made from Devonian sandstone and heavily patinated, were visible in front of their entrances. The Tihedine complex it aroused even greater interest for, just as in the Wadi Sughd, three shelters that were found which had no visible deposits (fig. 1.3b). However, here too it was possible to observe a heavy concentration of lithic artefacts in the open space in front of the shelters within sight of the wadi that represented the site catchment of the groups. In both cases, the typology of the lithic tools demanded detailed analysis, for the preliminary study suggested the presence of a more Epipaleothic archaic technique than had previously been known (7).

4. FUTURE RESEARCH PROSPECTS

The methodology described led to the definition of a reliable settlement system in the territory. A more extensive use of this research technique will undoubtedly result in more precise models. In any event, the results reported by B. Marcolongo provide significant information as to the distribution of the settlements and the location of the relative site catchment areas. It is a study that is, in my opinion, already represented in the results of the excavations.

---

6) For the possibility of developing models that might anticipate the location of prehistoric sites through the use of Landsat images, see J.F. Custer and T. Eveleigh (1983).

7) They will be the subject of a separate study.

Further efforts should be directed to the detection of more archaic settlement nuclei, of the type that came to light in the Wadis Tihedine and Sughd, always on the basis of specific environmental parameters. No lesser importance should be ascribed to the reconstruction of a distribution pattern of rock art that might indicate the topographic relationships with the settlement sites, preferred centres of habitation, and the grouping of stylistic phases.

The north-eastern area appeared to be rather poor in pictorial evidence but in a previous paper (Barich 1974: 8-14) I underlined a number of outstanding features from this part of the Acacus. The prevalence of engraved as opposed to painted specimens might depend on topography and be related to the location of this external area of the massif which was greatly affected by meteoric action; or it might be related to its peculiar economic capabilities and, thus representing a specific choice.

Within the framework of the scanty inventory, one may differentiate two themes: Hunting with a tendency towards schematic representations, and Pastoral. They correlate with the cultural processes reconstructed from the archaeology. As a general rule, the scarcity of pictorial evidence in this area is paralleled by scarcity of pastoral evidence compared with that of hunters and gatherers. It would almost seem an indication of the fact that people were little interested in any pictorial representation or in artistic work generally.

Quite a different situation is found in the Wadi Teshuinat and the system of wadis connected with it. The constantly recurring evidence of pastoral settlements is particularly noticeable, either close to or within shelters which were marked by the presence of outstanding art work. Therefore, one may ascribe a significant sensibility for forms of figurative art to the pastoral peoples, whatever the motivation for it might have been. This stresses the close relationship existing between the artistic inclination and the ethno-social situation of these groups, an explanation opposed to the prevailingly diachronic interpretations which classify art only in stylistic terms.

If the art evidence collected in the various areas of the Tadrart Acacus is representative of human settlement in the massif as a whole, it becomes possible to understand the type of development, for, on the basis of a well established style sequence, different associations may be suggested for a whole series of data: climatic, vegetational, faunal, and human group organization.

The distribution study should also include the megalithic structures found throughout the Tadrart Acacus and which are particularly frequent on the eastern slope. Gravegoods and other chronological indications are exceedingly rare and their association with occupation sites, and archaeologically documented groups, is problematical. A funerary use may be ascribed, in my opinion, to a part of them and I consider them related to a late period and associated with the movements of proto-Berberian nomads dating from 4000-3500 B.P. (Barich 1984c). This subject in the environs of Wadi Ti-n-Torha and the various branches of the Auis, is dealt with by M. Baistrocchi in Chapter 4.

This Volume is the evidence of an integrated research effort throughout the Tadrart Acacus. Within the context of this ecosystem, every attempt was made to identify elements of differentiation as well as of contact between the two micro-environments taken into consideration by the archaeological survey in the north-eastern and the central regions. B. Marcolongo undertook the geomorphological and the palaeoenvironmental study, A. Gautier carried out the faunal analyses, E. Schulz conducted pollen analyses, G. Belluomini coordinated the C14 dating programme, and K. Wasylikowa undertook the study of the botanical macro-samples. The results of this study will be published separately since, owing to the size of the collections, they could not be included in this Volume.

Moreover, I gratefully acknowledge the contributions made to this Volume by M. Baistrocchi (megalithic aspects of the Auis), A.E. Close (analysis of the lithic sequence of Ti-n-Torha facies), A.M. Palmieri (ceramics component analysis) and I. Caneva (study of the Uan Muhuggiag ceramic decorative motifs).

CHAPTER 2

THE TWO CAVES SHELTER: AN EARLY HOLOCENE SITE IN THE NORTH-EASTERN ACACUS

By Barbara E. BARICH

1. THE SITE AND THE STRATIGRAPHIC SEQUENCE

This shelter was found during 1972 in the wadi running parallel to the Ti-n-Torha. The location and the archaeological deposit allowed its immediate identification for it presented the same exposure towards the East as the East Shelter, with a similar structure consisting of a piece of sandstone blocks as an outer protection. At the time of discovery the boulders were almost completely covered by eolian soil and sediments (fig. 2.1*a*).

The presence of a man-made structure, however rudimentary, was in itself extremely significant, being symptomatic of an occupation of semi-permanent character, with obvious implications for the economic organisation achieved by the community.

This side wadi of the Ti-n-Torha runs in an E/W direction parallel to the previously surveyed one. Its north-eastern side presents a continuous ridge of ancient sandstone with a dome-like profile that, in any event prior to human occupation, gave way to landslides that left broad sections of smooth and sheer wall (fig. 1.2*b*). The opposite, south-western, side is pierced by a number of smaller wadis, delimited by rocky terraces of medium height alternated with higher peaks. These terraces are covered by sandstone debris that has turned black owing to weathering. Accumulations of large stones may be observed in various areas of the terraces.

The "Two Caves" name derives from the fact that the shelter consists of two communicating caves characterized by low ceilings and two openings onto the wadi side. The morphology of the massif, composed of extremely friable sandstones is easily subjected to erosion and seldom allowed the formation of actual caves. Therefore the identification of such a deep hollow that offered satisfactory protection to human groups represents an extremely interesting element. At present, the western opening is almost totally blocked by debris that, in relatively recent times, fell from the vault; the inner chamber contained a tumbled and extremely thick fill. In the adjoining chamber the archaeological deposit was covered by a 20-30 cm thick sand layer which preserved it. It is most likely that the activities of the occupation were predominantly carried on outside the caves, but the 'occupation' deposit was also present inside them. This was proved by the test trench opened to check the extent of that area (see below) (1).

1) The possibility of extending the excavation into the chambers would have meant clearing the western entrance and extracting the dust clouds that the excavation raised.

a

b

Fig. 2.1 - Two Caves, Wadi Ti-n-Torha. a: view of the rock shelter; b: the western cave.

The excavation in 1978 was on the terrace outside the caves where the archaeological deposit spread over about sixty square metres. It lay on rock approximately five metres above the wadi bottom. The whole SE front of the deposit had developed against an accumulation of boulders resting on bedrock (fig. 2.1b). Excavation showed that this mass extended inside the deposit, joining a second level of stones that were only marginally detectable in the first excavation stage.

The area excavated measured approximately thirty-four square metres; the point of maximum thickness of the deposit being in front of the western cave. The northern limit was a recent landslide while the southern side projected towards the wadi, beyond the massive blocks. The area was divided into three adjoining sectors for investigation: western, central and eastern, the whole area being covered by a layer of sand, varying in thickness from 2-5 cm at the eastern and western edges to a maximum of 20-22 cm in the central area (2).

The Stratigraphy of the Western Sector.

Level I: Soil having a varied composition, with superimposed lenses of fibrous material; dust mixed with ashes and charcoal; dust with fibres and seeds. Black colour (5YR 2.5/1) (3).

While this soil at the top was very compact, towards the bottom it was loose and the second level was marked by a change in colour and a compacted level with a concentration of charcoal. Leaving aside a marked doming, perceptible to the naked eye, in its upper part, it sloped to the West and East. At a depth of 30-35 cm from the surface, the stone accumulation in front acquired a clearer outline while the stones on the eastern side appeared as a proper wall in the deposit. In this way, the area was entirely delimited by stones.

Radiocarbon dating of this level: 8520±60 B.P. (R-1407) (4).

Level II: Loose soil with a significant component of ashes and charcoal, with a light grey colour (5YR 6/1). On the southern front, it lies below the first stone level, while on the eastern side it ajoins a large sandstone block (fig. 2.4a,b).

Radiocarbon dating of this level: 8450±60 B.P. (R-1408).

---

2) The preliminary study on Two Caves is in printing in Libya Antiqua, published by the General Direction of Antiquities, Tripoli (Libya). The major data of the stratigraphic sequence is reported below (cf.: Barich 1978 in press).

3) The soil colours reported here below refer to the MUNSELL'S SOIL COLOR CHARTS (1973).

4) All the datings obtained from the deposits, already published in Barich, Belluomini et Alii (1984), are summarized in Chapter 14 (this Volume). The carbon content of the samples collected from the Central Sector appeared to be too low to allow them to be dated.

a

b

Fig. 2.2 - Two Caves, Wadi Ti-n-Torha. a: the emerging of the stone floor in the Central Sector; b: a hearth in the Central Sector with *in situ* artifacts.

a

b

Fig. 2.3 - Two Caves, Wadi Ti-n-Torha. a: the stone floor of the Central Sector completely brought to light; b: detail of the stone floor.

Level III: The very friable sandstone of the eastern side was part of a landslide that had overrun most of the excavation area below the second level. The soil, resulting from the pulverisation of the sandstone was almost completely sterile and of a light reddish brown colouring (5YR 6/3).

Excavation in this section failed to uncover bedrock coming to a halt on the stones that announced it. This must be borne in mind when evaluating the few artefacts collected in this level.

Radiocarbon dating of this level: 8630 ± 50 B.P. (R-1404).

Central Sector.

Level I: The very compact blackish soil (5YR 2.5/1) showed an evident continuity with the western sector and became finer at its base, marking the transition to the subsequent level. Excavation revealed a stone floored hearth; large stone tools were present within the hearth (fig. 2.2a,b). The stone floor lay under a 20-30 cm thick, bowl-shaped sand accumulation that was obviously linked to the recently fallen boulders that obstructed the cave.

The stone floor which covered approximately 2/3 of the central sector, was joined to the small wall in the western section and was also linked to the outer massive blocks. The concentration of flaked artifacts suggested it was a tool-making area (fig. 2.3a,b).

Level II: The excavation area was limited to 2 x 1.50 m by stone constructions, especially in the West and South. The soil resting against the stone was very fine, with a grey (5YR 5/1) colouring and patches of sands and little charcoal.

Level IIa: At a depth of approximately 50 cm from the surface an arbitrary division was made (IIa). The composition of the soil was the same as in the upper part and hearths and ashes appeared quite frequently at the bottom (5).

Level III: A light reddish brown (5YR 6/4) soil, even finer than the preceding one, consisting of soil mixed with sand and very little charcoal. It was found beneath the stones and was thick. Patches of sand, that are a sign of the end of human occupation, appeared in the lower part. This was called Base and was a thin (5 cm thick) layer with an almost exclusively sandy composition, directly in contact with the Harlania bedrock.

In the direction of the wadi a N/S extension (1.50 m long; 1 m wide), was made close to the western baulk. This revealed that the wall-like structure, with an accumulation of stones, in two superimposed rows, occurred after the formation of Level III. The soil in the crevices has the same characteristics as that observed in Level I.

---

5) While being studied, the materials were considered altogether as Level II.

Eastern Sector.

Level I: Dark reddish brown (5YR 3/2), dusty and loose soil, mixed with sand resulting from the breaking up of sandstone and quite a number of rivers of charcoal resting on a crust of carbonate. It was to be connected with the base of Level I in the Western and Central Sectors. In this level, the excavation area lacked any structure marked limits.

Level II: The soil composition was similar to that of the preceding level but the stronger carbonate component gave it a lighter colouring tending towards a pinkish hue (5YR 7/8).

Radiocarbon datings of this level: 8620 ± 50 B.P. (R-1406);
8650 ± 105 B.P. (R-1409).

Level III: Very fine, light reddish brown (5YR 6/4) soil, with considerable quartzitic infiltrations resulting from the breaking up of the boulders which fell from the wall at the back of the shelter.

Radiocarbon dating of this level: 9350 ± 110 B.P. (R-1402).

Base Level: This level (maximum thickness 5 cm) mainly consisted of sand. It was directly in contact with the bottom clays and *Harlania*. In a few places, carbonate and iron oxide crusts (yellow ochres) were noted at the bottom.

A Test Trench (1.20 m long; 1 m wide), was dug within the inner part of the western cave, between the cave wall and the recent stone fall. The soil was quite undisturbed, unlike the visibly disturbed surface of the cave.

Level I: Very fine and dusty, dark reddish brown (5YR 3/3) soil, comparable with Level I of the Western Sector. Carbonate formations, tending to acquire a crust-like compactness, become evident at a depth of approximately 40-45 cm from the surface; beneath this point the soil is more coherent, with superimposed lenses.

The archaeological material present in the soil (lithic implements and a few ceramic fragments) decreased as the depth increased. After the carbonate formation, they became fewer, pointing to a very occasional occupation of the site.

The excavation stopped 65-70 cm below the surface (6).

Discussion:

The data may be interpreted as follows. The presence of the massive stone blocks on the south-eastern front and its continuation within the deposit through the transverse wall, resulted in the partitioning of the area in two chambers. The western chamber showed clearly defined boundaries while the eastern chamber - at least in the portion uncovered in excavation - did not appear to be so well defined. Therefore, even though

---

6) The stoppage was due to the rarity of archaeological evidence as well as to the growing difficulties in such a narrow area.

a

b

Fig. 2.4 - Two Caves, Wadi Ti-n-Torha. a: base of level II in the Western Sector; b: bone spatula *in situ* in the Western Sector.

a

b

Fig. 2.5 - Two Caves, Wadi Ti-n-Torha. a, b: the site at the end of the excavation; the internal wall has been completely brought to light.

the presence of a structure to make the shelter more effective is clearly evident, it did not have the regular shape observed in the Ti-n-Torha East shelter (fig. 2.5a,b) (7).

The partitioning of the inner space occurred during phases II and I, that is during the phases of greater use of the shelter. Similar to the findings at Ti-n-Torha East, it is clear that habitation had begun in Level III around the 10th millennium B.P.. Quite likely a group took advantage of the natural protection offered to the south-east by boulders, resting on bedrock. During the second habitation phase, the outer stone embankment was reinforced with further blocks and the addition of the transverse wall which defined an area (Western Sector) protected on all its sides. The wadi-projected extension in the Central Sector showed that the stones of the wall rested on Level III; they were added, therefore, from above during the second phase, even though the small wall had been strengthened several times during the first phase.

The great accumulation of occupation debris observed in the western part of the inhabited area during phase I has already been dealt with in the analytical account, and was also evident from the different cross-sections (fig. 2.6). This greater accumulation was associated with a wealth of finds in both the west and central areas. Its elaborate stratigraphy is a clear indication of the heavy utilization of this area which in Level I showed a certain similarity to the area inside the western cave. The carbonate formations - due to water dripping from the wall- found during the test excavation, one may hypothesize belonged to a moist period that caused the group to strengthen its walls and look more frequently for shelter inside the cave.

The remaining part of the living area, especially the Eastern Sector, was being occupied much less extensively as can be seen from the cross-section. The collapse of the large blocks that now close the western opening is, without doubt, a relatively recent event, that occurred after the abandonment of the cave. It caused a considerable accumulation of sand in the central area as well as subsidence under the pressure from above.

Leaving aside the cultural changes at the site, over at least a millennium, no stratigraphic break was reported in the series. The materials, analysed as a *continuum*, confirm the homogeneity of the sequence.

2. THE ARTIFACTS

In the preceding section it was pointed out that, in view of the stratigraphical evidence for continuity in the sequence, we should view the complex of artifacts as a single group. It consists of a lithic assemblage that, in addition to flaked, also includes a few polished, stone implements.

A limited though significant rôle is played by the few bone tools which show a high standard of manufacture. As for the ceramic sherds present, they were, for the most part, shown to be an intrusive element within a fundamentally aceramic context. This conclusion, based on the characteristics of these sherds as well as on their distribution within the

---

7) This site showed the presence of rows of stone forming semicircular dwelling areas ("huts") (cf.: Barich 1974).

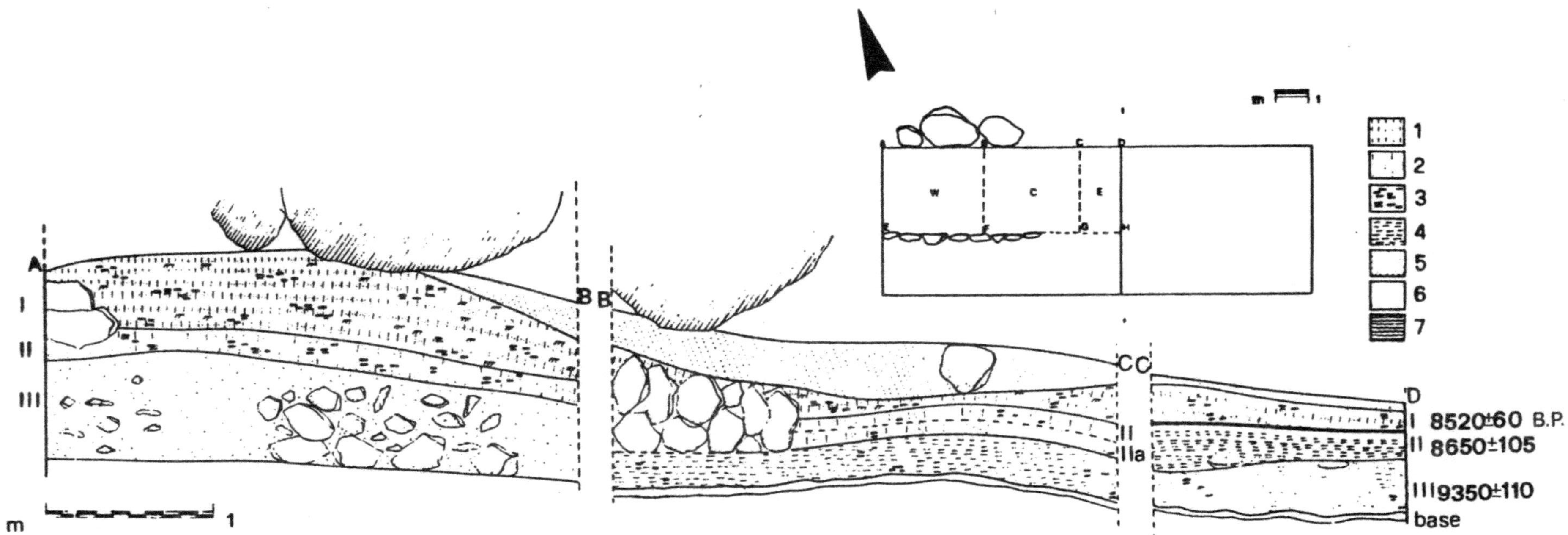

Fig. 2.6 - Two Caves, Wadi Ti-n-Torha. General stratigraphy. Key: 1) blackish soil; 2) grey soil; 3) coals; 4) light reddish brown soil; 5) sandstone; 6) sand; 7) clay.

settlement, will be discussed in the following paragraph.

The flaked stone was analysed with reference to the Ti-n-Torha East lithic assemblage (Barich 1974). It seemed advisable to refer to that report since it had endeavoured to provide a classification that might be applicable to other complexes present in the same area. This is absolutely necessary when having to deal, for the first time, with an Epipaleolithic lithic assemblage where the associations with ceramics did not allow us, at first, to match the type-list drawn up by Tixier (1963). Tixier's list appears to match the tool-classes in general but detailed agreement on specific types were rare and restricted to backed bladelets of La Mouillah and other types.

The typological structure of retouched tools - referred to in the present classification - may be outlined as follows:

1. End-scrapers
2. Burins
3. blunt unretouched blades
4. microbladelets
5. pointed unretouched blades
6. pointed backed blades
7. blunt-backed blades/flakes
8. Truncations
9. Geometric microliths
10. Microburins
11. Side-scrapers
12. Denticulates
13. Arrow-heads

The debitage analysis clearly distinguishes flakes <u>sensu stricto</u> (length/width ratio less than 2:1). Therefore, the different debitage-types described, are all flakes <u>s.s</u>, except for a few 'blade-like' flakes that have marked blade characteristics. The blades that are either unretouched or with scarce retouch (length/width ratio = 2:1) were included among the tools as a class, as in the Ti-n-Torha East classification. The concentration in a single class of the different blade-types is to stress the relative frequencies that are deemed relevant for the overall evaluation of the complex.

Having said that, the classification of artifacts is outlined in the reference table that allows us at a glance to grasp incidences and distributions (Tables 2.1-4, 8, 9). Within the text, cores/debitage and tools are commented upon by classes, with special reference to the following points that are deemed to be essential:

a) general definition of the core/debitage and tool types;

b) reference to particular items;

c) frequency variations in the stratigraphic sequence.

This form of comment is accompanied by illustrations of the types that appeared to be the most significant within each of the analysed classes.

The analysis of the raw materials used shows a similar artifact distribution to the nearby deposit of Torha East. The macrolithic implements (end-scrapers, denticulates, side-scrapers) were made on relatively fine grained Devonian sandstone and highly compact siltites. Backed tools and geometric microliths were made on quartzitic sandstone and quartz, as well as from a better quality sandstone. Small quartz pebbles, at times of a microlithic size, have often been observed in the debitage. Their frequency is constant throughout the sequence. Due relevance should be ascribed to the flint tools and debitage, including a reconstructable tabular pebble. Further details will be provided in cases of special utilization, as required.

Debitage

The incidence of the various debitage-types and their distribution throughout the sequence are shown in Table 2.8 and figures 2.7, 9, 13, 16.

Unretouched or partially retouched flakes total 1138, corresponding to 46.7% of the entire debitage. Primary flakes (234) were one of the most representative types; they tend to be larger as well as thicker than other flakes (8). The following varieties were observed: unbroken simple primary flakes with a dorsal flake-scar struck in the same direction as the flake itself; unidirectional primary flakes; bidirectional (opposite) primary flakes. The rest consisted of simple broken primary flakes. In all examples platform-types were prevalently cortex, although some faceted and many unidentifiable specimens were also present. In addition to the primary flakes s.s., another variety was also observed; primary blake-like flakes (seventy-three pieces). Although inconsistent with the conventional definition of blades, these had relevant blade characteristics (fig. 2. 16,1). Flakes from single platform cores were the commonest debitage-type (23.8% of the unretouched flake-class alone) but they were often broken (6.0% ca) (figs. 2.7, 6; 9, 2, 3; 13, 10; 16, 2, 3, 13). In the same way of the primary flakes, a blade-like variety (9.4%) was also identified in this type of debitage. They were very regularly shaped pieces - even though their lateral edges were not completely parallel or sub-parallel - with trapezoidal or triangular cross-sections (figs. 2. 7, 1, 2; 13, 3, 5, 7, 8; 16, 8). Excepting a few secondary items, most of them were tertiary. Platform-types were prevailingly lisse, with few faceted examples (less than 10%). Several tanged pieces (three from Level III, five from Level II and three from Level I) were also noted (figs. 2. 7, 12; 9, 9; 13, 1). Flakes from opposed platform cores were collected from Levels II and I; they were regularly shaped, either tertiary or secondary, with lisse platform (one dihedral). Flakes from ninety-degree cores were the third most frequent type (17.8%). They tended to be larger and shorter than the previous types and some of these specimens featured a use-retouch on the sides. Platform-types were lisse (65.0%) and faceted. Flakes from multiple platform cores were significantly shorter than any of the aforementioned types (fig. 2. 7, 13); they were all tertiary. Platform-types were faceted. In addition a number with denticulations were found, some with invasive bifacial retouches.

---

8) Primary flake: "an unbroken, unretouched flake of which the maximum dimension is not less than 15 mm and of which at least 50% of the dorsal surface is cortex or an identifiable fragment thereof" (cf.: Close, Wendorf, Schild 1979: 34).

The most common raw material was sandstone, often of very good quality. From Level I (Western Sector) some eleven flint leaves and ten from Level II. In the flakes s.s., lengths range from 21 to 35 mm; widths from 15 to 25 mm; thicknesses from 4 to 8 mm. Blade-like types can be divided into macrolithic pieces (lengths from 35 to 60 mm; widths from 15 to 40 mm) and diminutive ones, with length of less than 35 mm, width of about 15 mm and thickness of 4 mm.

The highest frequency of the class was reached in Level I, particularly in the Western Sector.

Core-trimming: (52, equal to 2.1% of the core/debitage group) were quite rare throughout the assemblage: only one core-trimming for every three cores. A very similar ratio had already been observed at Ti-n-Torha East. This means that rejuvenation or core preparation was not a basic element of the Wadi Ti-n-Torha technology. Pieces consist of undifferentiated core-trimmings, core-tablets and lames à crête (figs. 2. 7, 4; 9, 10, 11; 13, 2, 11; 16, 5, 6, 12). Sometimes the latter were transformed into hump-backed bladelets through backing-retouch. Most of the identifiable platforms were either lisse or faceted; few examples showed cortex. Core-trimmings were thick and generally short (on the average: length 23 mm; width 15 mm; thickness 6 mm). The percentage frequency of this class of debitage reached its highest values in Levels III and I.

Microlithic flakes (1088, equal to 44.7% of the cores/debitage group) had the following mean dimensions: length 10 mm; width $\leqslant$ 5 mm; thickness 3 mm (9). Their general index, showing a slightly lower value than that reported for macrolithic flakes, was in keeping with the general microlithic tradition of the Wadi Ti-n-Torha industry. In Ti-n-Torha East the percentage frequency was even higher (57.4%). In particular, micro-flakes were to be related to the production of micro-bladelets, backed bladelets (particularly those with a microlithic tendency) and geometrics. Specimens belonging to this class were always made on quartzite or small quartz or chert pebbles. Their presence was noted from the base of the deposit, where they reached their peak frequency, upwards - particularly in the Central Sector. A decrease was observed in Level I.

Cores

A total of 155 specimens were recovered from the various Levels. Table 2.1 shows the distribution as well as the absolute and percentage frequencies of the pieces throughout the sequence. A steady increase in their number was registered from Level III (twenty-three) to Level II (eighty). This is in keeping with the generally more stable utilization of the site in its later phase.

The base of the deposit and Level III yielded twelve single platform cores, four of which were fragments; two were ninety-degree cores with flaking-surfaces on the same side and ten were multiple platforms, both

9) They correspond to "chips" which other investigators enumerate as a class of debitage (Close, Wendorf, Schild 1979: 34-35; Close ed. 1980: 50).

patterned and unpatterned, with flaking-surfaces on adjacent and opposite sides. Four pieces were unidentifiable (one was a fragment). They generally had a globular shape and the platform-angles in the single-platform and in the ninety-degree examples were about 80°, while in the others they were generally less than 90°. Platform-types were _lisse_ or faceted and, in five specimens, cortical. A limited preparation was observed in single platform types. Ten were of quartzitic sandstone and the rest from sandstone. The mean piece dimensions were: length 40.5 mm; width 40.0 mm; thickness 32.0 mm.

Level II yielded forty-seven cores in addition to a tabular silt pebble with almost all its debitage (thirty-three pieces). Single platform cores were the commonest core-type (fifteen pieces) followed by the patterned, multiple platform variety (ten pieces) (fig. 2. 9, _4_, _7_, _8_). There were eight cases of unpatterned multiple platform (fig. 2. 13, _12_). In both varieties - patterned and unpatterned - flaking-surfaces were either on adjacent (five cases) or on opposite sides. There were five examples of the ninety-degree core and three of the opposed platform core with flaking-surfaces used to exploit the same face. Six examples were unidentifiable. Multiple, patterned and unpatterned specimens had irregular, tending towards globular, shapes; opposed platform core-types were prismatic. Platform-angles were always slightly less than 90°, while the type of platform varied between single and multiple platform core-types. It was mostly _lisse_ in the former while a higher frequency of cortex was noted in the latter. As in Level III, only the single platform variety presented a somewhat limited preparation of the core sides or back. All the specimens, except the flint item, consisted of quartzitic sandstone. Their mean dimensions were as follows: length 50 mm; width 42.0 mm; thickness 35 mm.

The highest absolute frequency of the various core-types was reported in Level I where, in addition to the aforementioned varieties, a discoidal specimen was also observed. The single platform core-type was still commonest (twenty-five, six of which were fragments). There were eight ninety-degree cores with flaking-surfaces exploiting single and adjacent faces; out of the nine opposed platform cores, three were fragments and the rest showed flaking-surfaces on the same side. In the twenty-four multiple platform cores (both patterned and unpatterned) the flaking-surfaces were often on adjacent sides (fifteen cases - on opposite sides in the rest). Ten specimens were unidentifiable and three were whole pebbles. The average platform-angle for these cores was about 74.5°, but the type of platform varied as in Level II. Single platform and ninety-degree types were mostly _lisse_, while the other types were faceted or dihedral. Five examples were cortical. Six cores showed signs of limited preparation of the back but the rest were unprepared. The favoured raw material was sandstone (thirty-five); there were three quartz pebbles and the rest consisted of quartzitic sandstone, except for one specimen that was made on silt. The mean specimen dimensions were: length 45.4 mm; width 38.1 mm; thickness 32.0 mm.

The presence of 125 small (fig. 2.7, _5_) pebbles of white quartz was quite interesting, although their figure could not be included in the calculation of the absolute frequencies since they were not considered true cores. Quite likely, many of the microlithic flakes (chips) observed in the debitage came from small pebbles of this kind. Their highest incidence was reported in Level III, while they decreased through Levels II and I.

Table 2.1 - Two Caves, Wadi Ti-n-Torha. Retouched tools from Level I.

	SECTOR western	central	eastern	Total	%
End-scrapers	5	2	2	9	2.5
Burins	5	-	1	6	1.6
Blades	112	143	41	296	80.4
Blunt unretouched	51	56	18	125	34.0
Microbladelets	30	23	3	56	15.2
Pointed unretouched	20	35	7	62	16.9
Pointed backed	6	16	12	34	9.2
Blunt backed blades/flakes	5	10	1	16	4.3
Truncations	-	3	-	3	0.8
Geometric microliths	2	12	1	15	4.0
Microburins	-	-	-	-	-
Side-Scrapers	5	6	-	11	3.0
Denticulates	14	17	-	31	8.5
Arrow-heads	-	-	-	-	-
Total	143	180	45	368	100.0

Table 2.2 - Two Caves, Wadi Ti-n-Torha. Retouched tools from Level II.

	SECTOR western	central	eastern	Total	%
End-scrapers	1	2	1	4	1.7
Burins	-	2	2	4	1.7
Blades	37	88	76	201	84.4
Blunt unretouched	11	34	33	78	32.7
Microbladelets	13	14	15	42	17.6
Pointed unretouched	4	11	7	22	9.2
Pointed backed	5	24	19	48	20.2
Blunt backed blades/flakes	4	5	2	11	4.7
Truncations	-	-	-	-	-
Geometric microliths	-	1	4	5	2.1
Microburins	-	2	3	5	2.1
Side-Scrapers	-	-	4	4	1.7
Denticulates	2	10	3	15	6.3
Arrow-heads	-	-	-	-	-
Total	40	105	93	238	100.0

Table 2.3 - Two Caves, Wadi Ti-n-Torha. Retouched tools from Level III.

	SECTOR western	central	eastern	Total	%
End-scrapers	-	1	2	3	4.1
Burins	-	2	-	2	2.8
Blades	3	40	14	57	79.2
Blunt unretouched	3	18	-	21	29.2
Microbladelets	-	16	4	20	27.8
Pointed unretouched	-	-	-	-	-
Pointed backed	-	5	8	13	18.1
Blunt backed blades/flakes	-	1	2	3	4.1
Truncations	-	-	-	-	-
Geometric microliths	-	1	1	2	2.8
Microburins	-	2	-	2	2.8
Side-Scrapers	-	2	-	2	2.8
Denticulates	-	4	-	4	5.5
Arrow-heads	-	-	-	-	-
Total	3	52	17	72	100.0

Table 2.4 - Two Caves, Wadi Ti-n-Torha. Retouched tools from the Base.

	SECTOR				
	western	central	eastern	Total	%
End-scrapers		-	-	-	-
Burins		-	-	-	-
<u>Blades</u>		<u>47</u>	<u>1</u>	<u>48</u>	<u>92.3</u>
Blunt unretouched		12	1	13	25.0
Microbladelets		5	-	5	9.6
Pointed unretouched		9	-	9	17.3
Pointed backed		15	-	15	28.9
Blunt backed blades/flakes		6	-	6	11.5
Truncations		-	-	-	-
Geometric microliths		1	-	1	1.9
Microburins		-	-	-	-
Side-Scrapers		-	-	-	-
Denticulates		3	-	3	5.8
Arrow-heads		-	-	-	-
Total		51	1	52	100.0

Retouched Tools

End-Scrapers: 16; general index 2.2%.

Figs. 2. 8, 16; 12, 2, 6; 14, 20; 20, 1.

End-scrapers on large tertiary flakes were the most common type since only five blades - with distal scraping-end - were collected from the different levels. Some end-scrapers on primary flakes retained the shape of the original blank from which they were obtained through an approximate retouch. Four sub-circular examples showed a more regular scraping-end, but the retouch did not involve their whole distal end. Two nucleiform, diminutive end-scrapers were collected from Level I, where a single example of denticulated end-scraper was also observed (fig. 2. 12, 6). The base Level and Level II yielded two dual purpose tools: both were backed and pointed bladelets with proximal end-scrapers (fig. 2. 8, 13). The mean class dimensions were: length 30 mm; width 15 mm; thickness 5 mm.

Except for the dual purpose tools, no end-scraper was identified at the base of the deposit. In Level III they had a 4.1% incidence (the highest in the whole complex) but their figure decreased moving towards the top of the series.

Burins: 12; general index 1.7%.

Figs. 2. 10, 7, 22, 25; 14, 23.

All the reported items were on tertiary blades with the working-edge on either the sinister or dexter side. Three specimens involved dihedral burins, of which two distal *déjeté* and one distal symmetrical on a backed blade. There were three nucleiform burins, both on secondary flakes, one of which was a multiple dihedral with an obverse sinister proximal truncation. The remaining six (from Levels II and III) were single angle burins on tertiary flakes, all with distal burin working-edges. The mean class dimensions were: length 35 mm; width 23 mm; thickness 4 mm.

Burins seemed to have a very low incidence in the complex. They were reported from Level III (2.8%) and their frequency greatly decreased towards the top of the stratigraphy.

Blades: 603; general index 82.6%.

The percentage frequency of blades defines the technological tradition of the Two Caves assemblage. The six classes included (Roche 1963) in the general blade group, confirm the pattern already observed in the Torha East assemblage. Except for micro-bladelets and pointed backed pieces, the distinctions made by Tixier (1963: 39) between blades and bladelets do not seem to apply. In term of size, the various classes (blunt unretouched, pointed unretouched, blunt backed blades/flakes) appear to form a *continuum*.

(a) Blunt unretouched: 237; general index 32.4%.

Figs. 2. 10, 24; 11, 3, 4, 6, 13; 14, 1, 2, 21; 15, 4, 11, 12; 17, 14; 18, 5, 6, 14.

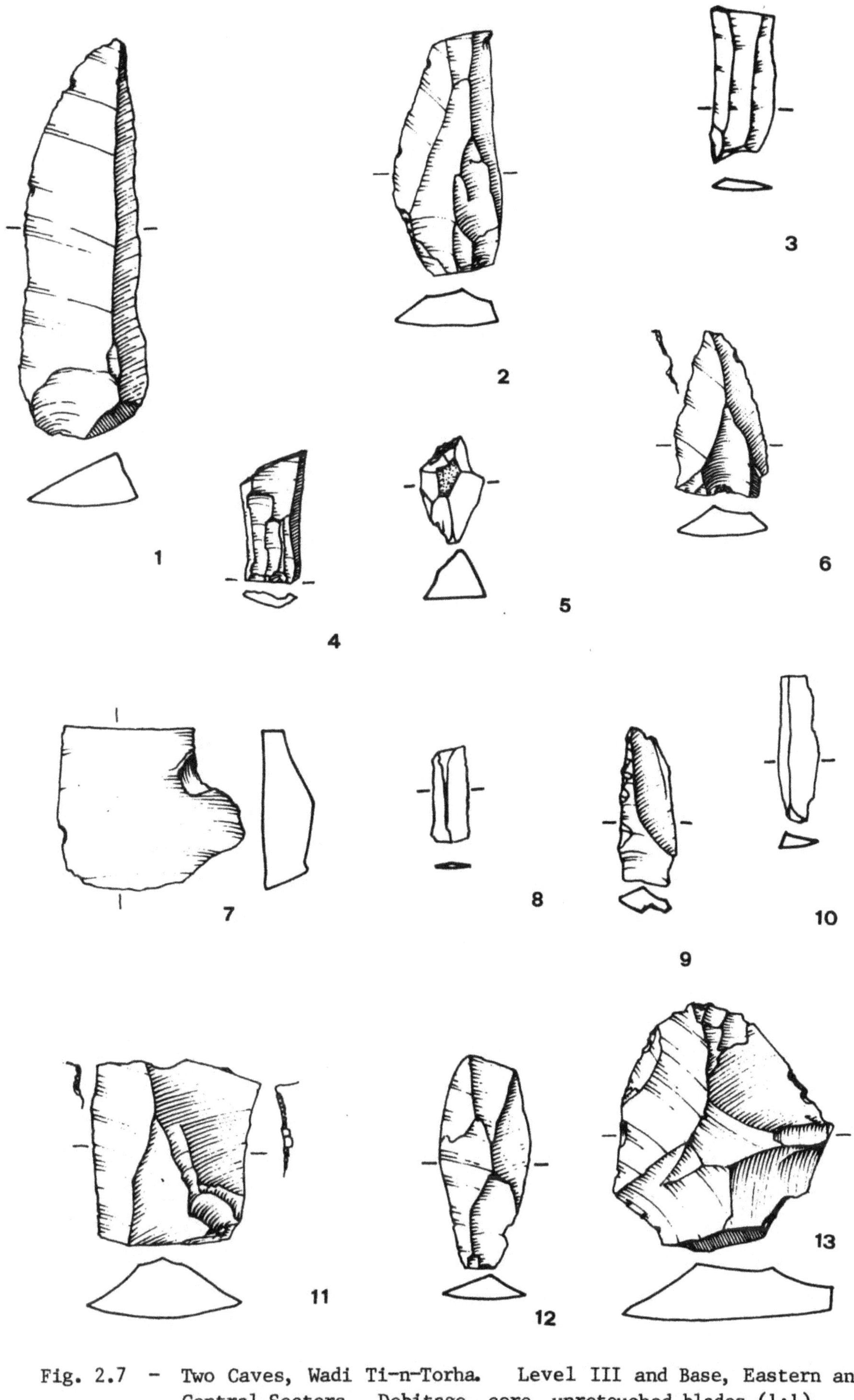

Fig. 2.7 - Two Caves, Wadi Ti-n-Torha. Level III and Base, Eastern and Central Sectors. Debitage, core, unretouched blades (1:1).

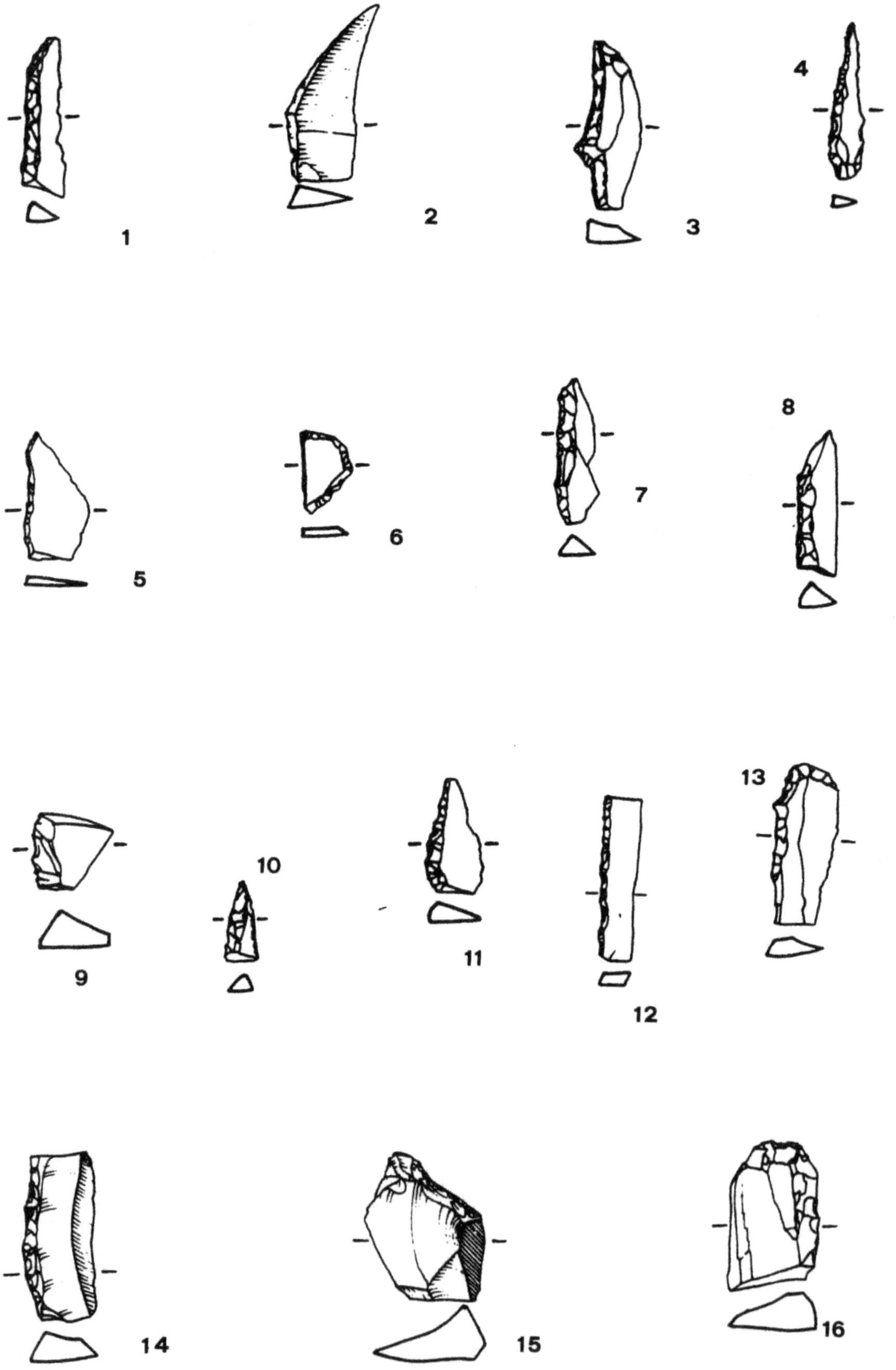

Fig. 2.8 - Two Caves, Wadi Ti-n-Torha. Level III and Base, Eastern and Central Sectors. Backed bladelets, backed blades/flakes, segment (1:1).

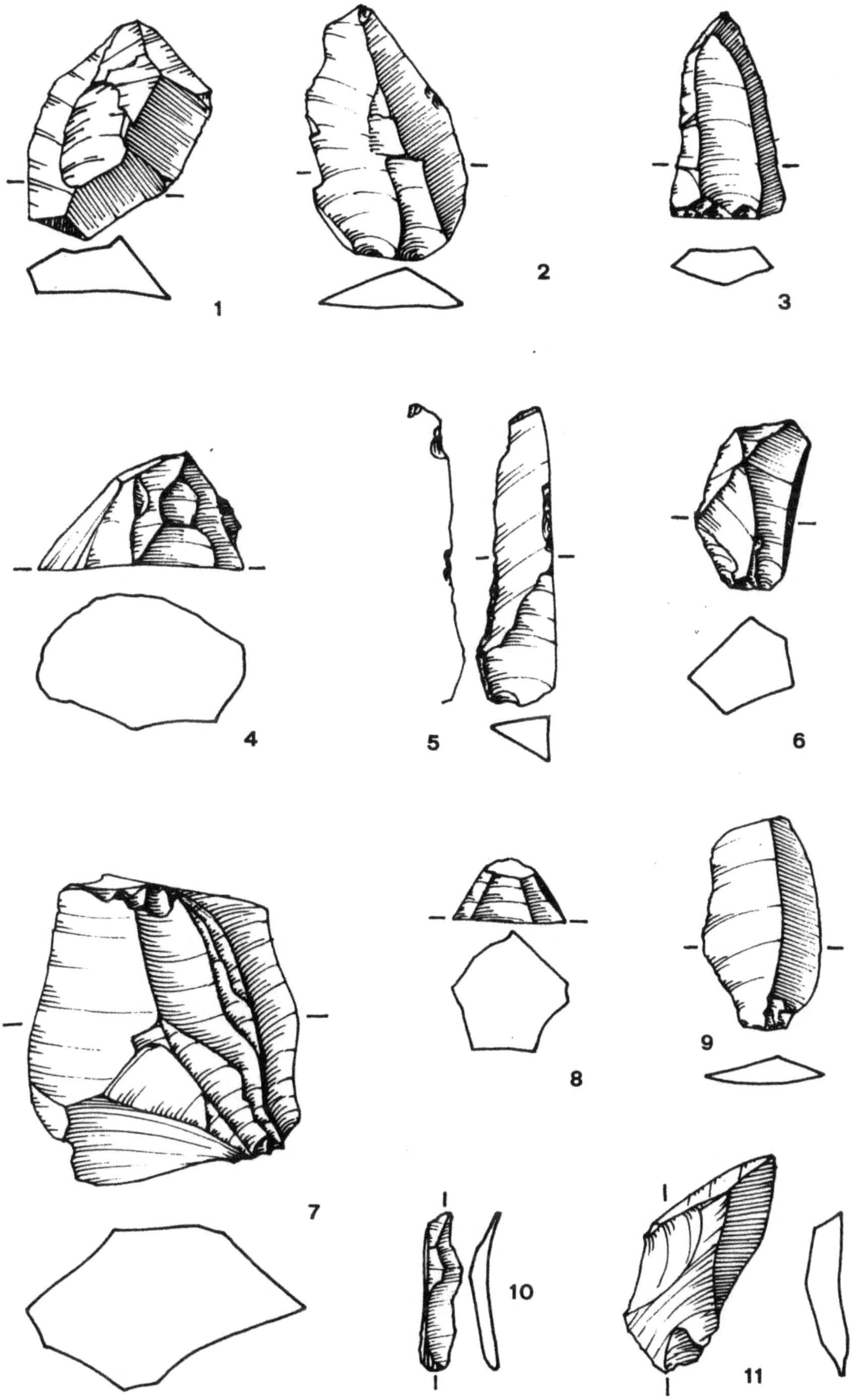

Fig. 2.9 - Two Caves, Wadi Ti-n-Torha. Level IIa, Central Sector. Debitage and cores (1:1).

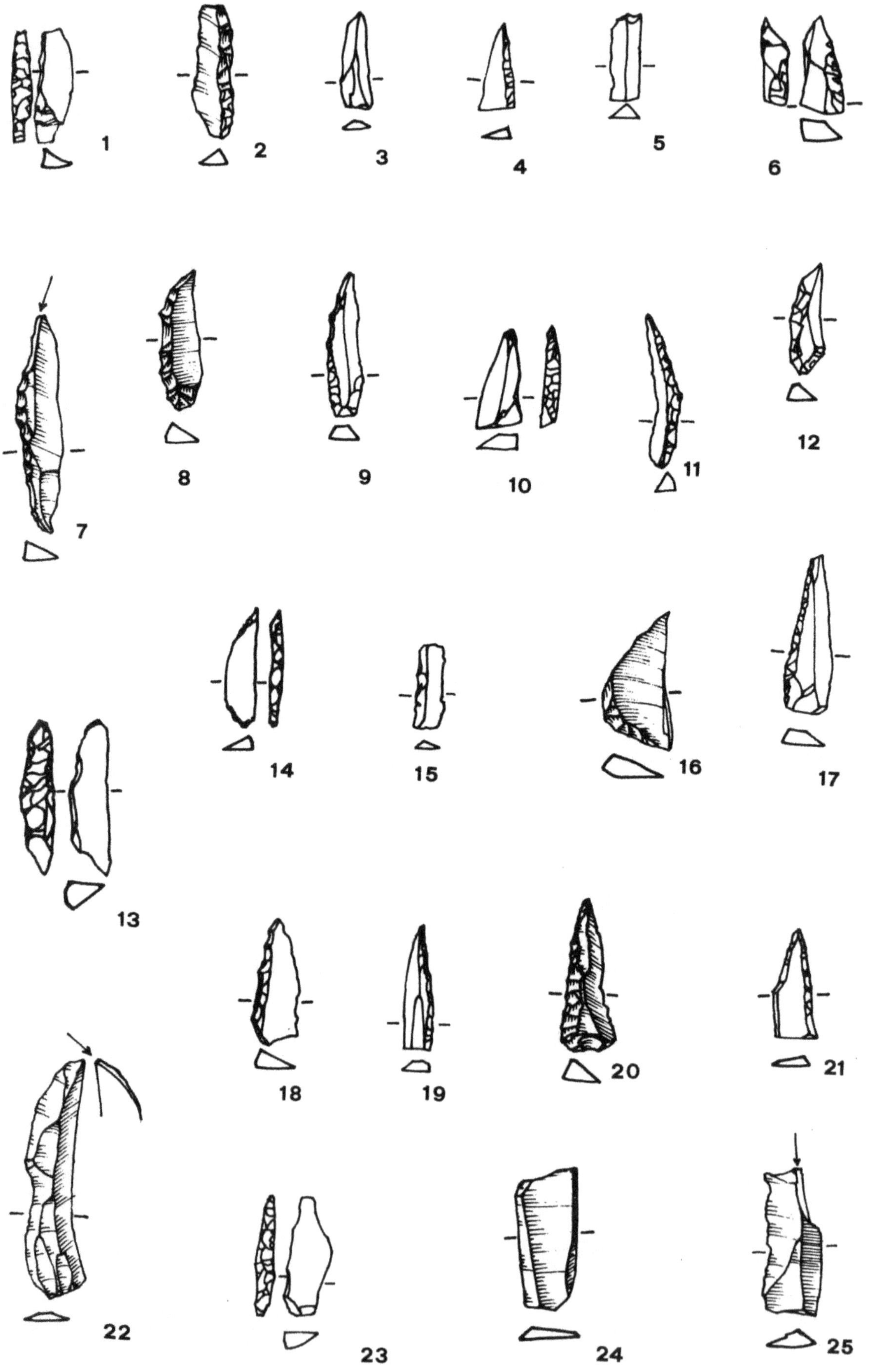

Fig. 2.10 - Two Caves, Wadi Ti-n-Torha. Level IIa, Central Sector. Burins, unretouched blade, microbladelets, backed bladelets, segment (1:).

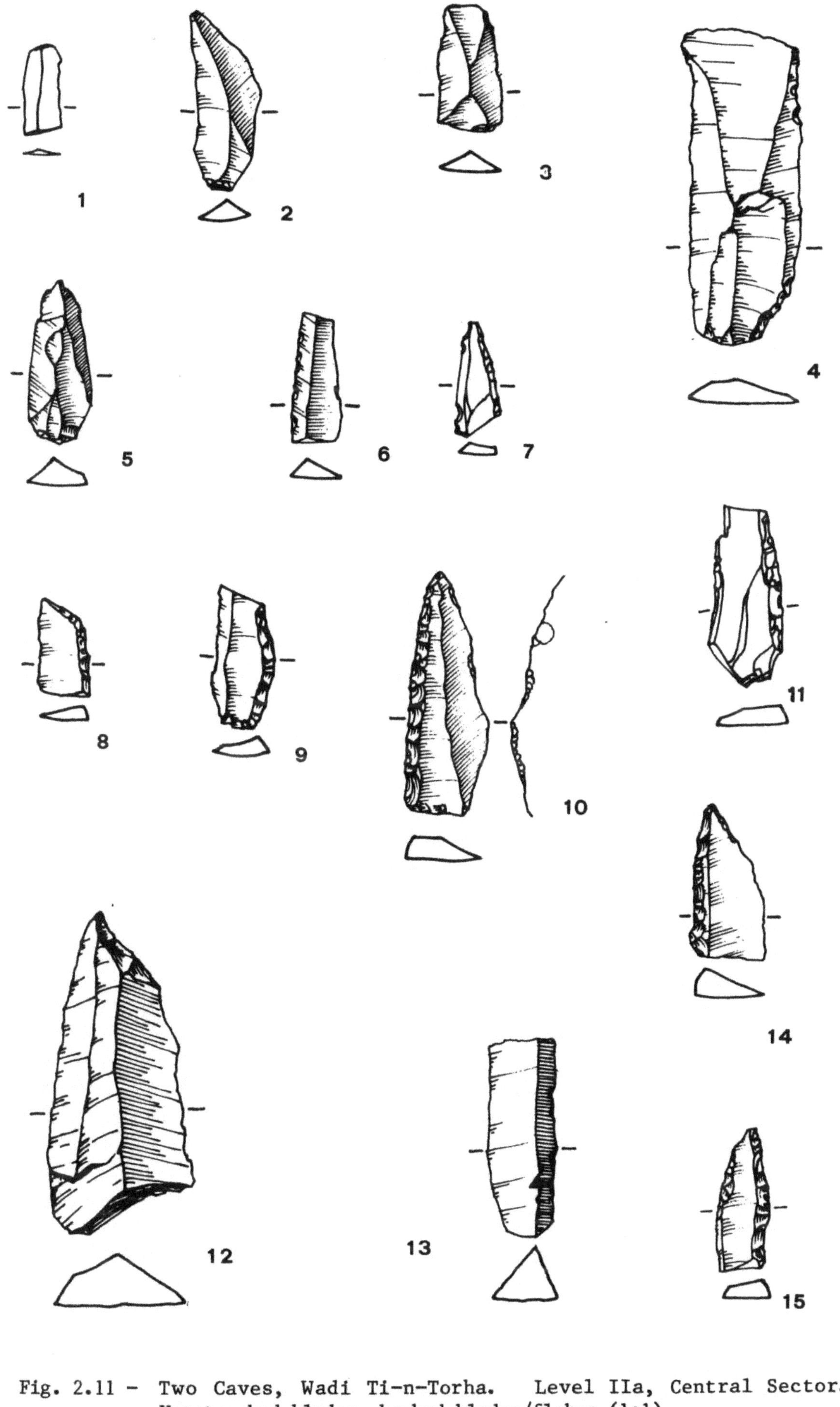

Fig. 2.11 - Two Caves, Wadi Ti-n-Torha. Level IIa, Central Sector.
Unretouched blades, backed blades/flakes (1:1).

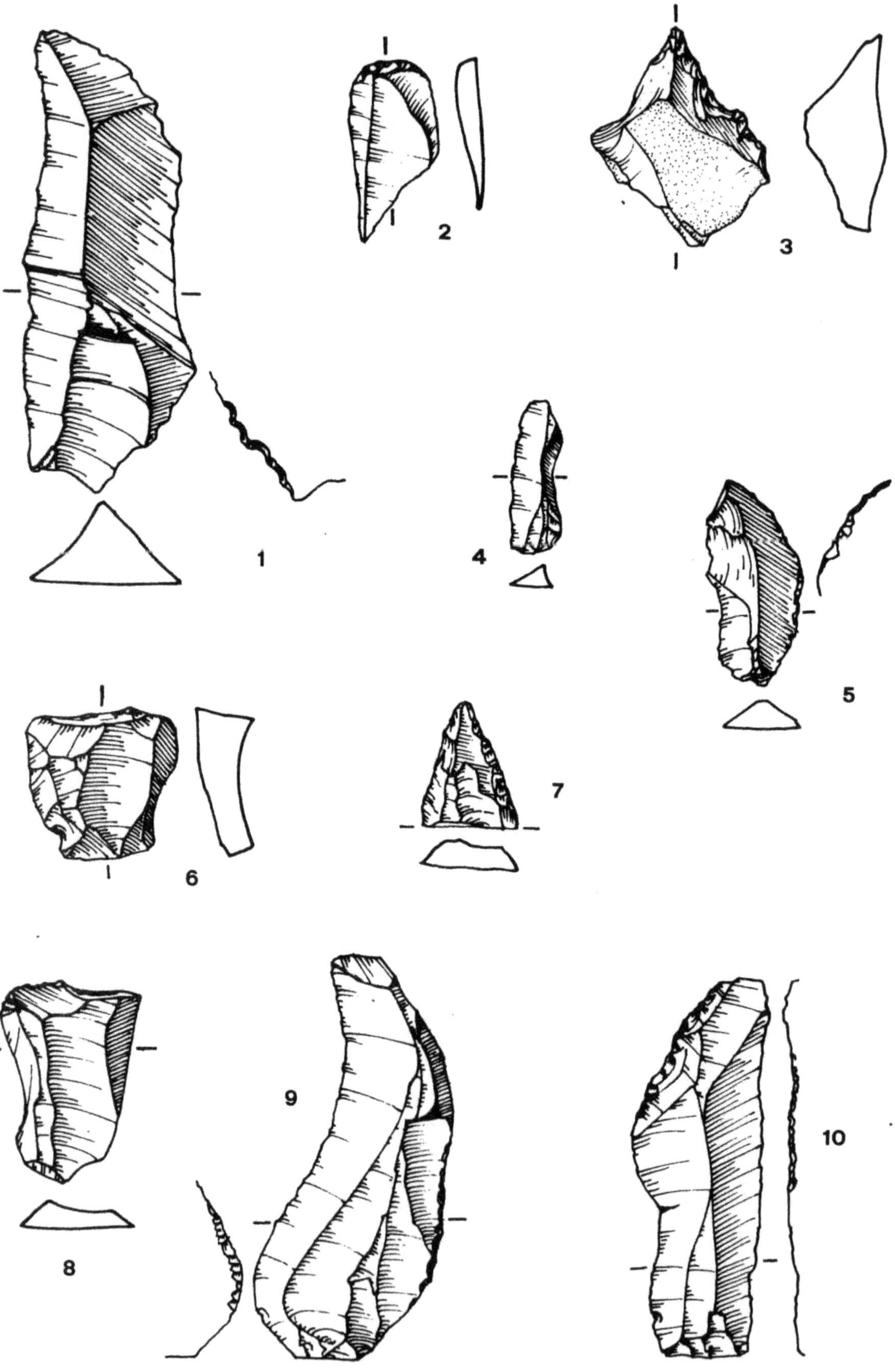

Fig. 2.12 - Two Caves, Wadi Ti-n-Torha. Level IIa, Central Sector. End-scraper, side-scrapers, denticulates (1:1).

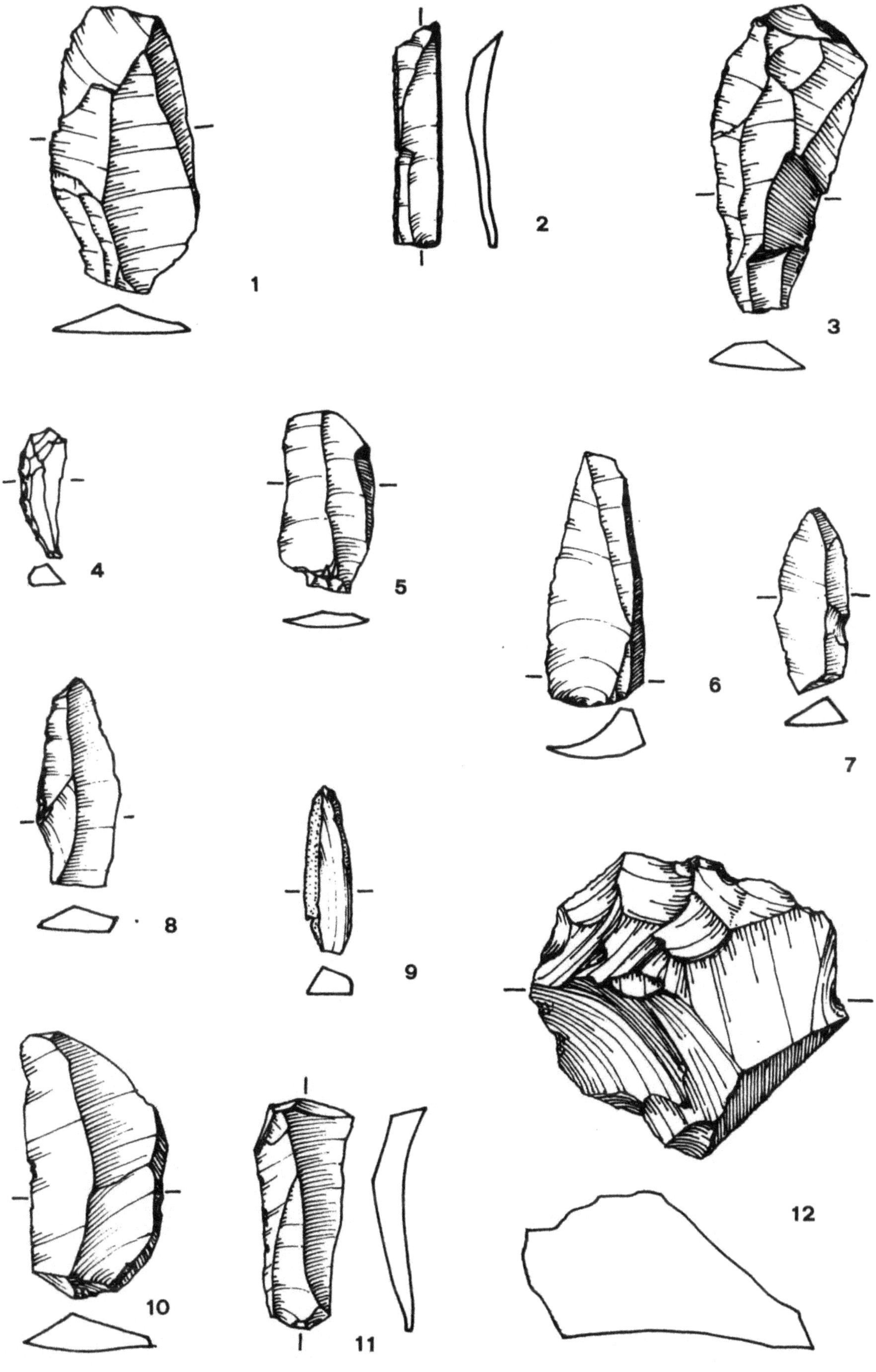

Fig. 2.13 - Two Caves, Wadi Ti-n-Torha. Level II. Western, Central and Eastern Sectors. Debitage and core (1:1).

Blunt unretouched blades from single platform cores were mainly tertiary or secondary, only a few specimens were primary. Pieces from opposed platform cores were also quite frequent. They had significantly shorter shapes with straight or slightly convex sides. Cross-sections were triangular and, in rare cases, trapezoidal, often showing rippled ventral surfaces with a slightly prominent bulb. Microbladelet-scars struck from the ventral surface at the point of its greatest thickness were occasionally detected (figs. 2. 14, 2; 15, 4; 18, 6). One specimen (Level II) was tanged. Another one (from Level I) was a lame outrepassée as defined by Tixier (fig. 2. 16, 6). A few pieces from different levels had poor use-retouch and the percentage of broken pieces was very high (50% *ca*). As for their dimensions, lengths ranged from 45 to 60 mm; widths from 20 to 25 mm; thicknesses were about 3 mm (measured over the entire pieces).

Within the blade group, the blunt unretouched class was the most common (32.4% of the whole lithic complex) and their incidence increased from the bottom towards the top of the stratigraphy.

(b) Microbladelets: 123; general index 16.9%.

Figs. 2. 10, 3, 5, 15; 11, 1, 8, 14, 16.

Following Tixier, conventional dimensions were established for microbladelets: mean length < 25 mm; mean width < 5 mm *ca*. All the specimens fell within these values. For the most part, they were unbroken pieces, always unretouched, with very regular shapes and ventral surfaces. For the most part they were made from quartzitic sandstone, but white quartz and chert were also well represented.

They were the second most frequent class within the blade group, with a general index of 16.9%. They decreased in number from Level III to Level I.

(c) Pointed unretouched: 94; general index 12.9%.

Figs. 2. 11, 2, 5; 14, 5; 17, 5, 6, 15, 16.

Tertiary or secondary blades, naturally pointed at distal ends, were the fourth commonest class within the blade group (and the lithic complex itself). They were generally unretouched, but some marginally (Ouchtata-type and, in rare cases, abruptly) retouched specimens were also detected (fig. 2. 17, 6, 16). The inner articulation of this class was shown in the section devoted to the typological definition of the Ti-n-Torha East complex. The number of items in that deposit allowed six types (or varieties) to be differentiated (Barich 1974: 118-20). Even though the sample from Two Caves was more limited, on the whole it was consistent with the observations that had already been made. In particular, the most common type consisted of unretouched blades from single platform cores with pointed oblique distal ends. The second most common group showed a Ouchtata-type retouch, together with some tanged pieces (that were not reported at Ti-n-Torha East). Nine tanged pieces were collected from Level I, four of which with Ouchtata-like retouches. The mean dimensions measured on the entire pieces were: length 35 mm; width 13 mm; thickness 4 mm.

The class is not found in Level III; twenty-two specimens came from Level II and the rest from Level I.

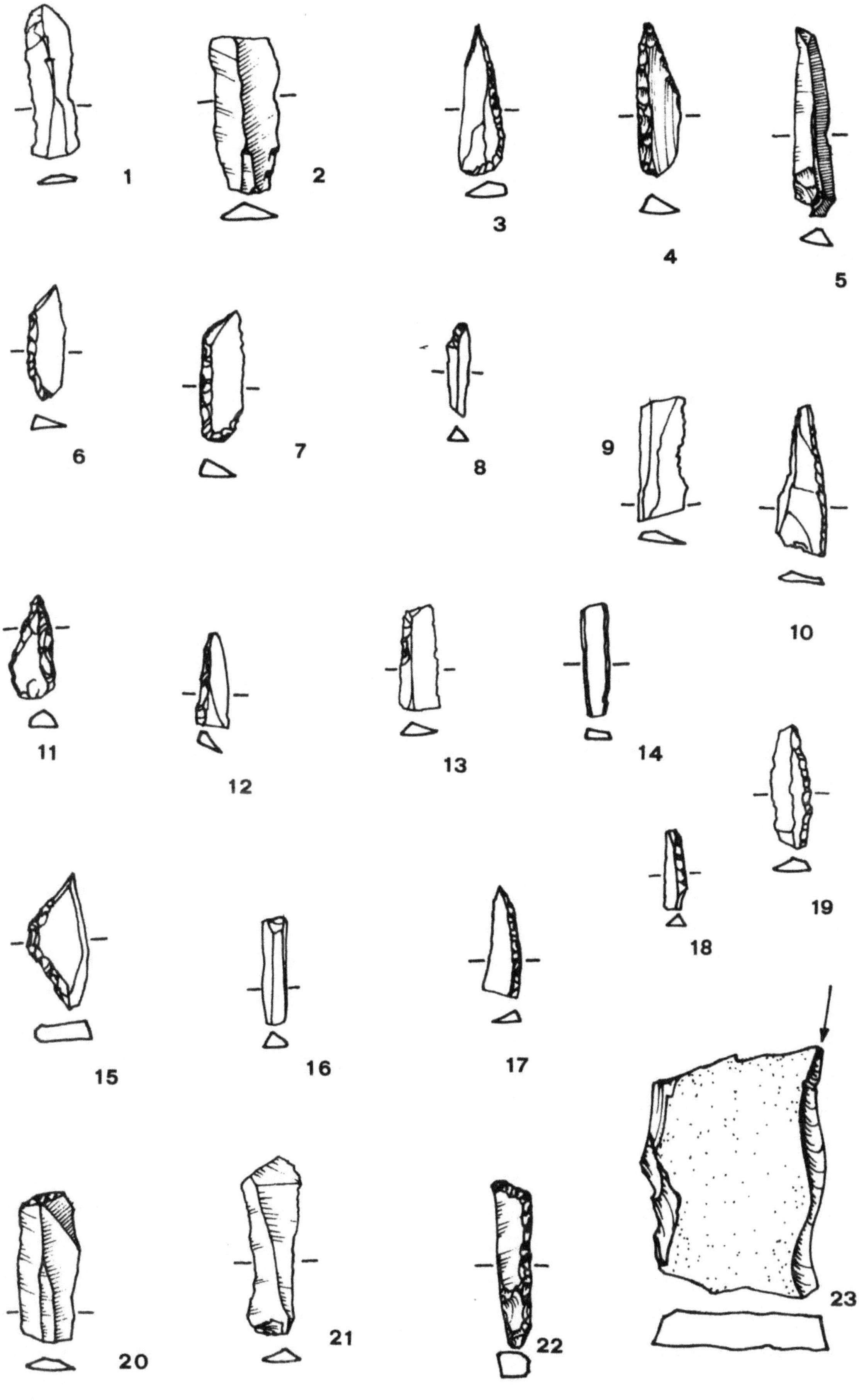

Fig. 2.14 - Two Caves, Wadi Ti-n-Torha. Level II. Western, Central and Eastern Sectors. Burin, unretouched blades, microbladelets, backed bladelets (1:1).

(d) Pointed backed bladelets: 110; general index 15.1%.

Figs. 2. 8, 1, 4, 7, 8, 10; 10, 1, 2, 4, 6, 8, 10-14, 18, 19, 23; 14, 3, 4, 6, 7, 12, 17, 18; 17, 1-4, 7, 9-13.

As previously pointed out, the distinction between blades and bladelets seems to apply to the pieces included in this class, having, on average, the following dimensions: length ⩽ 25 mm; width ⩽ 5 mm. A microlithic variety was also present. In the Ti-n-Torha East collection, six major types (or varieties) were differentiated. They were all comparable with the North African bladelet assemblage described by Tixier and, particularly, with those related to the Iberomaurusian technological tradition dating from the 10th to the 8th millennium B.P.: Mellalian, Hadjarian, Kristelian-Keremian, Columnatian-Elassolithic (Camps 1974a: 195-216). The Two Caves collection is not as rich as the Torha East complex and some of the previously established types were not reported.

Straight backed pointed bladelets were the first common type (figs. 2. 10, 2, 14, 17; 14, 4; 17, 1, 7) but specimens with modified-rounded (figs. 2. 8, 4; 14, 3; 17, 2, 4, 12) or simply retouched, base (fig. 2. 17, 11) were also reported. There were a few examples of arch-ended (figs. 2. 8, 1; 10, 6) or arch-backed bladelets (figs. 2. 10, 11, 19; 14, 17; 17, 3). Special relevance should be ascribed to the presence of five La Mouillah points showing microburins scars (figs. 2. 8, 8; 10, 12; 14, 6, 7). Two strangulated (Level II) and some hump-back (fig. 2. 8, 3) pieces were observed while a few specimens, with triangular cross-sections, obtained by _sur enclume_ backing, were also identified (fig. 2. 10, 1, 13, 23). Fragments of backed bladelets were very frequent representing 40% of the entire class. Simple obverse backing was the most common (45%), followed by _sur enclume_ (25% _ca_) and obverse-_sur enclume_ (15%) backings. Inverse backings seemed to be more rare. As for raw materials, backed bladelets were mostly obtained from quartzitic sandstone. A few specimens made on chert and, exceptionally, flint was also reported. In the various examples, lengths ranged from 25 to 35 mm; widths from 5 to 12 mm; and thicknesses from 2 to 3 mm.

Backed bladelets represented the third most important class by incidence; well documented from the very base of the settlement, their number rapidly increased towards the upper part of the series.

(e) Blunt backed blades/flakes: 36; general index 4.9%.

Figs. 2. 8, 5, 9-12; 11, 8-10, 14, 15; 14, 22; 15, 1, 2, 9; 18, 3, 4, 7-13.

This class includes secondary and tertiary blades, together with pieces that do not correspond strictly to the Tixier's conventional definition, even though they have a strong laminar shape. For the most part, they were obtained from single platform cores but examples from opposed and from multiple platform cores were also observed. The backing was the result of the classical obverse backing-abrupt retouch, struck from ventral to the dorsal side. An additional sinister distal and central denticulated retouch characterized one of the specimens (Level I: fig. 2. 18, 2). The mean class dimensions were: length 32 mm; width 14 mm; thickness 6 mm.

General index for the class was rather low and it showed a decrease

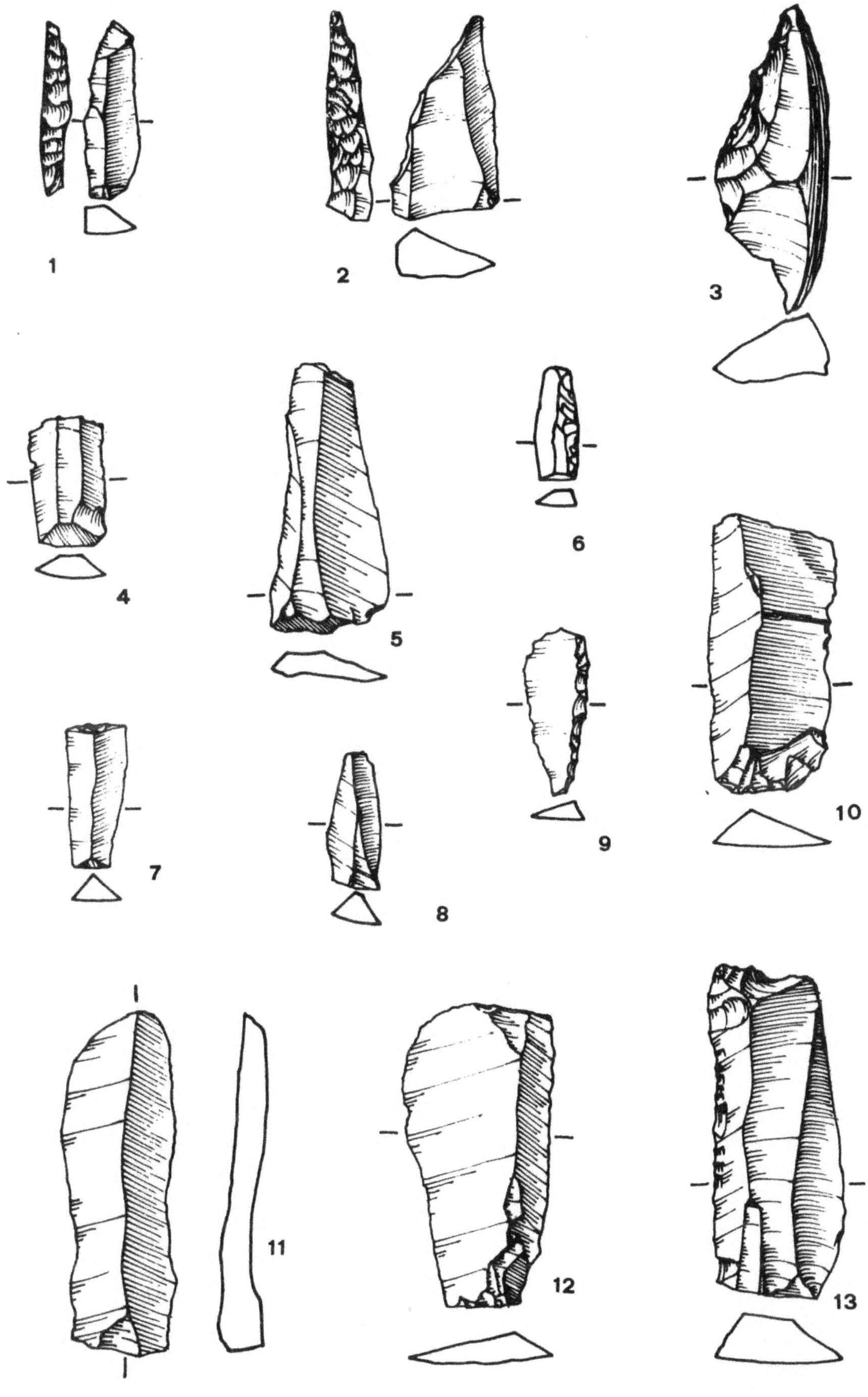

Fig. 2.15 - Two Caves, Wadi Ti-n-Torha. Level II. Western, Central and Eastern Sectors. End-scraper, unretouched blades, backed blades/flakes (1:1).

moving from the bottom to the top of the sequence.

(f) Truncations: 3; general index 0.4%.

Figs. 2. 19, 4; 20, 3.

The low incidence of this class within the blade group had already been noticed in the Ti-n-Torha East collection (Barich 1974: 129). In the Two Caves assemblage, it is even less representative: only three oblique, distal, specimens with obverse retouch were reported from Level I. Lengths ranged from 40 to 58 mm; widths from 18 to 22 mm; thicknesses from 3 to 6 mm.

Geometric Microliths: 23; general index 3.2%.

Figs. 2. 8, 2, 6; 10, 16; 14, 15; 17, 8.

Together with pointed backed bladelets, geometric microliths represented the most significant component of this complex. The higher incidence of segments (or lunates) with respect to other classes (triangles, trapezes, rhomboids) had already been noticed at Ti-n-Torha East. The Two Caves assemblage consists exclusively of segments. From a technological point of view, only three examples (figs. 2. 10, 16; 17, 8) comply with the definitions of "segment" provided by Tixier and Roche (Tixier 1963: 129-31; Roche 1963: 36 *Texte*), that is entire removal of the blank bulb in order to obtain a convex backing - dexter or sinister - opposite the other straight sharp side. The remainder, by far the majority (twenty pieces) were merely primary diminutive flakes struck from small discoidal chert or quartz pebbles from which they derived their characteristic "orange segment" shape (cf. fig. 2. 8, 2). None of these examples showed a true backing-retouch. Their mean dimensions were as follows. Length 20 mm; width 10 mm; thickness 3 mm.

The highest frequency of the class was reached in Level I: fifteen examples (thirteen of which were of the "orange segment" type).

Microburins: 7; general index 0.9%.

As showed by its percentage frequency, the microburin technique was not an important component of the technology. There were two tertiary distal specimens, a sinister and a dexter (from Levels III and II). The remainder were Krukowski microburins made from tertiary quartzitic sandstone with obverse backing. They were very well made with a mean width of 9 mm *ca.* The low frequency of microburins is matched by the absence of geometrics, such as trapezes and rhomboids, and by a very limited number of truncations. In the whole collection, the La Mouillah points were the only artifacts to be related to the microburin technique.

Denticulates: 52; general index 7.1%.

Figs. 2. 11, 12; 12, 3, 4, 8; 15, 3, 13; 18, 15; 19, 2, 5-7, 11, 12.

Only twelve examples were true notches on secondary and tertiary blades with obverse "single blow" or denticulated retouch (figs. 2. 15, 13; 19, 7, 11, 12). Four were sinister, eight were dexter and they all came from Level II and I. Pieces with denticulated retouch were the most common type: twenty-eight examples were on secondary and tertiary blades, while the rest (twelve pieces) were on flakes. Two specimens from Level II had been subjected to poor additional retouch to obtain distal end-scrapers. Another specimen from the same level consisted of a point with denticulated retouch. As it was observed in the collection from Torha East, denticulated tools were quite a significant component of this technology. The average size of denticulates were: length 50 mm; width 33 mm; thickness 6 mm.

Denticulates, which were collected from the bottom of the deposit, reached their highest frequency in Level I (8.5%).

Side-Scrapers: 17; general index 2.3%.

Figs. 2. 11, 4, 11; 12, 1, 5, 9, 10; 18, 1, 2; 19, 1, 3, 6, 8, 10.

Three pieces on tertiary blades were collected from Level I, while the remainder were on flakes, either tertiary or secondary. The various specimens lacked a precise typology and only a few examples showed a more accurate, either unilateral or bilateral scraping-retouch. They basically were flake tools with a partial invasive retouch along the edges. Their mean dimensions were: length 55 mm; width 25 mm; thickness 4 mm.

The percentage frequency of this class showed a steady increase from the bottom to the upper levels of the sequence.

Polished Stone Tools

Table 2.5 shows the extreme scarceness of polished implements. The six types that have been detected had already been found at Ti-n-Torha East where they were far commoner. The two circular hammerstones from Level II were made on limestone and had a diameter of approximately 70 mm. In the light of the brittleness of the raw material, they had most likely an ornamental function. The limestone and steatite tablets had a flattened trapezoidal shape (fig. 2. 21, 8). The single small ring made of limestone and the two small balls were the product of a natural polishing that man completed. The same can be said of the smoothing implements which retain the shape of pebbles from which they were obtained. The eight grinding-stones of sandstone might be classified as handstones, circular or sub-circular in shape. Their mean diameter was 85 mm *ca*, while their thickness ranged between 23 and 45 mm. Three examples were slightly flat on the grinding-faces, two were planoconvex while the remaining three were biconvex. These implements might be related to wild-grain milling. One should not overlook the presence of a number of yellow and red ochre fragments (twenty-eight) collected from Level III to Level I. Their presence means that the grinding-stones could have been used also for the preparation of the colouring substances used in art work. Even the limestone tablets, so brittle that their utilization as tool is quite problematic, might have been used in pictorial work.

Table 2.5 - Two Caves, Wadi Ti-n-Torha. Polished stone tools.

	I	II	III/Base	Total
Hammerstones	-	2	-	2
Smoothing Implements	4	-	-	4
Tablets	2	1	3	6
Rings and Small Balls	1	2	-	3
Grinding-Stones	7	-	1	8

Shell and Bone Industry

Ostrich eggshell bead production appeared to be carried on to a significant extent for eighty ostrich eggshell beads were collected from the different levels; unworked pieces had only a slightly higher total (eighty-two). The beads were found in practically all the stages of manufacture. Many fragments had been drilled but not rounded and many were burnt. The incidence of ostrich eggshell increased approaching Level I.

Polished bone implements are only represented by two specimens, both from Level I. One was a beautiful and accurately made spatula with curvilinear profile and serrated edges (dimensions: length 116 mm; width 19 mm; thickness 5 mm) (fig. 2. 21 3). It proves that, notwithstanding the few finds, bone working had reached a good technological level. The serrations along the edges suggest a utilization of the spatula in ceramic decoration but, as it will be later on specified, the deposit failed to provide much evidence of ceramics, especially of ceramics decorated with impressions in the Dotted Wavy Line style for which this spatula might have been designed. The only fragments which might have been decorated by it were three sherds with comb-type decorations detected in Level I of the Test Trench (see below).

Ceramics

The presence of ceramic sherds is a problem on account of their stratigraphical position and their technological and typological traits. With reference to the first point (see Table 2.7), they were missing in the Levels III and II and detected in Level I only in the middle-western section of the excavated area. A fragment without decoration was collected in the Western Sector and six fragments, only two of which are decorated, were collected in the Central Sector. It should be pointed out, however, that all six came from the depression, with marked infiltrations of soil and sand, resulting from the collapse of a large boulder. This being the case, the potsherds might have entered the deposit after the collapse. Eight more fragments, five with decoration, were collected from the surface.

The only well stratified specimens were the undecorated and, therefore,

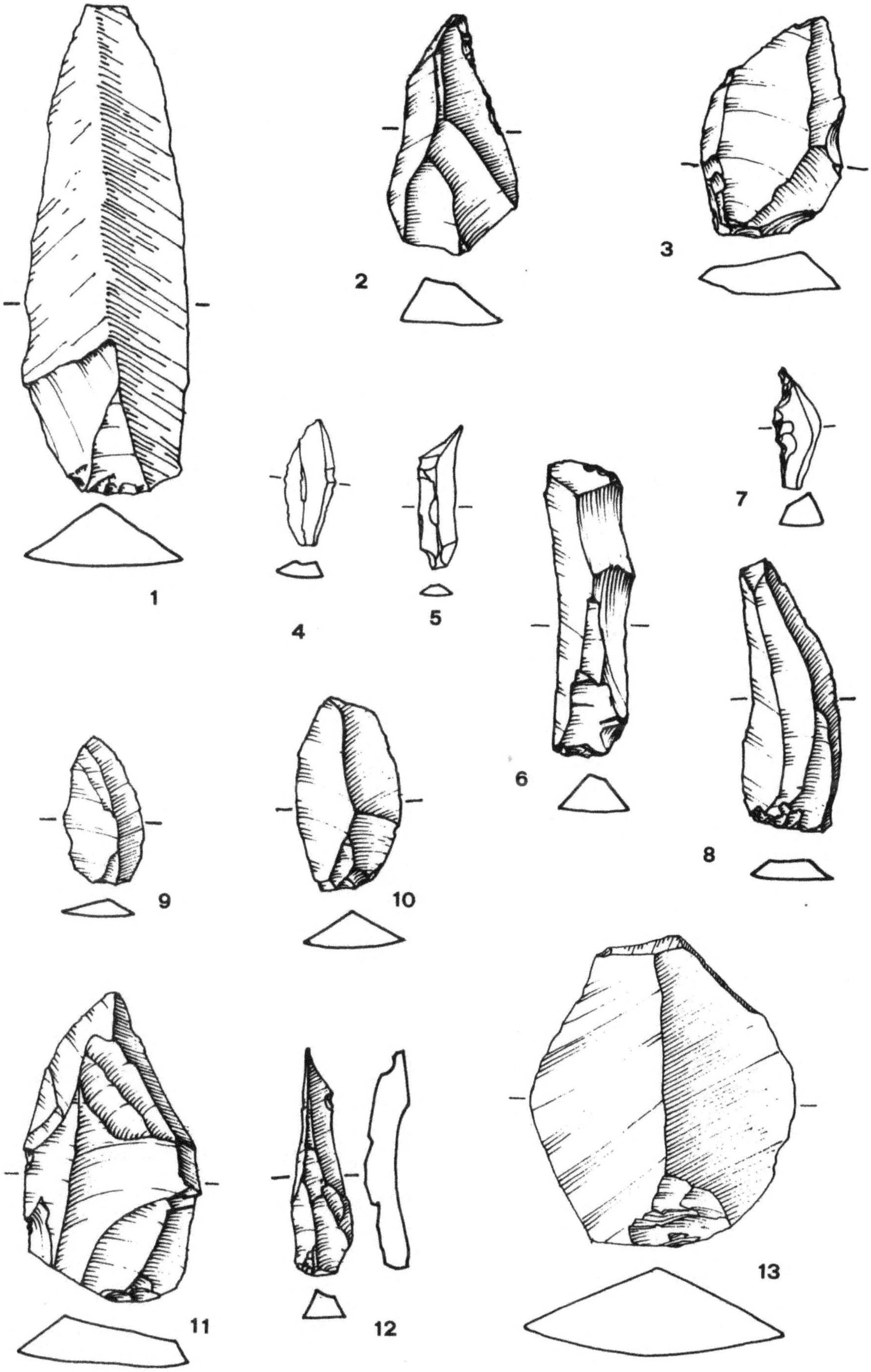

1 2 3 4 5 6 7 8 9 10 11 12 13

Fig. 2.16 - Two Caves, Wadi Ti-n-Torha. Level I, Western and Central Sectors. Debitage (1:1).

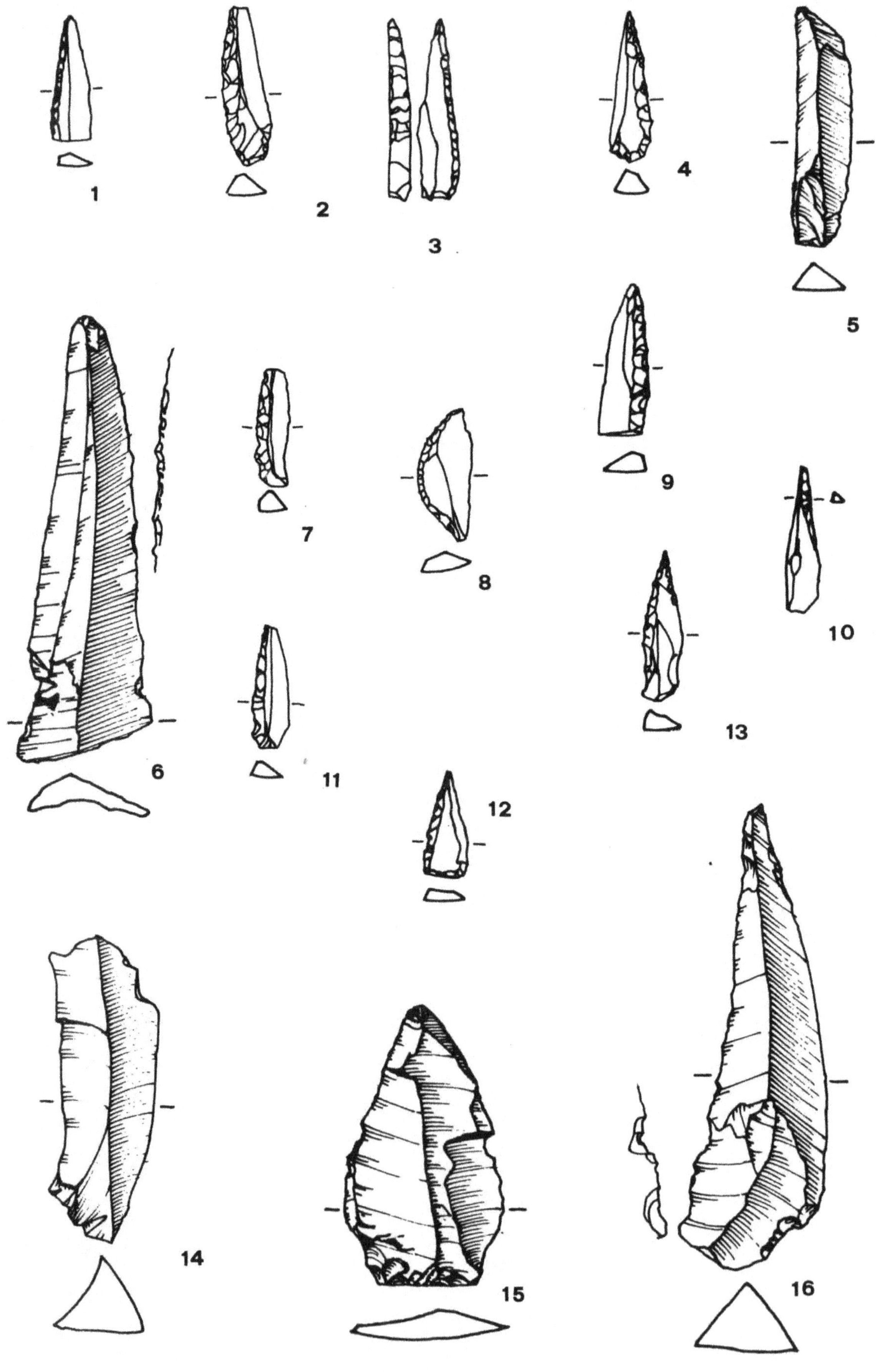

Fig. 2.17 - Two Caves, Wadi Ti-n-Torha. Level I, Western and Central Sectors. Blunt and pointed unretouched blades, backed bladelets, segment (1:1).

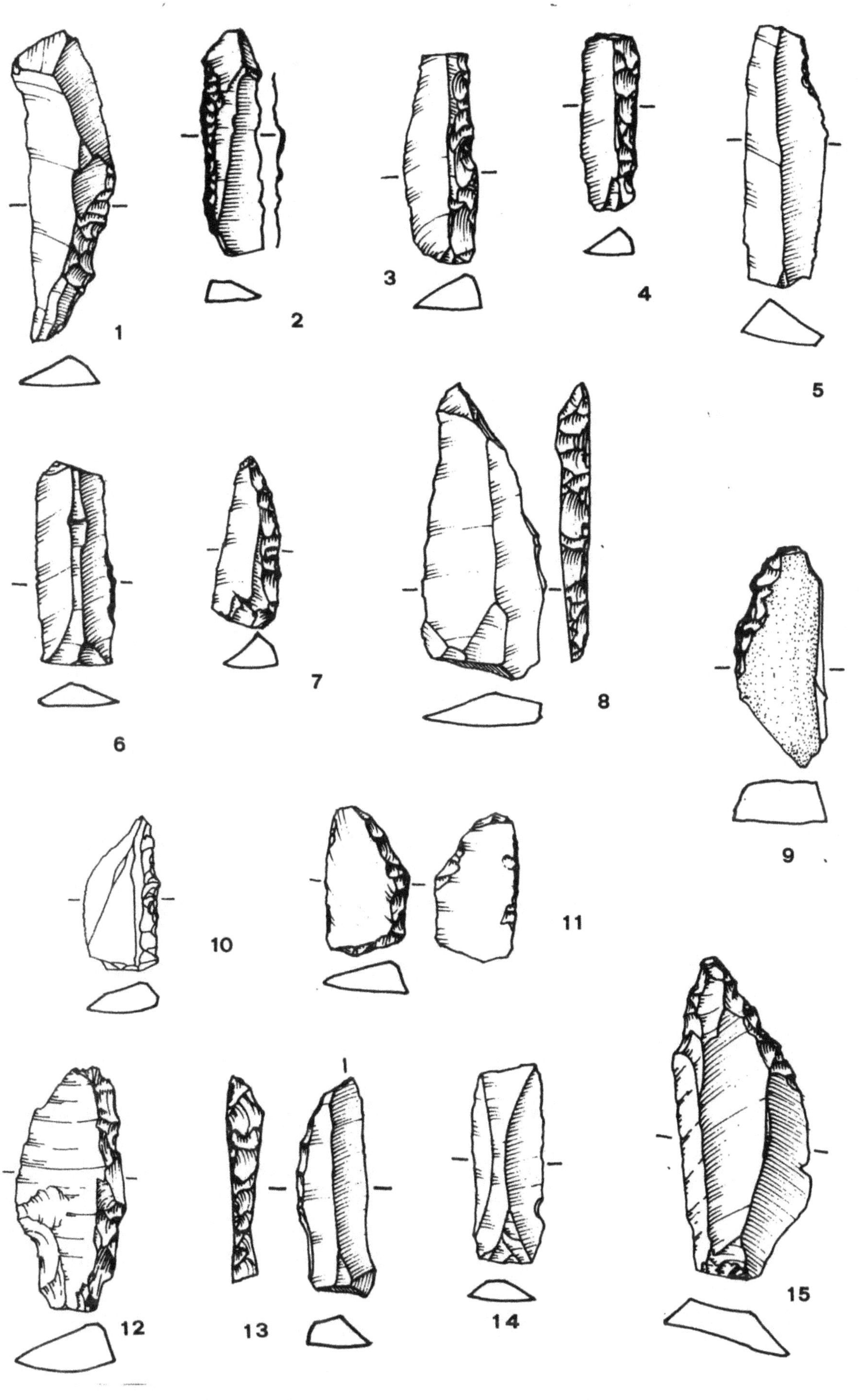

Fig. 2.18 - Two Caves, Wadi Ti-n-Torha. Level I, Western and Central Sectors. Blunt unretouched blades, backed blades/flakes (1:1).

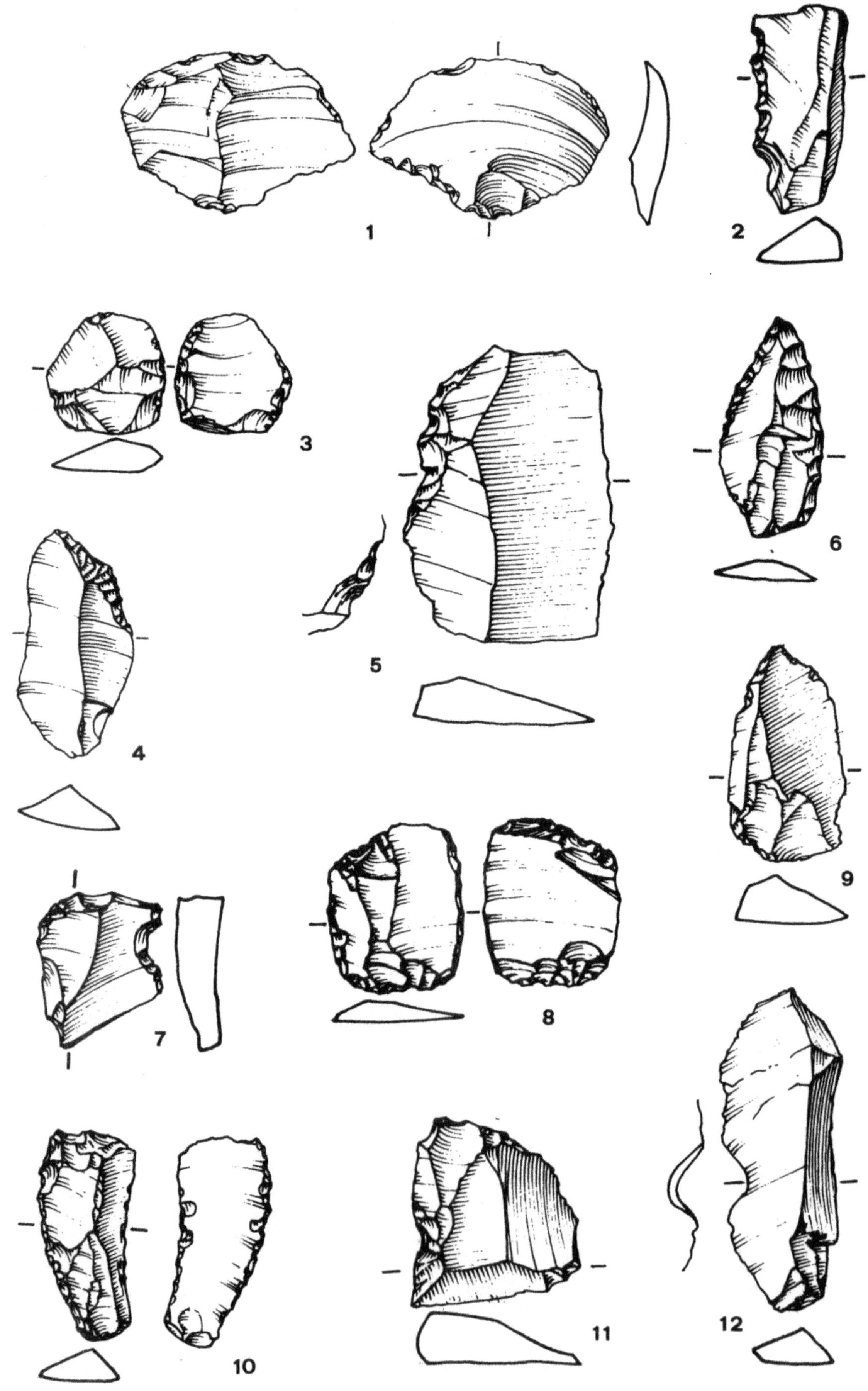

Fig. 2.19 - Two Caves, Wadi Ti-n-Torha. Level I, Western and Central Sectors. Truncations, side-scrapers, denticulates (1:1).

Fig. 2.20 - Two Caves, Wadi Ti-n-Torha. Level I, Central Sector (1-3): truncation, end-scraper, steatite tablet. Test Trench (4-10): debitage, backed bladelets, side-scraper (1:1).

poorly indicative sherds from Level I. Stratigraphically they coincided with four fragments collected in the Test Trench (see Table 2.6). The undecorated fragments have an extremely even and glossy surface, the paste is compact and the walls very thin. The decorative patterns are not at all in keeping with the "dotted wavy line" style, except for one fragment with punctuations in parallel lines (Central Sector: fig. 2. 21, 2). The other fragments come from the upper part of small vessels, with extremely thin walls, where the decoration, arranged around the rim, was obtained through the impressions of a quoin (Central Sector: fig. 2. 21, 5; Surface: fig. 2. 21, 7), and pivoting stamps (Surface: fig. 2. 21, 6). Three fragments were decorated by means of punch impressions (Central Sector: fig. 2. 21, 1), curvilinear or cross patterned incisions (Surface: fig. 2. 21, 4).

The technological characteristics of this small ceramic group conflict, both in the type of paste and the nature of the decorations with the antiquity of the deposit which, on the whole, corresponds to the earliest levels of Ti-n-Torha East.

TEST TRENCH

The small collection coming from this area comprises lithic, bone and ceramic materials, as shown in Table 2.6.

Table 2.6 - Two Caves, Wadi Ti-n-Torha. Artifacts from the Test Trench.

	Sub-Surface
DEBITAGE	
Flakes	3
Core-trimming el.	3
TOOLS	
Unretouched blades	2
Blunt backed blades	1
Pointed backed blades	3
Microburins	1
Side-scrapers	1
BONE	
Punch	1
Spatula	1
Ostrich Eggshell	4
CERAMICS	4

The lithic materials appear to be the major component of this collection: six debitage products and eight retouched tools. Special relevance should be ascribed to the presence of a distal microburin, a tool-type which is otherwise rather poorly represented. The three specimens of backed, pointed bladelets belong to the same tradition. The first piece is a straight backed pointed bladelet with rounded base (fig. 2.20, 6), comparable with the specimens recovered from Ti-n-Torha East (Levels RSup-CII). The second one is a backed pointed microlithic bladelet (fig. 2.20 7), while the third one is a triangular point made on a small and rather wide flake with obverse unifacial invasive retouch (fig. 2.20, 5). Even though it was not reported in the major excavation of Two Caves, it is a type present in the Ti-n-Torha East collection (especially in Levels CII-CI).

Recovery of two bone pieces is important given the general scarcity of bone artifacts; the same is true of the four ceramic sherds. The three decorated sherds (one of which is rather large) have impressions made, most likely, with a comb and feature traits that are definitely more archaic than those of the sherds coming from the area of the major excavation. These examples belong, totally, to the tradition of the Ti-n-Torha East ceramics.

The materials collected from the Test Trench are significant if they are related to the limited extent of the excavation. It may be inferred that the deposit inside the caves is very rich and keeps its original stratigraphy unchanged. The most interesting element is the ceramics, found in a primary position, which matches the sherd recovered in situ from Level I of the Western Sector. This suggests that at Two Caves, ceramics were a later acquisition attributable to the influence exerted by the community settled in the nearby site of Ti-n-Torha East.

3. TWO CAVES SHELTER: ITS PLACE IN THE NORTHERN ACACUS SEQUENCE

The stratigraphy proved to cover the span of the 9th millennium B.P. The base level of the Eastern Sector, dated to around the middle of the 10th millennium, may be indicative of the start of the occupation and, this being so, the strata of the Western Sector belong to an evolved phase of occupation. The date range characterizing this latter sector, shows that the accumulation occurred in a relatively short time and this is in keeping with the observations made during excavation. Subsequent to the enclosure of area with a small transverse inner wall, the second phase of inhabitation remains were less disposed and the accumulation more rapid.

Sedimentation in the lower levels characterized by a sizeable sand component and the lack of human constructions, except for the outer boulder wall, suggests that the occupation originated in a relatively dry period. The additional stone wall with the consequent concentration of the living area in the western-central section of the shelter, may be assigned to the second phase of occupation and the concentration of artifacts observed in the Central Sector (Levels II-I) shows it to have been a working area. Although this area lacks the huts observed at Ti-n-Torha East, the increased use of it suggests the attainment of a greater stability by the group and a more permanent use of the site. A moister period in the final phase of occupation may have prompted the group to strengthen its defences and to take greater advantage of the shelter offered by the inner caves.

The palynological study (Schulz, Chapter 13, in this Volume) of the samples collected from RBase and RInf Levels of Torha East, chronologically comparable to Two Caves (10) shows an open countryside with Gramineae and Cyperus present and a low tree and shrub component. Even though the reconstruction of a climatic evolution in terms of precipitation and temperature is a difficult task, the abundance of herbaceous vegetation witnesses a moist environment. The study of the groundwater distribution (Marcolongo, Chapter 11 in this Volume) points to an availability of water nearby (Tahala, Tasbät) so that the groups were not compelled to move long-distances for it.

The tool complex and, even more, the faunal remains (Gautier, Chapter 12) suggest a fishing, fowling and hunting (which included Barbary sheep) economy. We are dealing with a model of environmental exploitation that allowed the development of a sedentary life style. Reduced mobility allowed intensive exploitation and the selection of species for domestication to be undertaken.

The Polished Stone tools in the Two Caves techno-complex do not allow comparison with other areas since they lack specific characteristics although steatite tablets, used perhaps in preparing colouring substances, are analogous to the Uan Muhuggiag specimens (Chapter 7). Generally, polished stone tools were few, even at Ti-n-Torha East (cf. Barich 1974: 129, Tab. III), with the exception of grinding-stones which were very common there (546 specimens compared to the eight at Two Caves). Such a difference may indicate a different usage of the two sites; Torha East could be viewed as a 'proto-village' where gathering activities could be efficiently performed (11).

A difference between these two sites shows also the use of bone and shell. At Ti-n-Torha East, there is a much more diversified range of bone types (cf. Barich 1974: 129, Tab. IV), while at Two Caves the extremely well manufactured spatula shows that bone was used but, perhaps, represented a rarer and more precious material. Ostrich eggshell products were relatively more numerous (162 specimens including unworked elements and beads) at Two Caves, in line with what generally appears in Early Holocene Epipaleolithic complexes (12).

The technological observations made in paragraph 2 about the ceramics should be supplemented by a thorough analysis of its distribution within the settlement. Table 2.7 shows that the rôle played by ceramics in the Two Caves site appears to be limited. In the analytical description of the sherds, reference was made to the questionable stratigraphical position of

---

10) Only the samples collected during the early campaigns have as yet been analysed. The samples collected from Two Caves during 1978, were only returned by the laboratory of the Institute of Ecology of Parma University in 1985 and will be published separately.

11) Those activities might have been favoured by the predominantly herbaceous plant cover. The presence of prototypes of plants that might have been subsequently domesticated may be revealed by K. Wasylikowa's study of the botanical samples from Two Caves.

12) We may refer to similar associations in the Capsian sites (Camps-Fabrer 1975) and in the Early Neolithic sites of the Egyptian Sahara (Wendorf, Schild 1980; Close ed. 1984).

Fig. 2.21 - Two Caves, Wadi Ti-n-Torha. Steatite tablet (8), bone spatula (3) and ceramic sherds (1:1).

the examples coming from the Central Sector which appeared to be a later intrusion within the context. Therefore, only the fragment coming from Level I of the Western Sector and the four sub-Surface sherds from the Test Trench, with a stratigraphic position corresponding to Level I, may be considered as an integral part of the context.

Table 2.7 - Two Caves, Wadi Ti-n-Torha. Distribution of ceramic sherds within the excavated area.

		SURFACE	LEVEL I		
			W	C	TEST
Undecorated		3	1	3	1
Decorated					
	Comb	-	-	1	3
	punch	1	-	2	-
	spatula	1	-	-	-
	composite decoration	3	-	-	-

We may conclude that this justifies the definition of Two Caves as an "a-ceramic" complex. The ceramics fail to integrate with the other components and appear to be a later addition, most likely from the nearby site of Ti-n-Torha East.

If this is true then the most relevant elements in defining the settlement complex are the broadly differentiated flaked products. In the early utilization of the site the stone products were exceedingly poor. The Eastern Sector, excavated to bedrock, shows an early and discontinuous occupation. This changed very rapidly to a rich production of diverse types. From Level III, the industry presented features that remained constant, although with quantitative changes throughout the later occupation. A decisive increase in the quantity of material may be observed in the final phase (Level I). This agrees with the stratification and structures which confirm a form of semi-permanent occupation.

From the typological point of view, the Two Caves stones assemblage confirms the typological model inferred from the Ti-n-Torha East site. As previously mentioned, it agrees in general with the Tixier type-list but, as far as the individual types are concerned, both here and at Torha East there is a certain degree of variability.

A study of the debitage allows the following remarks; a predominance of blade-like oval flakes, a few of which have a short tang, was observed within the flake-class (Levels III, II, I). The rare use of rejuvenation techniques already observed at Torha East is echoed in this site by the limited ratio of core-trimmings to cores (ratio 1:3). A relatively greater frequency is instead reported for lames à crête as well as crescent ("orange segment") flakes. Table 2.8 shows the absolute and percentage frequency of microlithic flakes (or chips) and the high incidence of this class within the cores/debitage group (44.7%) is well in keeping with the microlithic and ultra-microlithic character of the collection as a whole. As far as cores are concerned (with a general index equal to 6.4%), the most common is the single platform type, from which numerous blades were struck. Both, patterned and upatterned, multiple platform types are also quite numerous and all of them are small. A few end-scrapers on core and nucleiform burins were also present.

The incidence of blades remains extremely high throughout the three levels (fig. 5.3) with a general index of 82.6%. It should be recalled that, following the criteria used in the Ti-n-Torha East classification, tools include also those blunt blades for which the criterion of differentiation from flakes (i.e.: length/width ratio = 2:1) seemed to apply, giving a general blade index in that site of 90.7%. A slight increase in geometric microliths (3.2% vs. 2.2%) and burins (1.7% vs. 0.1%) is reported at Two Caves. While the microburin technique remains of marginal relevance (0.9%), the recovered specimens have a high standard of manufacture. Arrow-heads are absent here, while at Torha East they were quite frequent in Level CI.

The following technological and stylistic constants may be detected; blunt unretouched blades frequently present a natural and often oblique scraping-end and are nearly always characterized by a triangular cross-section with predominance of very regular ventral surfaces with prominent bulbs. We are not dealing with an actual preparation technique although the Levallois ancestry is recognizable.

The distinction made by Tixier (1963:39) between blades and bladelets seems to apply to the pieces included in the pointed backed bladelet-class, the average dimensions of which are: length ≤ 25 mm; width ≤ 5 mm. Straight-backed pointed bladelets (Tixier's type 45), even with modified base (type 46), represent the most common types within the class. They are also sided by a few specimens of arch-ended (type 55) or arch-backed bladelets (type 56). Special relevance should be ascribed to the five specimens of La Mouillah points, while a few specimens of small bladelets with a triangular cross-section obtained through sur enclume backing of the type already observed at Ti-n-Torha East (Levels RSup - CII) have also been recovered. The triangular points with straight base, that had been observed at Ti-n-Torha East (especially in Levels CII-CI), were not reported in the major excavation although one specimen was recovered from the Test Trench. Among the burins, once may note a predominance of simple angle specimens. The two distal microburins may be related to the presence of La Mouillah points, while the five Krukowski-type specimens are characterized by a high standard of manufacture.

Reference should also be made to the tool/debitage ratio. At Two Caves, 728 tools form 23.0% of the whole assemblage while 2433 debitage pieces form 77.0% (ratio 1:3). The percentage difference decreases from the base section towards the top of the sequence. There is a similar situation at Torha East, where 7685 tools form 22.2% of the whole assemblage. Fluctuations of little consequence are reported throughout the sequence in the tool/debitage ratio, except for Level I (CI) where there is sharp

Table 2.8 - Two Caves, Wadi Ti-n-Torha. Absolute and percentage frequency of cores/debitage-classes.

	I	%	II	%	III/Base	%	Total	General index
Flakes	538	54.5	390	42.3	210	40.0	1138	46.7
Primary flakes	95		88		51		234	(20.6)
Primary blade-like flakes	35		20		18		73	(6.5)
Flakes from single platform core	110		95		66		271	(23.8)
Blade-like flakes from single plat. core	60		22		25		107	(9.4)
Flakes from opposed platform core	62		45		-		107	(9.4)
Flakes from 90% core	85		78		40		203	(17.8)
Flakes from multiple platform core	43		15		10		68	(5.9)
Unidentifiable flakes	48		27		-		75	(6.6)
Core-trimming elements	27	2.8	10	1.1	15	2.9	52	2.1
Microlithic flakes (chips)	341	34.6	475	51.5	272	51.8	1088	44.7
Cores	80	8.1	47	5.1	28	5.3	155	6.4
Total	986	100.0	922	100.0	525	100.0	2433	99.9

Table 2.9 - Two Caves, Wadi Ti-n-Torha. Absolute and percentage frequency of tool-classes.

	I	%	II	%	III/Base	%	Total	General index
End-Scrapers	9	2.5	4	1.7	3	2.5	16	2.2
Burins	6	1.6	4	1.7	2	1.6	12	1.7
Blades	296	80.4	201	84.4	106	85.4	603	82.6
Blunt unretouched	125	34.0	78	32.7	34	27.4	237	32.4
Microbladelets	56	15.2	42	17.6	25	20.1	123	16.9
Pointed unretouched	62	16.9	22	9.2	10	8.1	94	12.9
Pointed backed	34	9.2	48	20.2	28	22.5	110	15.1
Blunt backed blades/flakes	16	4.3	11	4.7	9	7.3	36	4.9
Truncations	3	0.8	-	-	-	-	3	0.4
Geometric microliths	15	4.0	5	2.1	3	2.5	23	3.2
Microburins	-	-	5	2.1	2	1.6	7	0.9
Side-Scrapers	11	3.0	4	1.7	2	1.6	17	2.3
Denticulates	31	8.5	15	6.3	6	4.9	52	7.1
Arrow-heads	-	-	-	-	-	-	-	-
Total e	368	100.0	238	100.0	124	100.1	730	100.0

increase in debitage that, in the end, makes up 82.3% of the whole lithic assemblage.

As for raw materials, a slight but significant increase in the utilization of flint may be observed if Two Caves is compared with Torha East (we may refer once again to the tabular flint specimen from Level I of the Western Sector). Quartzitic sandstone continues to be used for tools that require greater manufacturing accuracy, especially with backed pieces. Moreover, one should point to the high frequency of small, ultramicrolithic-type quartz pebbles which, as it was specified, were not included in the computation. These tiny pebbles (already observed at Torha East) are but impurities present in the rock walls but in certain cases might have been used to make some extremely small backed bladelets. It should be recalled that, at the time of the earliest habitation even the schists of bedrock were reportedly being utilized.

Notwithstanding the similar technological tradition of both Two Caves and Torha East, revealed by the cores and debitage, differences in specific tool-types allow us to refer Two Caves to an ancient phase of the Ti-n-Torha facies. This, together with the discrepancies in the frequency of grinding-implements, the types of bone tools and the nearly complete absence of ceramics, illustrates the difference between Two Caves and Torha East.

Whether the differentiation between the two sites should be interpreted as a normal variability between settlements belonging to the same tradition, chronologically close but with a different utilization, or whether this differentiation has a deeper meaning needs discussion. A.E. Close (Chapter 3 in this Volume) acknowledges at Two Caves the presence of more archaic lithic elements (in particular, backed bladelets, straight backed and pointed, often with modified base; microburin technique known but not common). The author compares the assemblage with the El Adam-type Early Neolithic in the Egyptian Sahara and she definitely sees Two Caves preceding Torha East. It represents a settlement from the time of the earliest occupation of the entire Ti-n-Torha area. It was a settlement with limited importance compared to Torha East, where a structure-building phase is more distinctly recognizable and better developed. The use of the habitation area to create huts, ascribes the dimension of a real village to Torha East. It is a village with its own inner development documented through progressive industrial change, particularly through ceramics (Barich 1974: 130-141). Two Caves, instead, shows a single period of occupation, and was rapidly deserted perhaps for a more favourable nearby location. It might represent the a-ceramic phase of the wadi occupation already suggested at Torha East, in Level RInf (Barich 1974:149). The absence of ceramics from Levels III-II, and its very limited role in Level I, poses a number of questions since the chronological series of the two settlements are contiguous and their respective values, spread over the 9th millennium B.P., overlap. How can we explain such a limited interest in ceramics by Two Caves dwellers when it is unquestionable that pottery had been plentiful in the same area since 8640 $\pm$ 70 ? (13). Whatever hypothesis might be suggested (i.e. alternative sites intended for different activities, with

13) Sample R-1035 α : Ti-n-Torha East, RInf.

Two Caves acting mainly as a workshop), the ways and means that led to the acquisition of ceramics by the groups need to be studied.

The lithic characteristics point to Epipaleolithic industries in different locations of North Africa (Maghreb, Southern Tunisia, Lower Sahara) that may be comparable in respect of some tools and have no knowledge of ceramics. A.E. Close, establishing a specific similarity with the El Adam-type Early Neolithic (in its turn, linked with the Nubian Shamarkian and the Arkinian), fails to provide an explanation for the absence at Two Caves of ceramics and domestic cattle that, although in limited quantities, were present elsewhere in the 9th millennium B.P. (Close ed. 1984: 7).

This dilemma also involves the difficulty of ascribing an origin to the domestication and other cultural phenomena. The diffusion theory appears to be increasingly unable to account for the different components. For instance, in the Acacus, having ascertained a number of affinities in the lithic technology, attempts to explain directions taken by diffusion processes fail to reconcile the spatial, chronological, economic factors. A survey of a lithic assemblage (carried out with statistical method and taking into account stylistic traits) should define the source where a number of related aspects were able to co-exist. Assuming a broad range of communities sharing a common, Late Paleolithic, substratum, one may explain the appearance of contemporary but different local details. The acquisition of ceramic technology may be accounted for in this way if detached from compulsory lines of descent, but exposed to different levels of interaction. If we accept the relationship of the Acacus to the Egyptian Sahara, as belonging to a common cultural background, the acquisition of ceramic technology emerges as a separate phenomenon that might have been independently invented or acquired by interactions with other regions; for example, the affinities with the Air-Ténéré area (cf. Roset 1983) (14).

In conclusion, if the cultural complex of Two Caves is an archaic aspect of the Ti-n-Torha facies, one should not overlook its relationship with the Torha East site. I am unwilling to consider Two Caves and Torha East as distinct cultural and chronological entities. In my opinion, they should be viewed as parallel manifestations of a common cultural context.

---

14) A more general level of reflection to be reached in the Conclusions.

CHAPTER 3

THE LITHIC SEQUENCE FROM THE WADI TI-N-TORHA (TADRART ACACUS)

By Angela E. CLOSE*

1. INTRODUCTION

Archaeological research in the Wadi Ti-n-Torha involved the excavation of three stratified rock-shelters: Ti-n-Torha Two Caves, Ti-n-Torha East and Ti-n-Torha North. Together, they provide a cultural sequence extending from the late tenth millennium B.P., thus covering almost all of the earlier Holocene occupation of the Sahara. The importance of this sequence for the Sahara as a whole is increased by its central location, and by the ever growing numbers of comparative data available from surrounding regions. The flaked stone artefacts from the excavations at Ti-n-Torha have already been published elsewhere (Barich 1974, 1978, this Volume Chapter 2). What follows, then, is not a full description of them, but rather an assessment of them, of their place in the later prehistory of North Africa and of the light they may shed upon the processes of colonisation and development in the Early Holocene Sahara Desert.

I was able to examine the collections from Wadi Ti-n-Torha which were temporarily housed for study in the Institute of Paleoethnology of the University of Rome. These were not the total collections, most of which are in Libya and were not available to me. The collections in Rome included almost no debitage or cores from any site, so that few, if any, technological observations can be added to those already published. There were also in Rome rather less than half of the retouched tools from Two Caves, probably about a fifth of those from Torha East and about a third of those from Torha North. These retouched tools from Wadi Ti-n-Torha were selected to form representative samples, therefore they cannot be considered random samples of the complete collections. However, the copious illustrations of the artefacts and comparisons of the complete published type-lists (Barich 1974: Tab. I and VI; 1978: Tab II) with those given below suggest that the examined portions are at least qualitative representative.

*Department of Anthropology, Southern Methodist University, Dallas, Texas 75275, U.S.A.

2. DESCRIPTION OF THE COLLECTIONS

Ti-n-Torha Two Caves

A total of 106 retouched tools were examined from Two Caves. The types (Tixier 1963) are given below in Table 3.1 and the tool-classes are summarised in Table 3.2. Of these tools, 25 are from the Eastern Sector (15 from Layer 2 and ten from Layer 3), 54 are from the Central Sector (22 from Layer 2, five from Layer 3 and 27 from the Base) and 27 are from the Western Sector (20 from Layer 1, six from Layer 2 and one from Layer 3). The stratigraphy and radiocarbon dates suggest that the later occupations of the site tended to be in the Western Sector. However, the dates from all sectors overlap considerably, with a maximal range of 9350 B.P. + 110 (R-1402) to 8400 B.P. $\pm$ 90 (R-1404 α), and there are almost no noticeable differences between layers or sectors in typology. In view of the small number of tools available, therefore, all of the examined material from Two Caves will be treated as a single unit.

End-scrapers are not common. They tend to be made on rather large tertiary flakes and, with the exception of the denticulated end-scraper, are very poorly made. The single double-backed perforator is atypical, and the collection examined included no burins.

Backed bladelets are the most important tool-class at Two Caves. They tend to be small, elongated and very finely made (fig. 3.1, 1-6, 9). Their mean length, width and thickness are 27.1 mm, 7.5 mm and 3.1 mm, respectively; the mean length:width ratio is 3.3. Two examples are noticeably wider and thicker than the remainder of the group, and could be considered blades rather than bladelets. Most of the backing is simple obverse backing (34 pieces), but *sur enclume* (20) and obverse-*sur enclume* (eight) backing are also common; inverse backing (one) and Ouchtata retouch (three) are rare. Backing is slightly more common on the left side (36 pieces) than on the right (28). Most of the backed pieces are very well made straight-backed and pointed bladelets, of which over a third have modified bases, usually rounded (fig. 3.1, 2-4). The arch-backed and arch-ended bladelets are less numerous and not quite so well made. The five La Mouillah points (fig. 3.1,5,6) are the only backed pieces on which traces of microburin scars could be identified. They are typical, well made and three of them are also rounded at the opposite end.

The notches and denticulates are large, broken, poorly made and quite unremarkable. The truncations are rare but typical, and the geometric microlith is a slightly broken, but very well made, segment with *sur enclume* backing. One of the simple microburins is atypical and dubious, but the Krukowski microburins show a real mastery of the technique. Most of them have *sur enclume* backing on the left side.

Scaled pieces are probably less important than the type-list suggests. Most of them (five) are from Layer 1 in the Western Sector and may represent some special activity in that restricted area (fig. 3.1,10). The pieces with continuous retouch are large and generally poorly retouched flakes. The *varia* include a naturally pointed blade with rounded base; a backed flake with rounded base, resembling more the triangular points of the later assemblages than a backed bladelet; and five otherwise unidentifiable retouched fragments. The possible triangular point, with bilateral convergent retouch, can only be considered such if the straight break across

Table 3.1 - Wadi Ti-n-Torha. Absolute frequencies of types of Retouched Tools at Two Caves, Torha East and Torha North.

	Tixier Types	Two Caves	Torha East				Torha North
			RInf	RSup	CII	CI	
1.	End-scraper on a flake	2	2	4	1	2	-
2.	End-scraper on a retouched flake	1	-	-	1	-	1
3.	Circular scraper	-	-	-	-	-	1
5.	Denticulated end-scraper	1	-	5	3	5	-
6.	Nosed or shouldered end-scraper	-	-	1	1	-	-
7.	Notched end-scraper	-	1	-	1	-	-
8.	End-scraper on a blade	-	-	-	3	1	-
11.	Double end-scraper	-	-	-	-	1	-
12.	Simple perforator	-	-	1	2	-	-
13.	Perforator on a backed bladelet	-	-	-	1	-	-
16.	Double-backed perforator	1	3	10	7	1	3
35.	Straight backed pointed blade	-	-	2	1	-	-
36.	Arch-ended blade	-	2	-	-	-	-
37.	Arch-backed blade	-	1	1	-	1	-
40.	Blunt backed blade	-	-	-	1	1	-

Table 3.1. continued

	Two Caves	Torha East				Torha North
		RInf	RSup	CII	CI	
41. Partially backed blade	-	1	-	4	2	-
42. Fragment of backed blade	-	3	2	2	-	-
45. Straight backed pointed bladelet	21	9	14	10	-	1
46. Straight backed pointed bladelet with rounded base	10	2	8	10	1	-
47. Straight backed pointed bladelet with truncated base	-	-	-	1	-	-
51. Straight backed pointed bladelet with retouched base	2	-	4	-	-	-
55. Arch-ended bladelet	1	-	-	-	-	-
56. Arch-backed bladelet	5	3	4	4	3	1
57. Arch-backed bladelet with rounded base	-	-	2	1	-	-
59. Arch-backed bladelet with retouched base	-	-	-	1	-	-
60. Gibbous backed bladelet	1	-	-	1	-	-
62. La Mouillah point	5	-	-	1	-	-
63. Partially backed bladelet	3	-	1	1	-	-
64. Shouldered bladelet	-	1	2	4	-	-
66. Fragment of backed bladelet	18	5	14	13	1	-
68. Scalene bladelet	-	3	-	-	-	-

Table 3.1. continued

	Two Caves	Torha East				Torha North
		RInf	RSup	CII	CI	
74. Notched flake	3	1	-	1	-	-
75. Denticulated flake	1	1	1	-	3	-
76. Notched blade	-	2	2	2	-	-
77. Denticulated blade	1	4	5	-	-	-
79. Notched or denticulated piece with retouch	-	-	-	3	2	-
80. Truncation	2	1	6	10	1	-
82. Segment	1	2	7	2	1	1
85. Trapeze-rectangle	-	1	-	-	-	-
89. Isosceles triangle	-	1	-	1	-	-
90. Scalene triangle	-	-	-	2	-	-
91. Triangle with one concave side	-	1	-	-	-	-
93. Triangle with one convex side	-	-	1	-	-	-
94. Elongated scalene triangle	-	-	-	1	-	-
95. Elongated scalene triangle with small short side	-	6	18	21	-	-
102. Microburin	2	-	-	-	-	-
103. Krukowski microburin	5	-	-	-	-	-
104. Scaled piece	8	1	1	-	-	1

Table 3.1. continued

	Two Caves	Torha East				Torha North
		RInf	RSup	CII	CI	
105. Piece with continuous retouch	4	5	2	4	–	–
106. Side-scraper	–	1	–	–	–	–
112. Varia	7	–	2	1	1	–
Triangular points	1	5	5	16	6	2
Bifacial points	–	–	1	1	6	8
Total	106	68	126	140	39	21

its base was deliberate (fig. 3.1,7). However, the illustrations indicate that a typical triangular point (fig.3.1, 8) exists among the collections housed in Libya.

Ti-n-Torha East

Most of the collections examined were from the excavation of Torha East. The deposits in this shelter were divided into four layers, of which the upper two were horizontally divided by a series of stone hut foundations. The lowermost layer, R Inf, was stratigraphically separable from the overlying R Sup by colour and consistency, and the three radiocarbon dates from it fall within the comparatively narrow range of 9080 B.P. ± 70 years (R-1036 α) to 8640 B.P. ± 70 years (R-1035 α). For these reasons and in spite of the rather small sample of artefacts, the R Inf material will be treated as one unit, falling probably a little later in time than that from Two Caves and offering an interesting contrast with it.

Except for the presence of the hut foundations in the upper layer, there is little stratigraphic difference between R Sup (lower) and CII (upper), the radiocarbon dates overlap completely, with both the oldest (8540 B.P. ± 140 years: R-1160 α) and the youngest (7330 B.P. ± 80 years: R-1158 α) being from R Sup, and the artefacts from the two layers are typologically indistinguishable (Tables 3.1 and 2). They will therefore be considered as a single unit.

The uppermost layer of Torha East, CI, is much poorer in artefacts than the underlying layers; only 39 retouched tools were found in the collections in Rome. Typological differences, however, argue against combining the CI collection either with the underlying material or with the later material from Torha North.

Ti-n-Torha East, R Inf

End-scrapers remain uncommon, large and not very well made. Perforators also are not common but are well made (fig. 3.1, 16); all three have one edge slightly convex and one particularly fine example resembles an elongated scalene triangle with all the sides retouched. There were, again, no burins among the material examined, although the debitage did include a burin spall.

The R Inf collection does contain a group of true backed blades, which can be separated from the bladelets in mean length (48.0 mm), width (17.0 mm) and, above all, thickness (6.7 mm), and which are not simply the larger end of a continuum. They are as elongated as the backed bladelets (their mean length:width ratio is 3.1) and, for the most part, equally well made (fig. 3.1, 19). Five are backed on the right side (three with obverse and two with obverse-*sur enclume* backing) and two on the left (both obverse). One of the arch-ended blades is rounded at the opposite end and the partially backed blade is, in fact, shouldered.

Backed bladelets are still the most important tool-class, although no longer so overwhelmingly as at Two Caves. They are still small, carefully made and relatively elongated. (Their mean length, width, and thickness

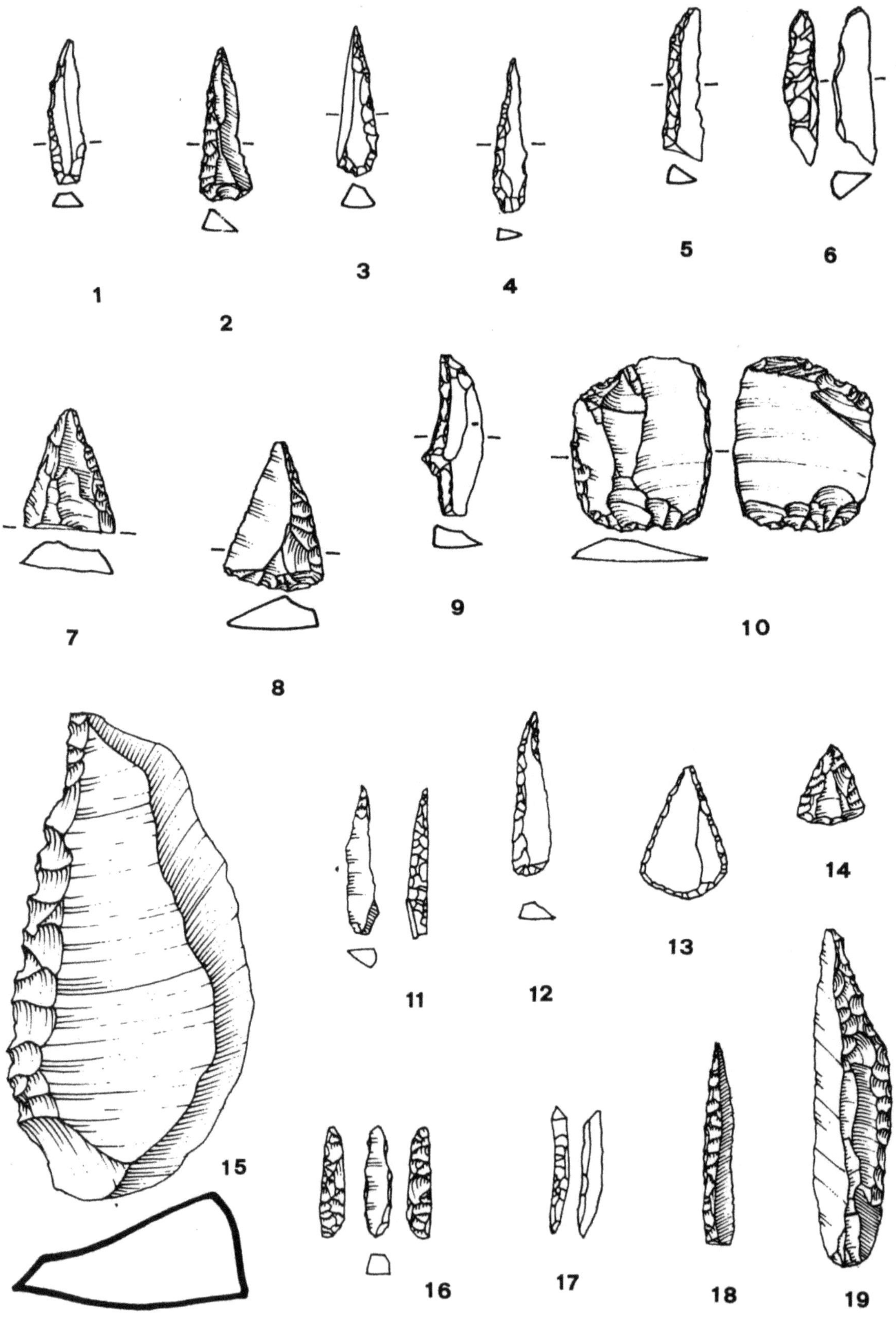

Fig. 3.1 - Wadi Ti-n-Torha. Retouched stone tools from Two Caves (1-10) and from Torha East, R Inf (11-19). 1-6, 9, 11, 12, 19, backed pieces; 7, 8, 13, 14, triangular points; 10, scaled piece; 15, denticulate; 16, perforator; 17, 18, triangles (1:1).

are 25.6 mm, 7.7 mm and 3.3 mm, respectively; their mean length: width ratio is 3.6). Only slightly more than half of the bladelets (13) are backed on the left side, and obverse backing (12 examples) is more common than *sur enclume* (six) and obverse-*sur enclume* (five) backing. Most of the backed bladelets are fine, straight-backed and pointed types (fig. 3.1, *11*, *12*). Modified bases are less common than at Two Caves (fig. 3.1, *12*), but most examples have additional retouch as well as the actual backing (fig. 3.1, *11*): on three examples, one edge has very fine denticulations, forming a very small backed saw. Of the arch-backed bladelets, two small examples may be broken segments while the third, with a wide, straight, *lisse* platform, rather resembles a triangular point. Two of the scalene bladelets are broken, but seem to be too wide and thick to be classified as triangles; the third example is the only backed piece to retain traces of a microburin scar.

The three notched pieces are rather poor, but the denticulates are much better made. The denticulations take up the whole of one or both lateral edges and are often formed by rather scraper-like retouch (fig. 3.1, *15*). Of the geometric microliths, the two segments are both well made with *sur enclume* backing; the trapeze-rectangle is also backed and resembles an elongated triangle (Type 95) with an extra proximal truncation; and the isosceles triangle and the triangle with two concave sides are small and very finely made pieces. The elongated scalene triangles with small short sides are a new and very characteristic type, which first appears locally at the base of Torha East. They are small (their mean dimensions are 23.0 x 5.7 x 3.0 mm), elongated (with a mean length: width ratio of 4.7) and exquisitely formed (fig. 3.1, *17*, *18*). The truncations usually show abrupt, obverse (five examples) or *sur enclume* retouch (four examples), obverse-*sur enclume* (two) and inverse (one) retouch being rare. The angle between the truncations may be as low as 90° and the mean value is only 110°. Most truncations (on four pieces out of six) are on the right side. Only one microburin scar was observed.

The two microburin scars noted above are the only evidence in this layer of the microburin technique. Such scars will rarely be detectable in this collection, since the retouch is almost always very abrupt and "strong", but the absence of microburins from the examined material and the indication that only three were found in the complete collection (Barich 1974: Table I) suggest that the technique was known but not important. The pieces with continuous retouch tend to be large, atypical and dubious, while the convex side-scraper is well made.

The triangular points are well made on short ($\bar{x}$ = 23.0 mm) and rather wide ($\bar{x}$ = 14.6 mm) flakes (fig. 3.1, *13*, *14*). One piece has a rounded base (with inverse basal thinning) and the others have normal basal truncations (three) or a wide straight platform (one). All have obverse semi-abrupt retouch along the left side, while the convergent right side may (three cases) or may not (two cases) be retouched.

Ti-n-Torha East, R Sup and CII

End-scrapers, as well as being more numerous than in earlier assemblages, are also better made. They are almost all on rather short wide and thick flakes, with mean length, width and thickness of 40.9 mm, 31.9 mm and 15.0 mm, respectively. Their mean length: width ratio is 1.2 and six of the 16 unbroken examples are wider than they are long. Most of the scraping ends are distal - only four are proximal - and the mean depth of scraper-retouch is 11.5 mm. One simple end-scraper on a flake is inverse; one end-scraper on a blade is carinated. The denticulated end-scrapers are particularly well made and the scraper-retouch tends to extend beyond the distal end of the flake and down the lateral edges; one denticulated example is double.

The perforators also are more numerous than before and are particularly finely made. Two of the simple perforators are isolated by opposed notches at the distal ends of large, wide flakes, but all the others are on bladelets. Double-backed perforators are the most characteristic type (fig. 3.2, 1, 2). One has a wide, flat platform and convergent sides, but the remainder have straight, parallel sides or one side slightly convex. Most (12) are double-pointed, one has a proximal point and two have rounded bases. The majority are formed by bilateral, obverse backing (12), two have *sur enclume* backing on one side, two have alternating backing on one side, and one has bilateral, *sur enclume* backing, giving the piece a triangular cross-section. The points on a few are also inversely thinned. The mean dimensions of the double-backed perforators are 22.8 x 6.5 x 3.8 mm, and their mean length: width ratio is 3.7.

These layers again contain a small group of true backed blades, separated from the bladelets principally by their thickness (their mean dimensions are 40.4 x 18.3 x 6.7 mm). They are generally as well made as the backed bladelets but, with a mean length: width ratio of only 2.3, are less elongated. One of the straight backed and pointed blades has a rounded base (as has also one of the fragments) and another has opposed basal retouch. Three of the partially backed blades are shouldered, one of them with both distal and proximal shoulders. The backing is usually obverse (six examples), other varieties being much less common (three are obverse-*sur enclume*, two are *sur enclume*, one is alternating and one alternate), and it occurs seven times on the right side, five times on the left and once bilateral.

Numerically, backed bladelets are not quite so important as before, but are exquisitely made (fig. 3.2, 3-6, 9-12). The backing is almost all obverse (45 pieces), *sur enclume* or obverse-*sur enclume* (21 each); Ouchtata retouch (five pieces), inverse and alternating backing (two each) are rare. Backing was applied rather more frequently to the left side (54 pieces) than to the right (38). The backed bladelets are small (with mean dimensions of 27.4 x 8.6 x 3.5 mm) and elongated (the mean length:width ratio is 3.2). The distribution of widths is bimodal, with a small, secondary mode at 5 mm. This is due to the presence of five small bladelets, 4-5 mm wide, with *sur enclume* backing extending from the ventral face to the dorsal ridge and giving them a triangular cross-section (fig. 3.2, 6). Four of these "trihedral rods" are straight backed and pointed and one is fragmentary. The beautifully made straight backed and pointed bladelets are the most common group (fig. 3.2, 3, 4), and are noteworthy for their frequency of additional retouch: almost half have modified bases - usually obverse rounding (fig. 3.2, 9-12) - and a third have other retouch elsewhere, frequently inverse thinning of the point(s). Arch-backed bladelets (fig. 3.2, 5) are much less common and not so well made, although they still tend to have additional retouch. Four double-pointed examples would be classified as segments but for their irregular or convex strings; one piece has traces of a microburin scar. One of the shouldered bladelets also has traces of a microburin scar, while a second has a rounded base. The La Mouillah point and four of the fragments have arched backs.

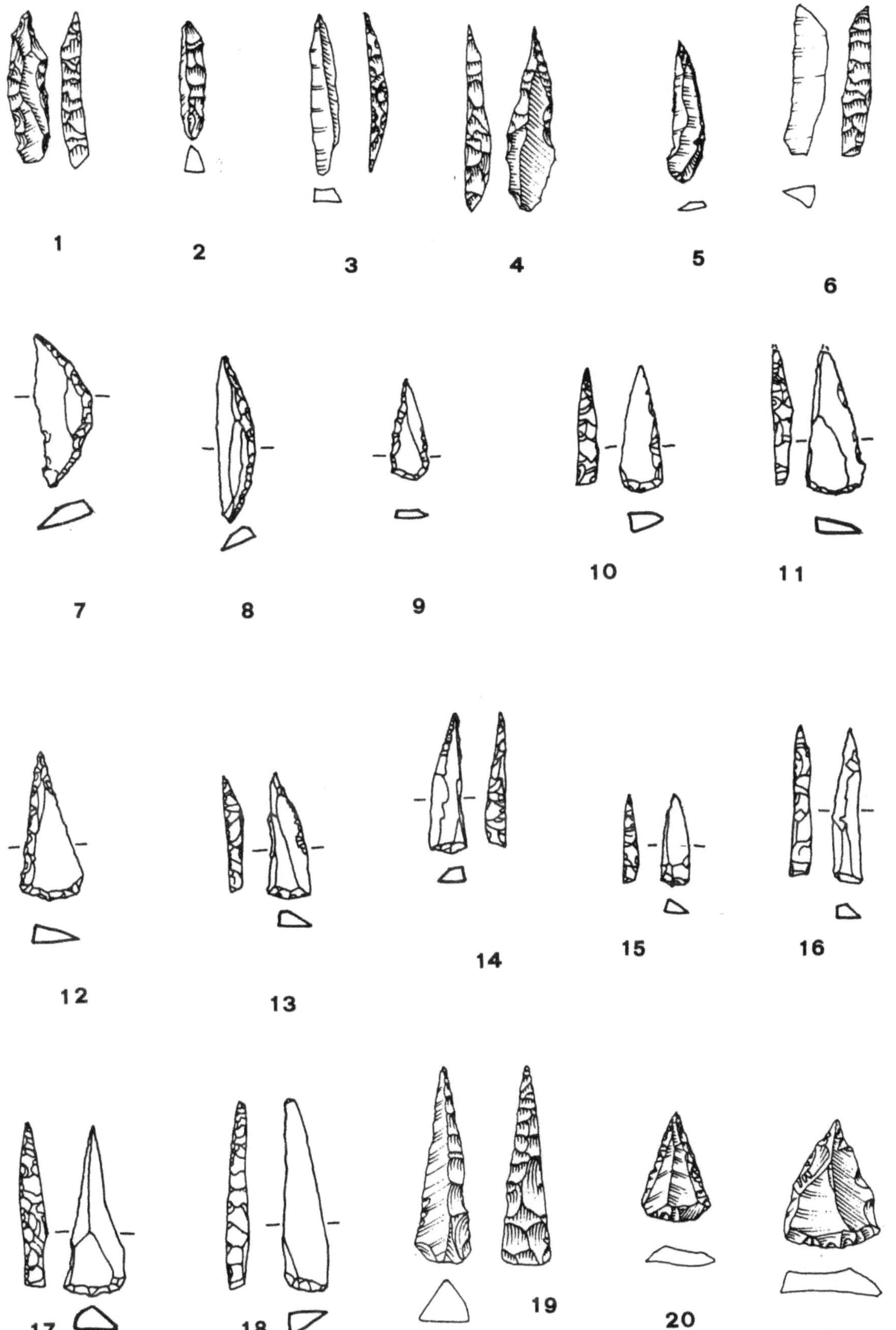

Fig. 3.2 - Wadi Ti-n-Torha. Retouched stone tools from Torha East, R Sup and CII. 1, 2, Perforators; 3-6, 9-12, backed bladelets; 7, 8, segments; 13-19, triangles; 20, 21, triangular points (1:1).

The notches and denticulates are made on large, thick flakes and blades. The denticulations are almost all single-blow and are generally better made than the notches. The only exception to this is a large blade and two wide, deep, opposed notches formed by the rather denticulated, scalar retouch which is typical of the El Nabta variety of Early Neolithic in the Eastern Sahara. The truncations tend to be made on rather short, wide blades (their mean width is 16.3 mm and mean length:width ratio 2.0), and are generally unremarkable. Two are double truncations, but not trapezes, and two have strong lateral truncations most reminiscent of those of the Sebilian in the Nile Valley; one resembles a triangular point and the other a blunt backed flake, but in neither case does the (distal) truncation reach the proximal end of the flake.

The segments are, without exception, very well made (fig. 3.2, 7, 8). Most (six) are backed on the left side; obverse and obverse-*sur enclume* backing are equally common (four each) and one piece has alternating backing. They are small (with mean dimensions of 22.7 x 8.0 x 2.7 mm) and quite elongated (their mean length:width ratio is 3.0). Of the triangles, only the isosceles triangle is not well made, but the elongated scalene triangles with small, short side are, again, by far the most numerous and characteristic type (fig. 3.2, 13-19). They are larger (the mean dimensions are 23.7 x 7.9 x 3.8 mm) and less elongated (with a mean length: width ratio of 3.14) than those of R Inf, but are still very well made. On most pieces, the short truncation is perpendicular to the long one and the most obtuse angle between them is only 108°. Most triangles are truncated on the left side (23; 14 are on the right and two are unidentifiable). The longer truncation is almost always straight (on two pieces it is convex and one one, concave), but over half (20) of the shorter truncations are convex. This latter feature means that a significant number of the triangles closely resemble straight backed and pointed bladelets with rounded bases (Type 46). The two types can normally be distinguished since, on the triangles, there is a sharp angle between the two truncations, but the backing and rounding of a bladelet follow the same unbroken line. The two types do, however, grade into each other to a surprising degree. (Compare, for example, fig. 3.2, 9-12 with fig. 3.2, 13-18). Most triangles have one obverse and one obverse-*sur enclume* truncation (14) or two obverse truncations (13); in either case, the shorter truncation is obverse. Other combinations of retouch are much less frequent: obverse/*sur enclume* - 3; *sur enclume*/obverse-*sur enclume* - 1; obverse/inverse - 1; obverse-*sur enclume*/inverse - 1; obverse/alternating - 1; *sur enclume*/*sur enclume* - 5. The last group includes three examples which, since they also taper from the proximal to the distal end, are triangular in all cross-sections (fig. 3.2, 19). None of the geometric microliths retains any trace of a microburin scar.

The single scaled piece is well made and typical. The pieces with continuous retouch are large and unremarkable. The *varia* include a proximal fragment of a bladelet with a long curved tang (or perforator-bit).

Although also present in the underlying R Inf, triangular points are most characteristic of R Sup and, especially, CII (fig. 3.2, 20, 21). They are made on small and rather wide flakes, with mean dimensions of 23.6 x 15.6 x 4.0 mm, and a mean length:width ratio of 1.6. The two sides converge towards the distal end and are almost always straight (three were noticeably arched). Usually both edges are fully (eight examples) or partially (seven) retouched; four examples have unilateral retouch and one is unretouched (and one piece is broken). The retouch is obverse, and more often semi-invasive to semi-abrupt than abrupt. Two points have rather denticulated retouch, so much so on one as to form "spines" like those on some of the bifacial points. Two points have bases formed by a wide, flat platform, but the remainder are retouched; ten are truncated (five straight and five convex) and nine are rounded bases, where the base and the sides form an unbroken line.

Finally, the collections included two well made and completely bifacial points. One is leaf-shaped with a rounded base and the second has a short expanding stem.

Ti-n-Torha East, CI

The end-scrapers are well made, particularly the denticulated examples. The double end-scraper is denticulated at one end (fig. 3.3, 5). The perforator has a triangular cross-section and sur enclume backing on one edge.

The backed blades (again, wider and thicker than the bladelets) are quite well made. The blunt backed blade is gibbous and is made on a blade which also has bifacial scaling at each end. The backed bladelets (fig. 3.3, 4) are rather better made and are mainly arch-backed pieces, two of them with traces of microburin scars. The straight backed, pointed bladelet with rounded base is perfectly typical.

The denticulates are better made than the notched pieces, the truncation is very fragmentary, and the segment, the only geometric microlith, is very well made (fig. 3.3, 2).

The single varium is an inverted triangular point (with blunt distal and pointed proximal end), with semi-abrupt, bilateral, obverse retouch, and invasive retouch over the entire ventral face; it may be unfinished. The triangular points are essentially similar to those in the underlying layers (fig. 3.3, 3), but only one has a truncated base (two are broken and three are wide platforms). One example has retouched serrations along both edges.

The bifacially retouched points are what most separate CI from the earlier material. They are fully bifacial and very well made. Two are simple leaf-shaped points (one is a willow-leaf); one is a leaf-point with serrations all round the perimeter (fig. 3.3, 1); two have ogival bases and incipient barbs and one has a squared base with incipient barbs.

The elements of continuity from the earlier parts of the Torha East sequence to CI are principally in the denticulated end-scrapers and the triangular points. The major differences are the virtual disappearance of straight backed, pointed bladelets, the complete disappearance of elongated triangles and the appearance or bifacial points. Clearly, CI is different and it is extremely unfortunate that the sample of tools is so limited. The two radiocarbon dates from CI (8440 B.P. ± 50 years (R-1162) and 8460 B.P. ± 50 years (R-1161 α) indicate an age of about 8400 years, but are older than two of the dates from the underlying layers. Such an age would also be typologically unexpected, since bifacial points do not appear elsewhere in North Africa until at least a thousand years later. The CI collection may, therefore, include some later material and, since the

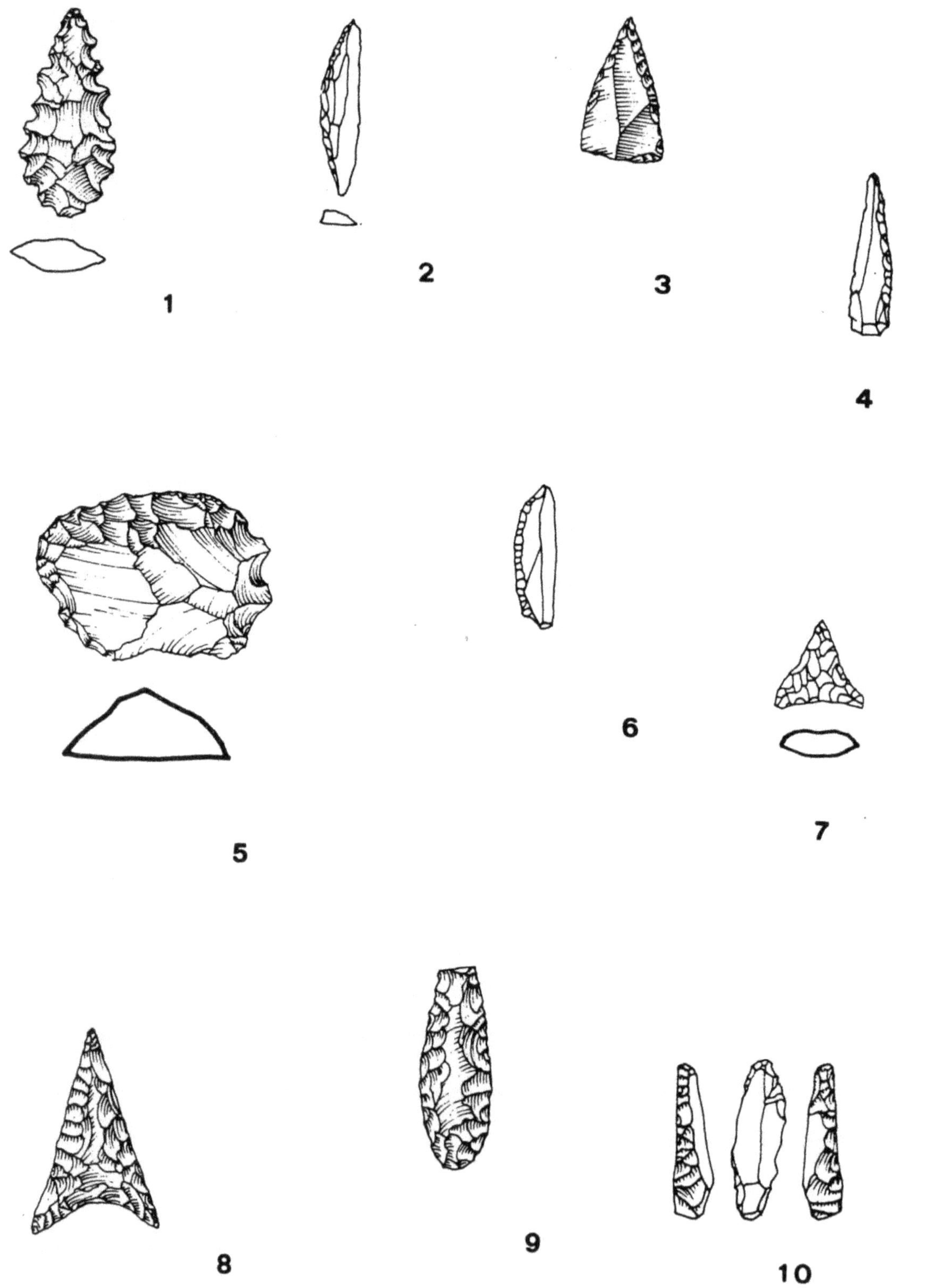

Fig. 3.3 - Wadi Ti-n-Torha. Retouched stone tools from Torha East CI (1-5) and from Torha North (6-10). 1, 7-9, Bifacial points; 2, segment; 3, triangular point; 4, 6, backed bladelets; 5, end-scraper; 10, perforator (1:1).

examined collection is in any case so small, it seems wisest not to place too much reliance on it.

Ti-n-Torha North

Ti-n-Torha North is the latest part of the prehistoric sequence from the Wadi Ti-n-Torha, covering the period from ca 7000 B.P. to ca 5300 B.P. Five stratigraphic layers were recognised during excavation, but all of them were extremely poor in artefacts and only 21 of the retouched tools were available for study. They will therefore be grouped as one unit.

The two end-scrapers are very well made, even though neither is denticulated. The circular end-scraper has an alternating scraper-edge. The double-backed perforators are rather irregular in shape (fig. 3.3, 10) and two of them have invasive retouch over the entire ventral surface. The backed bladelets (fig. 3.3, 6) and the segment are quite well made, while the notched pieces and the scaled piece are atypical and dubious. One of the triangular points has a rounded distal end and pointed base, the other has a pointed tip and straight proximal truncation; both are well made.

Bifacial points are the most characteristic (and most frequent) element of the Torha North collection, and all are of a very high standard of manufacture (fig. 3.3, 7-9). Two are elongated leaf-points (one willow-leaf) with rounded bases (fig. 3.3, 9), and two are short leaf-points, each with a straight proximal break that may be deliberate. Of the triangular pieces, one is a small equilateral triangle with three concave sides (fig. 3.3, 7), one is a small isosceles triangle with a slightly concave base, one is an elongated triangle with incipient barbs and a rounded base and the last is a large wide triangle with a hollow base (fig. 3.3, 8).

3. WADI TI-N-TORHA AND THE NORTH AFRICAN CONTEXT

The relative frequencies of the various classes of retouched tools in the six units defined here for the Ti-n-Torha sequence are summarised in Table 3.2. It is unfortunate that the collections on which this study is based were not true random samples, since this precludes the use of statistical techniques in comparing the assemblages with each other or with material from elsewhere in North Africa. We are therefore forced to discuss the artefacts principally on a qualitative and rather subjective basis. It may be noted that such an approach would not, in any case, be completely avoidable, since the quality of information from North African sites is extremely variable.

One of the most striking aspects of the Ti-n-Torha collections as a whole is the very high standard of manufacture and, particularly, the prominence of blades and bladelets. This led Barich (1974: 120) originally to compare the material from Torha East with the Northwest African Epipalaeolithic industries, which arise out of the Iberomaurusian and seem to be independent of the Capsian. A developed bladelet technology is certainly common to all of these occurrences, and the earlier material from Wadi Ti-n-Torha would conventionally be considered "Epipalaeolithic". The resemblances are, however, only very general. After ca 25-20000 B.P. the Late Palaeolithic industries throughout occupied North Africa (the Maghreb, the Mediterranean littoral and the Nile Valley) were all essentially backed bladelet industries, like the Iberomaurusian of the Maghreb. Holocene industries throughout the area are therefore all based upon very similar foundations and tend to look somewhat alike; linkage between them can be justifiably proposed only in the case of temporal continuity, spatial continuity or specific artefactual analogies.

The Epipalaeolithic industries of the Maghreb proper - the Keremian or Kristelian (Tixier 1967), the Columnatian (Cadenat 1970; Roubet 1968) and the "Elassolithic" (Camps 1974) - may be comparable in age to the Ti-n-Torha material but have few detailed similarities, are far removed geographically and, above all, are Maghrebi, rather than Saharan, phenomena. Slightly closer in space are the southern Tunisian bladelet industries, particularly those of the Oued Akarit (Gobert, Howe 1955). These have not been dated to at least the mid-ninth millennium B.P., and possibly even as early as 9200 B.P. (Lubell, Sheppard, Jackes 1984). Straight backed and pointed bladelets, many of them with additional retouch or inverse thinning (cf. Torha East R Sup and CII), are very important, but the numerous triangles are short and microburins are very common. Oued Akarit also lies beyond the Sahara proper.

Within the Western Sahara, the "Northwest Saharan Epipalaeolithic" (or Mellalian), known from Hassi Mouilah (at about 8600 B.P.), El Hadjar and El Haouita (Aumassip 1972), has a number of more specific resemblances to the earlier Ti-n-Torha material. Straight backed and pointed bladelets, often with basal retouch, are important, denticulates are better made than notches and Hassi Mouilah yielded some small saws. These features recall Two Caves and Torha East R Inf, but, again, microburins and, above all, notches and denticulates are much more important in the northwestern Sahara. The Hadjarian, from the Oued Mya, is generally comparable but, with an estimated range of 7-5000 B.P., post-dates most of the Wadi-n-Torha sequence (Aumassip 1973). Farther south, the Epipalaeolithic of Taoudenni (Petit-Maire, Riser 1983) dates to at least 7500 B.P., but the industry, composed mostly of scrapers and segments, has little in common with that of Wadi Ti-n-Torha, nor does the even more remote Epipalaeolithic of the Saoura (Chavaillon 1964). Along the southern fringes of the western and central Sahara, the Ounanian is characterised principally, it seems, by Ounan points, of which none was found at Ti-n-Torha. The Kiffian, from Adrar Bous (Smith 1976), seems to be late (an age of 7500-6000 B.P. is estimated on the basis of stratigraphy and one radiocarbon date on carbonates) and, with its emphasis on segments, again has nothing specific in common with the Ti-n-Torha material.

Further to the east, Cyrenaican Libya is of some interest, although beyond the desert proper, since it has yielded an unbroken sequence covering the same period as that of the Wadi Ti-n-Torha. The Libyco-Capsian of the Haua Fteah (McBurney 1967) corresponds chronologically to Ti-n-Torha, is of generally similar Epipalaeolithic character and has a few points of more specific resemblance. It includes a number of trihedral rods identical to those of Torha East R Sup and CII (based upon personal examination of the Haua collections; see also McBurney 1967; Figure VIII.19: 1-2, 4-5), and the backed bladelets, although not straight backed, frequently have rounded or otherwise modified bases. The Haua Fteah Late and Epipalaeolithic sequence as a whole also shows a tendency for the backed and truncated bladelets, or scalene bladelets, the most important tool-type, to become smaller through time and to develop towards elongated scalene triangles with small short

Table 3.2 - Wadi Ti-n-Torha. Percentage frequencies of classes of Retouched Tools at Two Caves, Torha East and Torha North.

	Two Caves	Torha East RInf	Torha East RSup	Torha East CII	Torha East CI	Torha North
End-Scrapers	3.8	4.4	7.9	7.1	23.1	9.5
Perforators	0.9	4.4	8.7	7.9	2.6	14.3
Burins	-	-	-	-	-	-
Backed Pieces	62.3	44.1	42.9	38.6	23.1	9.5
Notches & Denticulates	4.7	11.8	6.3	4.3	12.8	9.5
Truncations	1.9	1.5	4.8	7.1	2.6	-
Segments	0.9	2.9	5.6	1.4	2.6	4.8
Trapezes	-	1.5	-	-	-	-
Triangles	-	11.8	15.1	17.9	-	-
Microburin Technique	6.6	-	-	-	-	-
Varia	17.9	10.3	4.0	3.6	2.6	4.8
Triangular Points	0.9	7.4	4.0	11.4	15.4	9.5
Bifacial Points	-	-	0.8	0.7	15.4	38.1

sides; a gradation between backed bladelets and triangles was noted above for the Ti-n-Torha sequence.

In Egypt, Epipalaeolithic material from Siwa is generally similar to the earlier Ti-n-Torha assemblages and has been compared to the Libyco-Capsian (Hassan 1976); unfortunately, detailed descriptions of the material are not available to me. South of Siwa, along the extreme west of Egypt and northern Sudan, considerable Epipalaeolithic or Early Neolithic material is coming to light (R. Kuper, pers. comm.), but that published thus far (Kuper 1981) post-dates most of the Ti-n-Torha sequence (the tenth millennium date from the Gilf el Kebir is not associated with cultural remains) and has little in common with it.

Farther east, many data are now available from the Egyptian Sahara, and the Neolithic sequence established there shows remarkable correspondences to that from Wadi Ti-n-Torha (Wendorf, Schild 1980; Wendorf, Schild, Close 1984; Wendorf, Close, Schild 1985; Banks 1984). The Early Neolithic covers a period from about 9500 B.P. to about 7900 B.P. and is therefore equivalent to the Two Caves and Torha East assemblages. Part of the sequence was formerly called Terminal Palaeolithic but, since pottery and domestic animals (cattle) are present throughout, it is now all regarded as Early Neolithic.

The Early Neolithic of the Eastern Sahara is divided into four stages, which are chronologically separate and typologically distinct. The first, the Early Neolithic of el Adam type, dates from about 9500 to 9000 B.P. and is characterised by a high frequency of backed bladelets, especially straight backed and pointed bladelets, often with modified (usually rounded) bases. The microburin technique was known but not commonly used. End-scrapers, burins or notches and denticulates may be quite numerous, but vary from site to site. In chronology and overall typology, this first stage of the Early Neolithic is closely similar to the collection from Two Caves.

The second stage of the Eastern Saharan sequence is the Early Neolithic of el Kortein type, which dates from ca 8800 to ca 8500 B.P. The overall typology is similar to that of el Adam, predominantly of straight backed and pointed bladelets, but with the addition of a very distinctive type of stemmed point, the Ounan-Harif point, which is perhaps closer to the Harif point of Sinai and the Negev (Marks, Scott 1976) than to the Ounan point of the southern Sahara (Tixier 1963). Ounan-Harif points are not represented in the Wadi Ti-n-Torha, although perfectly typical examples are known from Shati in central Libya (Petit-Maire 1982; Ounan-Harif points are not illustrated in the publication but were observed during personal examination of the material from Shati, for which my thanks are due to Dr. Petit-Maire).

The Early Neolithic of el Ghorab type, the third stage of the Eastern Saharan sequence, dates from about 8500 to 8200 B.P. Ounan-Harif points are now replaced by elongated triangles with small short sides. Backed bladelets remain common, and retouched pieces and notches and denticulates may also be numerous. This clearly corresponds to and coincides with the change from Two Caves to Torha East R Sup and CII, with Torha East R Inf fulfilling an essentially transitional role between the two.

The fourth stage in the Eastern Sahara is the Early Neolithic of el Nabta type, which dates between ca 8100 and ca 7900 B.P.; this is less easy to relate precisely to the Wadi Ti-n-Torha. It is characterised in Egypt by finely made, double-backed perforators, which become relatively numerous in Torha East R Sup and CII, but also by burins, which are virtually absent from Ti-n-Torha, and by retouched pieces, notches and denticulates made on large blades by semi-abrupt scalar retouch; a strangled blade, absolutely typical of this stage, was noted above for R Sup CII. The correspondence between the Eastern Sahara and the Tadrart Acacus is no longer, then, quite so close as it was in earlier phases of the Neolithic, although the overall trends still seem to match.

The diagnostic elements of the Middle Neolithic of the Eastern Sahara are segments and triangular points (the unifacial, proximally truncated points of Banks 1984). Identical pieces are common at Torha East in R Sup and, particularly, CII, which also overlap in age with the Eastern Saharan Middle Neolithic (ca 7700-6200 B.P.). This would suggest that the occupations represented by Torha East R Sup and CII may, indeed, extend from the mid-ninth millennium B.P. of el Ghorab type of Early Neolithic to the late eighth millennium of the Middle Neolithic, which accords with the overall range of the radiocarbon dates from those two layers (ca 8500-7300 B.P.). The stratigraphic inversion of some of the radiocarbon dates, and what would, from an Egyptian perspective, be regarded as a mixture of artefacts from three stages of the Neolithic, may reflect disturbance inevitably caused by the construction of the stone-lined huts. (Stone-lined huts are another feature common to Ti-n-Torha and to the Eastern Saharan Middle Neolithic). The two bifacial points from R Sup and CII might be intrusive from CI, but both are rather simple forms and simple bifacial points do occur in the Egyptian Middle Neolithic.

The presence of triangular points and, especially, segments in R Sup and CII link the Wadi Ti-n-Torha not only with the Egyptian Neolithic, but also with the Neolithic of the Central Sahara. Segments are very common in the Central Saharan Epipalaeolithic and Neolithic (the Saharo-Sudanese Neolithic of Camps 1974 and 1975), and both segments and triangular points occur in the middle layer of Amekni (Camps 1968), which probably dates after 7000 B.P. This linkage to the west becomes much more pronounced at Torha North, where segments and triangular points are present but the many and varied bifacial arrow-heads are the most numerous and characteristic group. Bifacial arrow-heads are almost ubiquitous in the later Neolithic of the central Sahara and can be very numerous. While it is possible that they are present in the Kiffian at 7500 B.P., this is based only upon surface sites; they are known by about 7000 B.P. in the Hadjarian and they occur in the Hoggar by 6000 B.P. The Neolithic sequence from Hassi Mouilah (Marmier, Tixier, Trécolle 1978-79) would indicate that bifacial arrow-heads become more common through time and, certainly, they are numerous in the Ténéréan (mid-sixth to mid-fifth millennia B.P.; Tixier 1962) and at Méniet at about 5400 B.P. (Hugot 1963).

The material from Torha North is not, however, completely Central Saharan in character, and the difference lies in the continued high standard of manufacture of the retouched tools. The tools of the later Central Saharan Neolithic, with the notable exception of the bifacial arrow-heads, tend to be unstandardised and rough. Maître (1974), for example, in synthesising the Neolithic of the Hoggar, felt obliged to construct a new typology to accommodate the ill-defined and often *ad hoc* assemblage. The non-bifacial tools from Torha North, however, still fall into the types defined for the Epipalaeolithic by Tixier (1963), and the group of large and crude tools is completely absent. Instead, the unifacial tools show a significant continuity with the types present in earlier assemblages from the wadi.

The affinities of the lithic artefacts from the Wadi Ti-n-Torha may cast some light upon the Early Holocene colonisation of this area, and upon the later prehistoric networks of which the wadi's inhabitants were a part. The earliest Ti-n-Torha artefacts, from Two Caves, are closest in both chronology and typology to the earliest Neolithic of the Egyptian Sahara. The latter, in turn, seems to be linked to the Arkinian and Shamarkian industries of the tenth and ninth millennia in Nubia (Wendorf, Schild, Close 1984: 410-411). This implies that Wadi Ti-n-Torha was colonised by movement across the desert from the east. Two Caves thus far lacks the pottery and domestic cattle which occur with the Early Neolithic of el Adam type in Egypt. However, both pottery and cattle bones are <u>extremely</u> rare even in the Egyptian sites, and the evidence from Torha East (Barich 1974; Gautier, van Neer 1977-82) shows that both were known in the wadi at least soon after the occupation of Two Caves.

While the Wadi Ti-n-Torha may have been colonised from the east, thereafter it was not, in any sense, a mere province of the Eastern Sahara. The next variety of Early Neolithic in the Eastern Sahara, el Kortein type, is known to have extended as far west as central Libya but never affected Ti-n-Torha. Instead, Torha East R Inf continues the lithic tradition of Two Caves, with the addition of a few of the elongated triangles which were later to become the characteristic tool-type in both Ti-n-Torha and the Eastern Sahara. R Inf also shows notable, and perhaps significant, similarities to the contemporaneous Northwest Saharan Epipalaeolithic.

The Early Neolithic of el Ghorab type is the last Eastern Saharan tradition to be clearly visible at Ti-n-Torha. In Torha East R Sup and CII, there are strong indications of contacts with both the Nabta-type Early Neolithic and the Middle Neolithic groups to the east, but the correspondences are no longer exact. Instead, there are now elements typical of the central Sahara, which, by the time of Torha North, have become dominant in the collections.

It was noted above that even Torha North, with a full suite of central Saharan bifacial arrow-heads, still shows significant continuity with the earlier assemblages from the wadi. This continuity is a final and striking element of the lithic sequence from the Wadi Ti-n-Torha. Changes which occur in the sequence are gradual: new types first appear in small quantities and established types decrease in frequency before disappearing. This might be due in part to the mixing of archaeological layers, which is always a hazard in caves and rock-shelters with multiple occupations, but mixture cannot be a complete explanation. The sequence is based on three different shelters between which mixing is impossible, while the presence of one or two triangular points and no other types at Two Caves and the complete absence of elongated triangles from Torha East CI would argue against a great deal of mixture within each shelter. Additional evidence of regional cultural continuity is provided by the discovery of at least one elongated triangle at the base of Uan Muhuggiag.

The continuity seen in the Wadi Ti-n-Torha sequence is, then, real and stands in contrast to the breaks evident in the Neolithic sequence of the Eastern Sahara. This is a reflection of ecological difference. In the

Eastern Sahara, the four stages of the Early Neolithic correspond to wetter periods and are separated by intervals when the desert was arid and uninhabited. At Wadi Ti-n-Torha, on the other hand, water was consistently available and there are no apparent hiatus in human occupation. The evidence of the lithic sequence implies that the wadi was occupied by the same diachronic social group throughout the first half of the Holocene. This contradicts traditional notions of wandering and rootless prehistoric hunters, but indications of very long-term persistence of social territories in the Palaeolithic and Epipalaeolithic do occur in other stratified cave sequences elsewhere in North Africa (Close 1977, 1980-81). At the Wadi Ti-n-Torha, therefore, for some four millennia, we are able to trace the maintenance by a single group of a stable and successful pattern of settlement, of a developing process of adaptation and of a complex and changing network of social relations.

Acknowledgements

I wish to thank Dr. Barbara E. Barich for inviting me to study the lithic collections from Ti-n-Torha, and for her unfailing kindness and hospitality during may stay in Rome. Professor Fred Wendorf helped with the study of the artefacts and commented upon versions of this paper, both of which are much appreciated. The analysis of the artefacts was partially supported by a grant from the Institute for the Study of Earth and Man of Southern Methodist University, for which my thanks also are due.

REFERENCES

AUMASSIP G. 1972 - Civilisations prénéolithiques des régions sahariennes, Sixième Congrès Panafricain de Préhistoire, Dakar: 273-278.

AUMASSIP G. 1973 - Néolithique sans Poterie de la Région de l'Oued Mya (Bas-Sahara), Mémoire du C.R.A.P.E., 20 (S.N.E.D.: Algiers), 235p.

BANKS K.M. 1984 - Climates, Cultures and Cattle: the Holocene Archaeology of the Eastern Sahara, Department of Anthropology, Institute for the Study of Earth and Man, Southern Methodist University: Dallas.

BARICH B.E. 1974 - La serie stratigrafica dell'Uadi Ti-n-Torha (Acacus, Libia), Origini, VIII: 7-184.

BARICH B.E. 1978 - Nuove evidenze nell'area del Tadrart Acacus (Fezzan), Libya Antiqua, XV, in press.

CADENAT P. 1970 - Le Columnatien, industrie épipaléolithique d'Algérie, Bulletin de la Société d'Etudes et de Recherches Préhistoriques, XX: 40-50.

CAMPS G. 1968 - Amekni, Néolithique Ancien du Hoggar, Mémoire du C.R.A.P.E., 10 (A.M.G.: Paris), 230p.

CAMPS G. 1974 - Les Civilisations préhistoriques de l'Afrique du Nord et du Sahara, Doin: Paris.

CAMPS G. 1975 - Nouvelles remarques sur le Néolithique du Sahara central et méridional, Libyca, XXIII: 123-132.

CHAVAILLON N. 1964 - Une industrie post-atérienne de la vallée de la Saoura, Bulletin de la Société Préhistorique Française, LXI: 84-104.

CLOSE A.E. 1977 - The Identification of Style in Lithic Artefacts from Northeast Africa, Mémoire de l'Institut d'Egypte, 61 (S.O.P. Press: Cairo), 314p.

CLOSE A.E. 1980-81 - The Iberomaurusian sequence from Tamar Hat, Libyca, XXVIII-XXIX: 69-104.

GAUTIER A., VAN NEER W. 1977-82 - Prehistoric fauna from Ti-n-Torha (Tadrart Acacus, Libya), Origini, XI: 87-127.

GOBERT E.G., HOWE B. 1955 - L'Ibéromaurusien de l'oued el-Akarit (Tunisie), Actes du Deuxième Congrès Panafricain de Préhistoire, Algiers: 575-594.

HASSAN F.A. 1976 - Prehistoric studies of the Siwa Oasis region, northwestern Eygpt, Nyame Akuma, IX: 18-34.

HUGOT H.J. 1963 - Recherches Préhistoriques dans l'Ahaggar Nord-Occidental, 1950-1957, Mémoire du C.R.A.P.E., 1(A.M.G.: Paris), 206p.

KUPER R. 1981 - Untersuchungen zur Besiedlungsgeschichte des östlichen Sahara. Vorbericht über die Expedition 1980, Beiträge zur allgemeinen und vergleichenden Archäologie, III: 215-275.

LUBELL D., SHEPPARD P. and JACKES M. 1984 - Continuity in the Epipaleolithic of northern Africa with emphasis on the Maghreb, in Advances in World Archaeology, Volume 3, eds. Wendorf F., Close A.E., Academic Press: New York, pp. 143-191.

MAITRE J.P. 1974 - Nouvelles perspectives sur la préhistoire récente de l'Ahaggar, Libyca, XXII: 93-143.

MARKS A.E., SCOTT T.R. 1976 - Abu Salem: type-site of the Harifian industry of the southern Levant, Journal of Field Archaeology, III: 43-60.

MARMIER F., TIXIER J., TRECOLLE G. 1978-79 - Etude de l'industrie du gisement d'Hassi Mouillah. Région de Ouargla. Sahara algérien. III L'industrie lithique néolithique, Libyca, XXVI-XXVII: 81-115.

MCBURNEY C.B.M. 1967 - The Haua Fteah (Cyrenaica) and the Stone Age of the South East Mediterranean, Cambridge University Press: Cambridge.

PETIT-MAIRE N. ed. 1982 - Le Shati. Lac pléistocène du Fezzan (Libye), Editions du C.N.R.S.: Paris.

PETIT-MAIRE N., RISER J. eds. 1983 - Sahara ou Sahel? Quaternaire récent du Bassin de Taoudenni (Mali), C.N.R.S. and Ministère des Relations extérieures: Paris.

ROUBET C. 1968 - Nouvelles observations sur l'Epipaléolithique d'Algérie orientale. Le gisement de Koudiat Kifèn Lahda, Libyca, XVI: 55-101.

SMITH A.B. 1976 - A microlithic industry from Adrar Bous, Ténéré Desert, Niger, Panafrican Congress of Prehistory and Quaternary Studies, VIIth Session, Addis Ababa: 181-196.

TIXIER J. 1962 - Le "Ténéréen" de L'Adrar Bous III, in Mission Berliet Ténéré-Chad, A.M.G.: Paris, pp. 333-348.

TIXIER J. 1963 - Typologie de l'Epipaléolithique du Maghreb, Mémoire du C.R.A.P.E., 2(A.M.G.: Paris), 209p.

TIXIER J. 1967 - Procédés d'analyse et questions de terminologie concernant l'étude des ensembles industriels du Paléolithique récent et de l'Epipaléolithique dans l'Afrique du Nord-Ouest, in Background to Evolution in Africa, eds. Bishop W.W., Clark J.D., University of Chicago Press: Chicago, pp. 771-820.

WENDORF F., SCHILD R. 1980 - Prehistory of the Eastern Sahara, Academic Press: New York.

WENDORF F., CLOSE A.E. and SCHILD R. 1985 - Prehistoric settlements in the Nubian Desert, American Scientist, LXXIII: 132-141.

WENDORF F., SCHILD, R. assemblers, CLOSE A.E. ed. 1984 - Cattle-Keepers of the Eastern Sahara: the Neolithic of Bir Kiseiba, Department of Anthropology, Institute for the Study of Earth and Man, Southern Methodist University: Dallas.

CHAPTER 4

PRE-ISLAMIC MEGALITHIC MONUMENTS OF THE NORTHERN TADRART ACACUS

By Massimo BAISTROCCHI*

In this brief study the pre-Islamic megalithic monuments present in the northern area of the Tadrart Acacus and, in particular, those at Jebel Ghelou and Auis (Wadi Tasbät and Ti-n-Eggiart) that have been partly excavated (Barich 1974), will be examined on the basis of the typology formulated by the author for the central saharan area. The peculiar aspects which set them apart from the funerary monuments of the Germa area (Wadi El Agial) will be analysed. The author analyses several types, the V-shaped tumulus identified at Jebel Ghelou appearing to be the most interesting. This type of monument, a tumulus with corridor-like access wings, is compared with similar monuments of central Sahara. The author rules out any possible relation of these monuments with the ancient peopling of the area aend underlines the difficulty of providing specific chronological dating for them in the absence of any specific archaeological evidence.

In this regard, it should be immediately pointed out that, while the tumuli in this region come under the classification of "pre-Islamic funerary monuments", typical of central Sahara, they differ from the megalithic typologies of other regions of Fezzan, as for example the necropolises of Germa and Wadi El Agial where the Pace, Caputo and Sergi Mission found (between the wells of Ti-n-Abunda and El Abiodh) about 45000 pre-Islamic tombs (Caputo <u>et Alii</u> 1951); Reygasse, professed to never having seen "such a great quantity of tumuli" in any other Saharan region (Reygasse 1950:32). It was there that Caputo distinguished various tumuli: - cumuli, pseudo-craters, drum-shaped, with "false vaults", round, with pseudo-domes - and structures with one or two, horn- or hand-shaped (either with two or four fingers) stelae. In the Northern Acacus and in particular in the Auis/Ti-n-Torha region, although only a short distance away from the above mentioned Garamantic necropolises, the Joint Mission found the following structures: slabs of sandstone, cairns, conical and truncated-conical tumuli, flat tumuli and "dallages" as well as "V" shaped monuments (Barich 1974: 14-15).

Apart from the different nomenclatures used, the types noted by Barich are much the same as the monuments most commonly found in the Sahara, with

* Ministero degli Affari Esteri, Rome (Italy).

the exception perhaps of the Jebel Interiki structure (slabs of sandstone) (1).

Considering the structures in the northern Acacus, it is worthwhile remembering that the slabs of sandstone were generally found at the bottom of dried wadi beds, and this could be an indication of a relatively recent construction - that is either contemporaneous with or subsequent to the final drying-up of the region. In the northern Acacus, this type of structure is frequently found at Jebel Interiki.

At the same time, the small cairns, like those found in the Sughd region, were at the bottom of dry wadis, either isolated or grouped together, in line or scattered. One of the cairns in the Ti-n-Torha region was excavated without result (Barich 1974: 15). Similar cairns were found both at Tibesti (where in one case they were arranged in neat lines covering an area of more than 23 square metres) and many hundreds of kilometres further West, as at Adrar des Ifoghas (Hugot 1974: 314). It should also be added that similar structures exist in many other regions of the Sahara and they are not related to burial customs but rather to propitiation rites; every traveller who passes by must leave a stone on top of the pile either as a sign of devotion or to protect himself against the spirit of the place (Djenoun). This custom - still observed today (Lhote 1976: 52) - is therefore a confusing element in the study of pre-Islamic funerary structures.

The above classification is 'open' (see Baistrocchi 1984: 597); it takes into consideration the architectural structure of pre-Islamic monuments

---

(1) For greater convenience in dealing with the subject matter, I provide a brief description of the typologies of the pre-Islamic funerary monuments of Central Sahara while, for a detailed account, I refer you to the more systematic study contained in Atti del II Colloquio Euro-Africano, Parigi 1983, on the theme "Il passato dell' Africa e delle zone limitrofe dai Garamanti al Medio Evo" (Baistrocchi 1984).

a) simple slabs of sandstone, small cumuli of stones;
b) common tumuli in conical and truncated-conical shapes, of varying heights and dimensions;
c) crater-shaped or 'cromlech'-type tumuli, or tumuli with stone at the centre;
d) 'chouchet'-type, either simple or with steps;
e) 'niche'- or 'chapel'-type tumuli;
f) oval composite tumuli, tumuli in 'key-shape' (also called 'ideban à enclos' by French scholars) or surrounded by concentric circles, often of considerable size;
g) 'winged' tumuli, consisting in a central unit and two extremities which can measure as much as dozens of meters across; the V-shaped tumuli belong to the same family just as the 'fusiform-type' which has arms running perpendicular to the tumulus;
h) canal-type tumuli, composed of a central tumulus and a stone ditch;
i) platform tumuli, known as 'dallages';
j) 'blockhouse'-type tumuli;
k) 'horse-shoe' and 'Fatima's tent' styled tumuli;
l) circles of stone without a tumulus.

a

b

Fig. 4.1 - a: Tadrart Acacus (Libya). Bazina in the Wadi Tasbät; b: Ghat (Libya). Bazina of the El Barket necropolis.

rather than their territorial distribution. This is not to say that various types of structures cannot be contemporary with other models within the same regions. It is also possible that, in certain sites, the pre-Islamic tumuli might have been reused by Islamic nomads of the Tuareg tribes.

The tumuli characterized by a conical and a truncated-conical shape are the most common and widespread pre-Islamic burial monuments. They can be found from the westernmost Mauritania to Nubia, from the Moroccan Tafilalt to the buttresses of the Air Mountains, and from Tibesti to the Ennedi. Like all the constructions in the Sahara, they are piles of loosely assembled and uncemented stones that are arranged according to the simple structure described above. In the northern Acacus, and indeed everywhere in the Sahara, this type of funeral monument is usually found on crests and summits. This is the case, for example, of the tumuli at Wadi Ti-n-Eggiart, which average eight meters in diameter (Barich 1974: figs. 1, 2).

The "bazinas" of Wadi Tasbät and Ti-n-Eggiart (fig. 4.1a) are larger (more than 10 metres diameter) and their circular structure, supported by stones fixed into the ground, is well marked by slabs of stone placed onto the summit. Even the bazinas - both in their simpler forms (like the one mentioned above) and in composite models - are some of the most common structures found in the Sahara from Maghreb (Camps 1952: fig. 50-57; Camps 1980: fig. 34) to Air (Beltrami 1979: 419-423).

Of all the pre-Islamic monuments found in the northern Acacus, the most interesting is the V-shaped tumulus sited at the starting point of Wadi Ti-n-Torha (fig. 4.2a,b) to the West of Jebel Ghelou; an analogous structure was also found in the Wadi Tasbät region. The first of these two tumuli has been excavated. The main body consists of huge, rectangular sandstone slabs, set vertically. Two corridors, at right angles to each other and constructed with double rows of stone slabs, branch off from the central structure. The excavation (Barich 1974: 18) involved only the central part of the tumulus, which was reassembled in its original form after the excavation, and retrieved part of a skeleton buried at a 70 cm depth in a fetal position, with the lower limbs contracted and folded against the head. The skeletal remains (about half of the cranium, bones from the thorax and lower limbs) were deposited in the prehistoric section of the National Archaeological Museum of Tripoli where they are awaiting further study. Funeral monuments with "arms" or V-shapes have also been found, though in different forms, both North of the Acacus, at Ti-n-Abunda, and further West in the Ghat area, at El Barket (Caputo et Alii 1951: 385) (fig. 4.1b). The tumuli found on the Ténéré border at the Falezlez well (In Afellahlah) and in the Tefasasset basin (Hugot 1962: 152 fig. 2; 155 fig. 3) are similarly shaped (fig. 4.3). V- and arm-shaped tumuli have also been found in many other regions of central Sahara, even though their structure may vary from region to region, as for example the tumuli at Toufine (Hugot 1974: 306) and those at In Ekker, Tan Ainesnis and Gara Tesmad (Baistrocchi 1984: 599).

Having described the phenomena, a few general remarks on the tumuli and their purpose may be made.

Until now, the pre-Islamic monuments have only sporadically aroused the interest of the few specialists who are primarily interested in the study of Saharan rock art and excavations were limited and hurriedly conducted. Consequently, the little data available to us is varying and inconsistent, and does not allow us to construct a framework of reference and comparison.

a

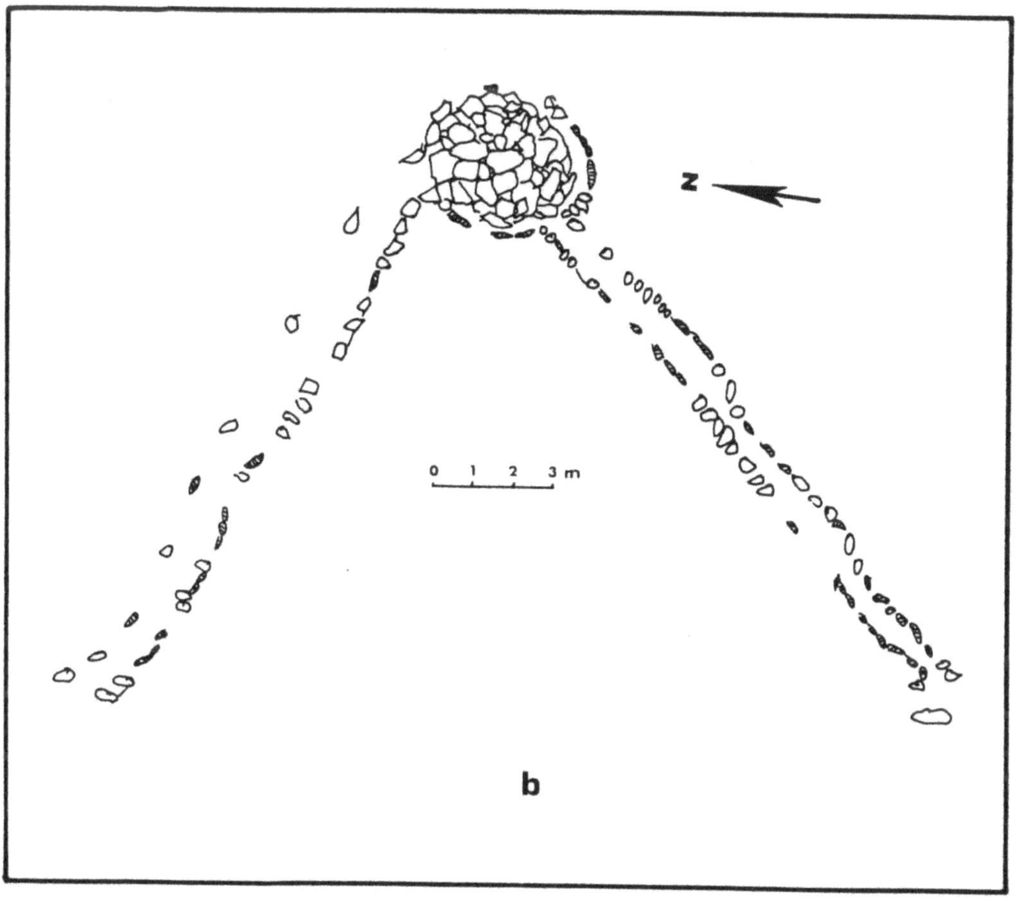

b

Fig. 4.2 - Tadrart Acacus (Libya). a,b: V-shaped tumulus at Jebel Ghelou.

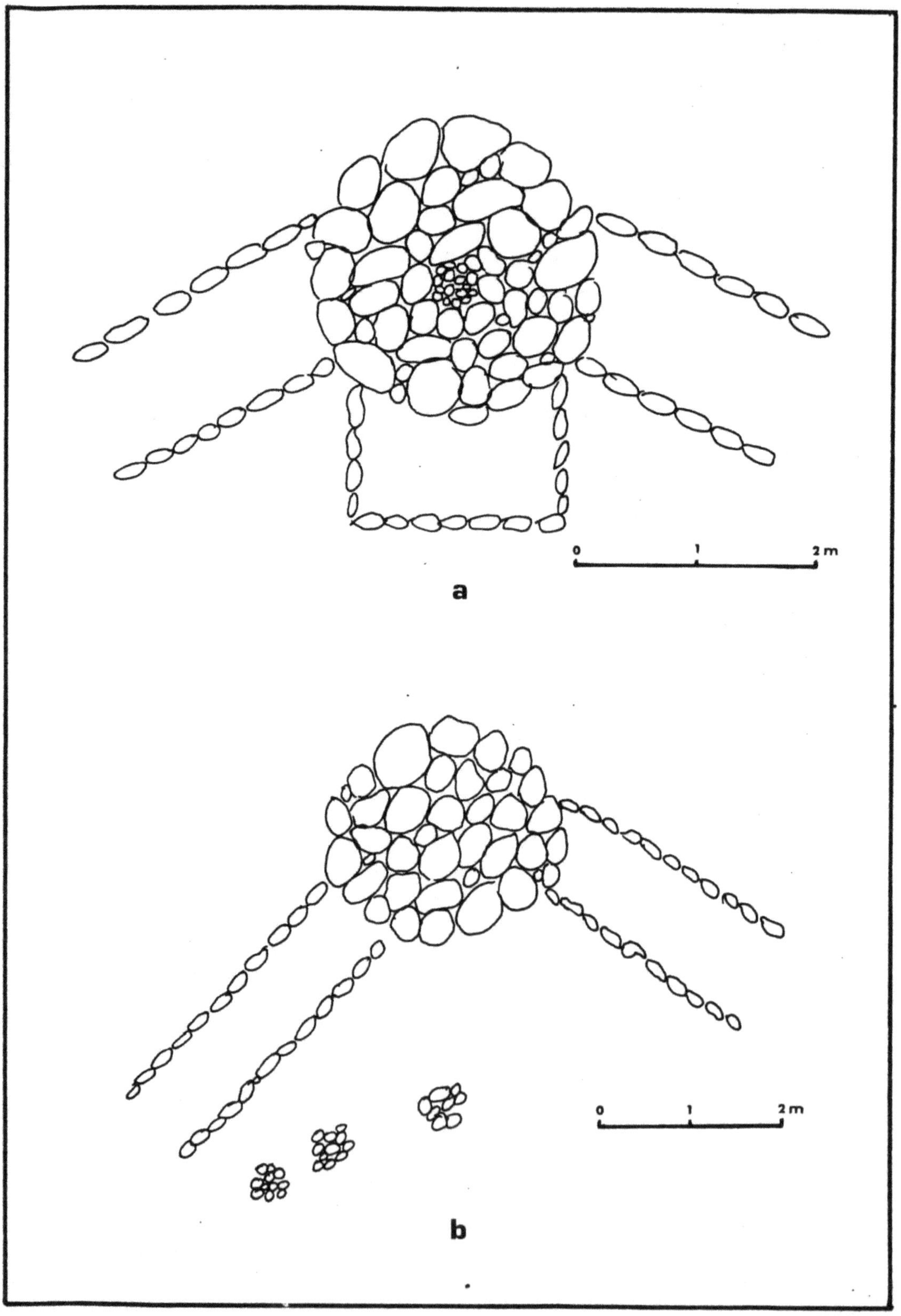

Fig. 4.3 - Ténéré (Chad). a: V-shaped tumulus at Falezlez (In Afellahlah); b: V-shaped tumulus in the Tefasasset basin. (After: Hugot 1962).

We lack the elements to formulate a general theory on the significance and the purposes of the monuments, only based on their shape and structure. Even the material that has been retrieved within the excavated monuments does not allow us to make comparisons or direct links with the data we were able to obtain from the excavation of deposits in shelters at the base of walls bearing engravings or paintings. Up to now, the dating of pre-Islamic tumuli has been imprecise: the oldest structures go back to the end of the Neolithic period which in this region may be between the first and second millennium B.C., and the youngest to the complete Islamization of the Tuareg population, since from that time onwards the funerary structures take on characteristics peculiar to Islam.

Therefore, one may say that the builders of the tumuli have nothing in common with the people who populated the northern shelter of Ti-n-Torha which dates from 5260 ± 130 B.P. to 7070 ± 60 B.P. to 9080 ± 70 B.P. (Barich 1974).

The funeral objects on their part, are of little help in providing information, except in the case of the Ti-n-Hinan tumulus at Abalessa, where a dark clay Roman lamp with a trophy of victory in relief, dated from about the 4th Century (Reygasse 1950: 100). Funeral objects are generally very poor and usually consist of ostrich eggshells for necklaces, glass or stone beads, an occasional bronze bracelet and pottery. For example, the V-shaped mound at Jebel Ghelou produced no funeral materials except for a few fragments of ostrich eggshell.

It should also be said that in the Acacus there are numerous and high quality rock drawings but the funerary structures are quite scarce. The same phenomenon was noted by H. Lhote in the Tassili-n-Ajjer which overlooks the Acacus; this fact led him to affirm that there should be no direct link between the presence of tumuli and rock art so that "the method by which the people of the 'Round Heads' period and the 'bovidian' herders buried their dead remains a complete mystery" (Lhote 1976: 59). Furthermore, it should be added that other regions which have little rock, such as the Ténéré surrounding areas, are rich in many thousands of tumuli.

In conclusion, if scholars have been able to establish relative chronologies for rock art in the Saharan regions, and for the central Acacus itself an absolute chronology (Mori 1965: 233-241), the pre-Islamic funerary monuments have rarely provided any firm indication of date. Indeed, it is that these structures are linked to burial rites the significance of which is not yet fully understood. It is not yet established whether the burials were laid with an intentional orientation, although the V-shaped tumulus of Jebel Ghelou is perfectly positioned facing West (fig. 4.2<u>b</u>) and indeed its two arms run in the direction of the setting sun. On the other hand, some of the most characteristic tumuli of the Sahara - the "key-shaped" ones - face the opposite direction. They are particularly numerous in the Fadnoun region (Savary 1966), but they can also be found in other surrounding areas (Oued Djerat, Tefedest, Tassili-n-Ajjer). According to some scholars, the corridor of stones leading to the central tumulus could have been positioned by the builders in the direction of the rising sun (for the most part, they face East) on the day when the death occurred (Milburn 1979: 59). This is an interesting hypothesis even though, until now, it had no scientific proof (Baistrocchi 1984: 610).

Therefore, only a systematic survey, although difficult to accomplish due to the vastness of the area, would allow us to collect homogeneous data for a scientific interpretation of the phenomena. Such a study is all the more urgently needed, since in some regions environmental degradation is already under way and, in some cases, in an advanced stage.

REFERENCES

BAISTROCCHI M. 1984 - Monumenti funerari preislamici del Sahara (Atti del II Congresso Euro-Africano, Parigi 1983), L'Universo, 5: 589-613; 655-658.

BARICH B.E. 1974 - La serie stratigrafica dell'Uadi Ti-n-Torha (Acacus, Libia). Per una interpretazione delle facies a ceramica saharo-sudanesi, Origini, VIII: 7-184.

BELTRAMI V. 1979 - Repertorio dei monumenti cosiddetti "preislamici" presenti nel territorio dell'Air ed aree limitrofe, Africa, XXXIV/4: 417-423.

CAMPS G. 1952 - Monuments et rites funéraires protohistoriques - Aux origines de la Berbérie, A.M.G.: Paris.

CAMPS G. 1980 - Berbéres - Aux marges de l'histoire, Ed. des Hésperides: Toulouse.

CAPUTO G., PACE B. e SERGI S. 1951 - Scavi Sahariani. Ricerche nell'Uadi El-Agial e nell'oasi di Ghat, Monumenti antichi, XLI, Accademia dei Lincei: Roma.

HUGOT H.J. 1962 - premier apercu sur la preéhistoire du Ténéré du Tefasasset, in Mission Berliet Ténéré-Tchad - Documents scientifiques, A.M.G.: Paris, pp. 149-178.

HUGOT H.J. 1974 - Le Sahara avant le désert, Ed. des Hésperides: Toulouse.

LHOTE H. 1976 - Vers d'autres Tassilis, Arthaud: Paris.

MILBURN M. 1979 - Secrets of South Sahara, Vintage Press: New York.

MORI F. 1965 - Tadrart Acacus, Einaudi: Torino.

REYGASSE M. 1950 - Monuments funéraires préislamiques de l'Afrique du Nord, A.M.G.: Paris.

SAVARY J.P. 1966 - Monuments en pierres seches du Fadnoun, A.M.G.: Paris.

CHAPTER 5

THE WADI TI-N-TORHA FACIES

By Barbara E. BARICH

Excavations and thorough ecological studies allowed us to identify at least two successive phases of development in the Tadrart Acacus regional sequence. This resulted in the recognition of two aspects of food production dynamics which, in the Ti-n-Torha area and, more generally, the Tadrart Acacus, were in keeping with our knowledge of development in the Saharan area as a whole - a conclusion to be dealt with in the final chapter.

The two deposits of Torha East and Torha North (fig. 5.1) had already suggested the type of definition to be given to the Ti-n-Torha groups according to an anthropological model rather than an exclusively typological one (Barich 1974, 1980a, 1984a, b). Taking into account all the available data, in particular that relative to the type of economy that resulted from its "ergologic" character (1), I dismissed the definition of the Ti-n-Torha society as Neolithic, notwithstanding the presence of ceramics in the Eastern site. This represented only one side of the coin, the other being represented by an equally significant lithic production, particularly important on account of the possible links it implied for earlier communities. The contexts of the stone artifacts allowed us to identify major constants that deserve to be outlined in the light of what was observed also at Two Caves Shelter (figs. 5.2, 3).

The Lithic Industry

The backed blades and bladelets which make up the richest and most representative class of the lithic industry are particularly significant in their morphological and technological correspondence to the Late Paleolithic and Epipaleolithic industries. A distinction can be drawn between (a) unretouched tools, or tools with distal, proximal or central retouch of the simple or abrupt type, and (b) tools with abrupt marginal or deep retouch made with the backing technique, on one side, on the proximal end and one side, or on both sides. Unlike other classes, where a certain degree of decline may be observed, backed bladelets show a very high standard of manufacture throughout. The index of the whole class within the lithic complexes is very high (12% and 15.1% are the general indexes for Torha East and Two Caves respectively) and quite significant in itself. A comparison was made (Barich 1974: 120) with the blade complexes found between the Lower Sahara and coastal Algeria which, originating from the Iberomaurusian

1) In the text the term "ergology" is used with reference to the material and technical aspects of culture.

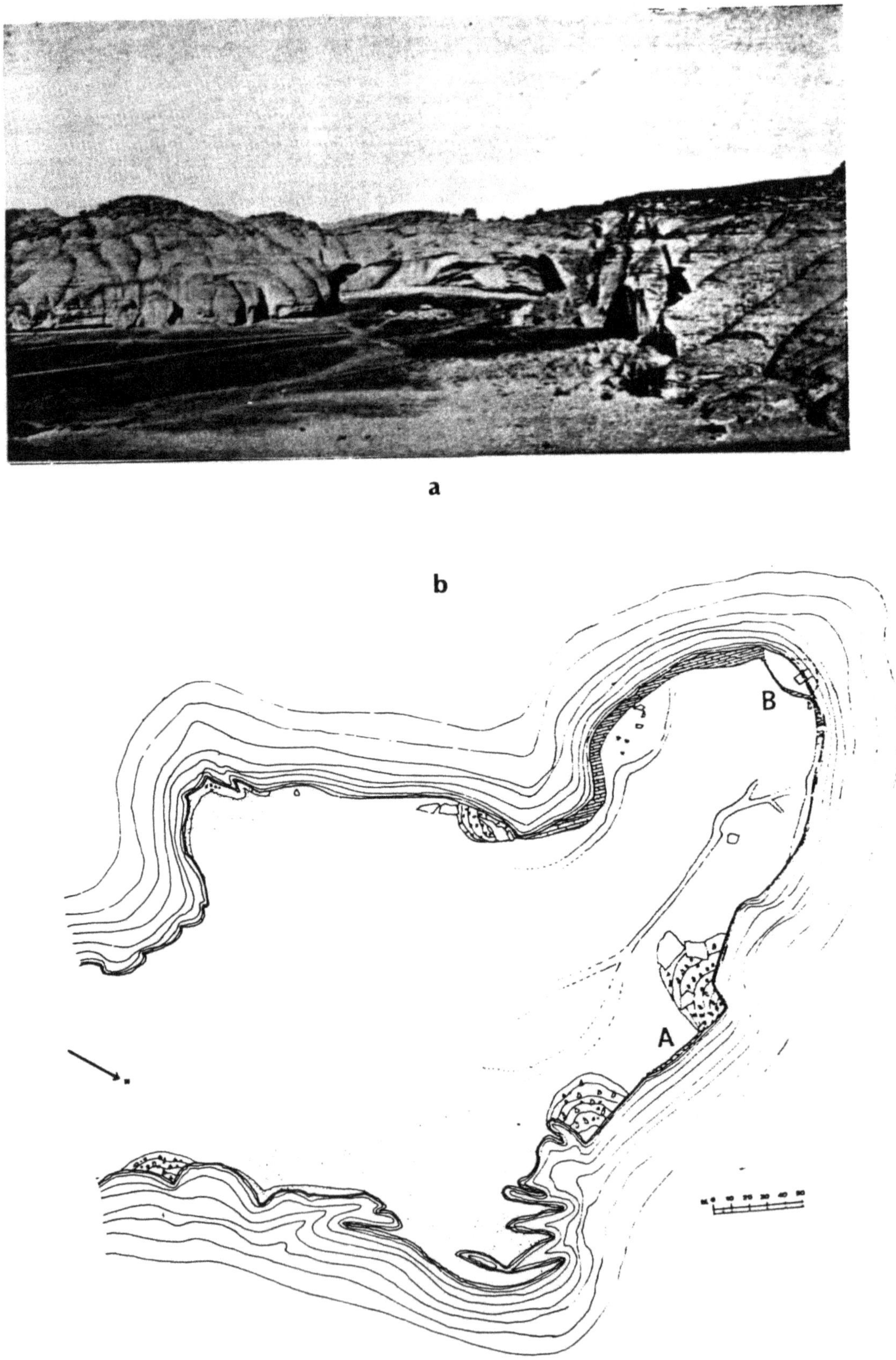

Fig. 5.1 - Wadi Ti-n-Torha. a: general view of the wadi valley; b: map of the valley with localisation of East (A) and North (B) shelters.

tradition, came to be placed between this and the Capsian tradition (2). The chronological evidence traces these industries from the 10th to the close of the 8th millennium B.P. and as they become younger in date so we can see a progressive lowering of the percentage index of backed bladelets.

It is common knowledge that blade segments - an element already significant in the Hagfet-Et-Tera complex (McBurney 1960: 192-93, fig. 26) - are persistently present even in the Epipaleolithic bladelet-industries of coastal Algeria (Columnatian, Keremian-Kristelian). Conversely, the complexes originating from a similar tradition in Southern Tunisia develop by preference triangles and trapezes (Camps 1974a: 195-201).

This aspect of geometric microlithic production appears as a common element of the Epipaleolithic structure (leaving aside local preferences for one or other of the known types) and is taken up by the ceramic context which preserves the same structure. For example an extremely rich frame of reference is given by the pottery of Khartoum (Arkell 1949: Pls. 10, 11/3, 12, 13, 14/4) and Es Shaheinab (Arkell 1953: Pls. 7, 10).

The segments or lunates found in the Ti-n-Torha industry, correspond only in part to the definition given by Tixier (1963: 129-31) and Roche (1963: 36, Texte), for few, although having two of the requirements - bulb removal and a straight sharp side - have a regularly convex backing. They are, therefore, defined as "segment-like" artifacts. In any event, the bulb removal leaves no choice and they must be considered segments rather than pointed bladelets with arch backing. There are also a few specimens with humped backs that are transitional to triangles.

Burins and the microburin technique are a marginal component of the Ti-n-Torha industry. The rare examples found at Torha East (ten burins corresponding to 0.1% and three microburins, equal to 0.03%) and Two Caves (twelve (1.7%) and seven (0.9%) respectively) show that although the technique was known it was employed to a very limited extent. A few specimens of high-quality Krukowski microburins are known and even some microburin scars have been found.

The presence of core-trimming elements, although in a low percentage, is symptomatic of a specific technological orientation of the industry. A few of these, with triangular cross-section and arched ventral surfaces, are quite close to the "lames outrepassées" defined by Tixier (1963: 43). In a number of cases it was evident that these blades, through subsequent retouches, were adjusted and turned into definite tools, such as denticulates or backed tools (particularly triangular pointed backed bladelets). Where the blow was struck transversely with respect to the platform, we may instead note a special type of rejuvenating flake, with a wide dorsal face showing the signs of the preceding blows and detachments

---

2) In Chapter 2 I have referred to the Mellalian, Hadjarian, Kristelian-Keremian and Columnatian-Elassolithic environments as described by G. Camps (1974a:195-216). For Magreb, in particular, a more recent synthesis is given by D. Lubell (Lubell et Alii 1984).

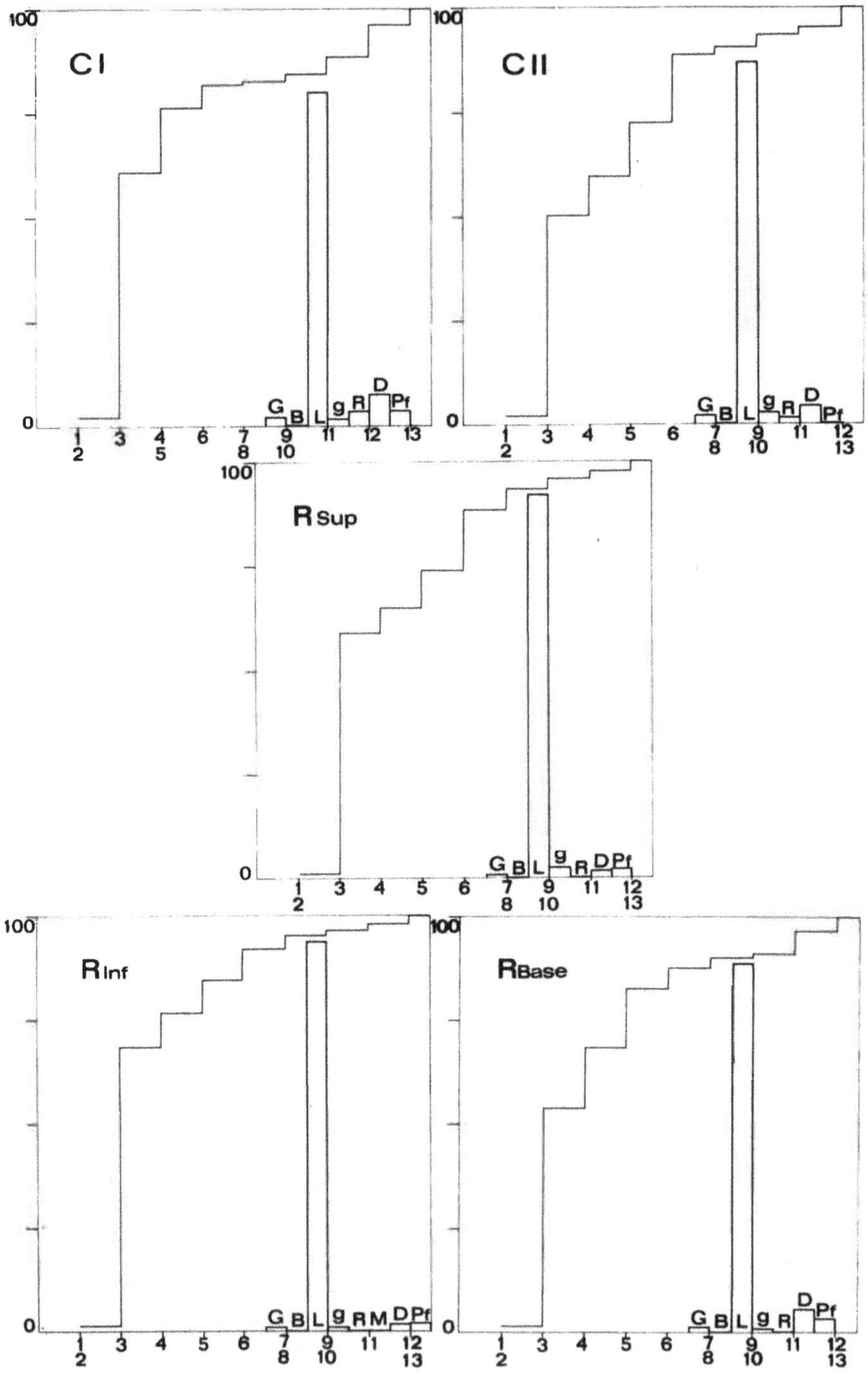

Fig. 5.2 - Wadi Ti-n-Torha. Lithic industry from Ti-n-Torha East (Tools). Key: G=end-scrapers; B=burins; L=blades; g=geometrics; R=side-scrapers; D=denticulates; Pf=arrow-heads. The numbers correspond to the different tool-types of the Ti-n-Torha type-list (cf. Chapter 2).

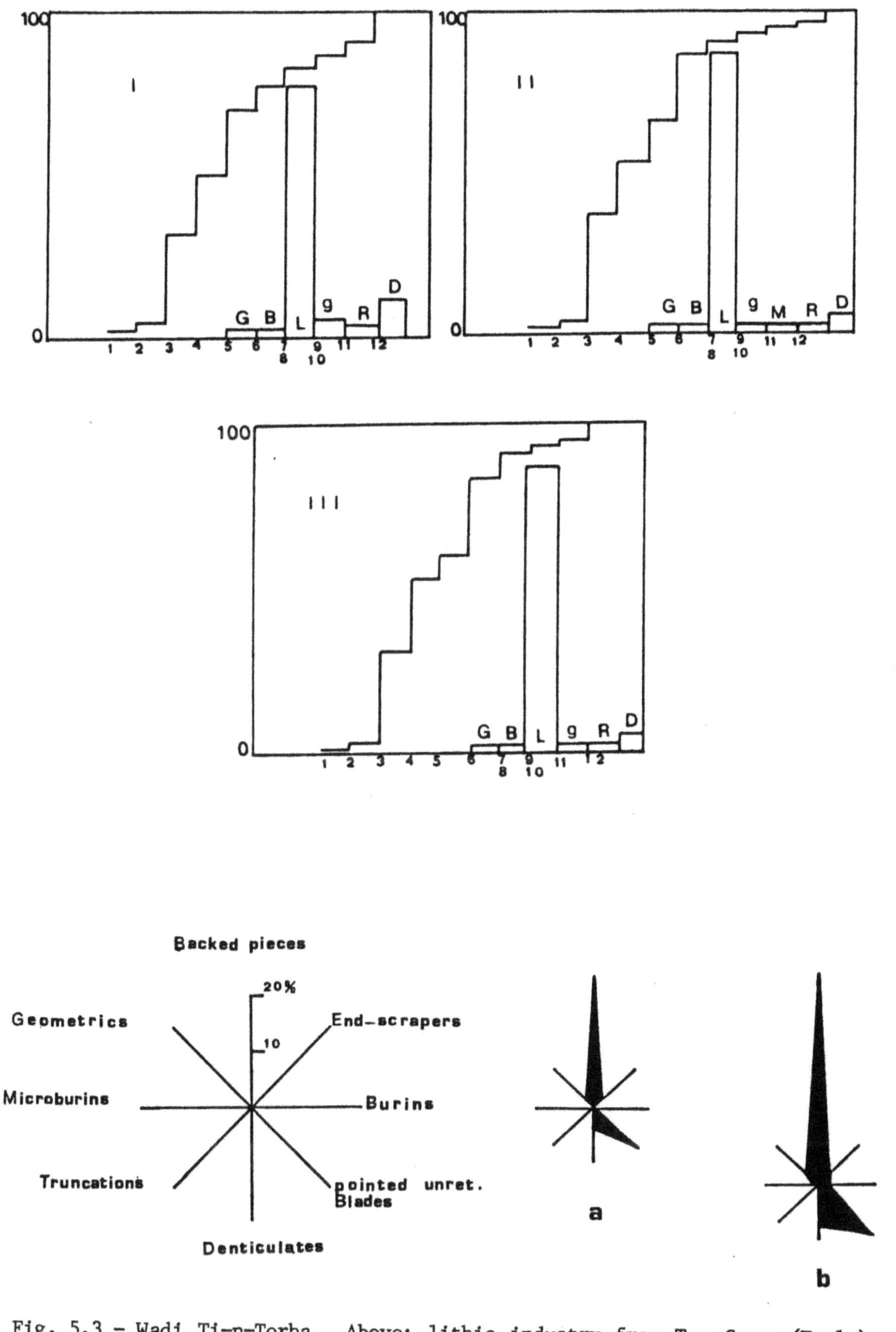

Fig. 5.3 - Wadi Ti-n-Torha. Above: lithic industry from Two Caves (Tools). Below: star-diagrams of the lithic industry from Ti-n-Torha East (a) and Two Caves (b).

from the core (3).

Unretouched or partially retouched pieces frequently belong to a type of blade-like flake from single platform cores. Although these pieces do not correspond to the conventional definition of blade, they have a marked blade-type characterization with trapezoidal or triangular cross-section and a regular ventral surface. Blade-like flakes pointed at distal ends obtained with the para-Levallois flaking technique may be observed quite frequently.

In general, cores preserve the shape of the pebbles they were taken from. This applies in particular to the initially struck cores of brown-reddish siltite in which the platform is cortex with a very deep patina. The most usual type in Levels CII-CI at Torha East is a discoidal multiple unpatterned core, often with flat cross-section. In the base levels of both deposits - Two Caves and Torha East - cores are smaller and more regularly shaped. They are prismatic artifacts of the single platform type with laminar strucks. Nucleiform burins (Tixier 1963:79, figs. 23, 24) are present.

The Two Caves assemblage presents no marked statistical differences with respect to Torha East. The stylistic approach used by A.E. Close has pointed to significant elements of distinction between the two sites, whereby Two Caves evidences a definitely more archaic aspect. This leads us to regard Two Caves as the first inhabitation of the wadi, more circumscribed than Torha East that comes to be regarded as a true proto-village.

The beginning of the Eastern Site (fig. 5.4), dating from the close of the 10th millennium B.P., entails a settlement with a microlithic (and ultra microlithic) type industry with a high blade index (88.8%). The context witnesses the introduction of ceramics which, missing in the base layer (from 9080 $\pm$ 70 B.P., R-1036 α), is unquestionably present in that dating from 8640 $\pm$ 70 B.P. (R-1035 α), with ceramics in the wavy line style.

The flourishing of the settlement is between the levels of 8640 and 7990 $\pm$ 70 (R-1034 α) and covers the intermediate period between Horizon R and Horizon C. The change made in the primary settlement, which had additional stone elements, demonstrates that the attainment of an optimum environmental condition gave impetus to the development of the settlement (fig. 5.5) (4). Although no break can be detected in the sequence, one can observe a qualitative enrichment of the ceramics and a different percentage ratio of types in the associated stone industry, where the progressively lower incidence of microlithic elements is countered by the development of macro-implements and arrow-heads. We are facing a technological

---

3) These could approach the "burin resharpening flakes" reported by McBurney (1960: fig. 20/10-17).

4) As I will clarify later on, the transformation of the site coincided with a relatively climatic drier phase which ended the seasonal abandonment of the wadi, not easily practicable during the summer rainy periods. In fact our available datings correspond to the dry period that J. Maley (1981) posed from 8500 to 8200 _ca_ B.P. when comparing the sedimentary series of the Tibesti and Nabta (Egyptian Sahara).

Fig. 5.4 - Wadi Ti-n-Torha. The East Shelter before the excavations.

a

b

Fig. 5.5 - Wadi Ti-n-Torha. a:"huts" in the East Shelter; b: sub-structure levels in the East Shelter.

advancement defined by less recourse to retouch and backing techniques and their replacement by a bifacial technique.

Level I shows signs of decline in the lithic industry after the preceding development: a decrease in all the lithic implements except arrowheads, that are prevalently fusiform and based on the blade technique. A more uniform ceramic decoration, oriented towards repetitive patterns is apparent. Was this aspect to be related to the alternation of moist and dry periods disclosed by the palaeoecological reconstruction (Bonadonna 1974), it could be ascribed to the beginning of an adverse climatic phase, negatively affecting the development of the industries.

Levels V-IV of the North Shelter witness the further impoverishment of the facies betweeen the 8th and the 7th millennia with a generalized contraction of the tool assemblage and, particularly, a decline of the blade-oriented and a consolidation of the flake-oriented, technique. The climatic change intensified until it caused the abandonment of the shelter, unusable due to the thermoclastic collapse of the roof as shown by Level III of the sequence.

The Ceramic Evidence

The earliest ceramic evidence (dating from 8640 ± 70 B.P.) consisted of fine, reddish pottery with small shapes and decoration mainly ascribable to two-toothed implements, arranged in zones and concentrated in the upper part of the vessels. Among the patterns adopted, special relevance is ascribed to the 'pseudo-dotted wavy line' pattern obtained through small cord-wrapped (or "filetés") combs. This pottery differs from that characterizing the climax of the settlement phase, being thicker and having wider spherical shapes decorated with broad-spanning side combs or spatulae used with the rocker-stamp technique. This period marks the appearance of the real dotted pattern', obtained with various types of combs that, to all appearances, are also used with rocker impressions (see I. Caneva, in this Volume) (5). In the terminal level (CI), even though the dotted wavy line pattern is still being used, there is a marked tendency towards a simpler and standardized decoration, once again with the predominance of two-toothed implements used for the impression (fig. 5.6).

The association of this early ceramic horizon of Ti-n-Torha with the wavy and dotted-wavy line tradition is not only ascribable to the presence of this special pattern and the whole series of adopted fictile techniques and themes, but also and especially to the ergologic characterization of the group as a whole. From a stratigraphic point of view, the time when the Torha East settlement was at its height (infra-structural levels from 8640 ± 70 to 7990 ± 70 B.P.), may approximate to the lower horizon of Amekni (8670

---

5) J. Hugot (1963:142) suggests a different technique based on the use of combs with one curved end in the "alternate and jointive" impression. It is well known the broad diffusion of the dotted pattern which is present in cultural contexts scattered at a long distance from each other. Cf.: Amekni (Camps 1968: Pl. XVI/1-3;XIX,6); Méniet (Hugot 1963:141, fig.54/5; 143, fig. 55/8; 145, fig.56/7); Delebo II (Bailloud 1969:37, fig.3/B); Wanianga (Arkell 1964:Pl.31; Courtin 1969: fig. 21/1,2,4,5); Shaheinab (Arkell 1953:Pls.29/1;38/3); Khartoum (Arkell 1949: Pl.72); El Qoz (Arkell 1953:Pl.38/3).

± 150/8050 ± 80 B.P.) that G. Camps traces back to the ancient phase of the "Saharan-Sudanese Neolithic", according to his latest definition (Camps 1974a: 24; 1982: 559 et seq.). The affinities with Ti-n-Torha witnessed by their respective products, both lithic and ceramic, are quite evident, even though the stone industry of Amekni was already simplified compared with Ti-n-Torha. On the other hand, the pre-structural levels of Torha East preceded Amekni and witness an older phase, at the climax of the Epipaleolithic tradition, with a geometric microlithic industry and pottery in its earliest stages, which is not at all matched at Amekni.

Discussion

As previously pointed out, many studies succeeded in giving a specific chronological and historical setting to the bladelet-industries in coastal Algeria around the 10-9th millennia B.P. They have aspects that different authors (6) relate to the Iberomaurusian tradition, leaving aside the local adjustments found. A high percentage of backed bladelets also characterizes the complexes of Southern Tunisia (Akarit, Menchia, Lalla) and of the Souf (El Oued) area. Further south, similar features are found in the Mellalian environment of the Ouargla region (Sebkhra Mellala) (Marmier, Trecolle 1968) as well as in the Hadjarian environment of the Wadi Mya (Aumassip 1973).

It was indeed daring to mention these environments when constructing a reference framework for the industry and the economic model at Ti-n-Torha East (Barich 1974: 121, 149-51). One of the guiding criteria was the willingness to emphasize the non Neolithic character - from an anthropological standpoint - of the Torha East complex, as well as its lack of relationship to the Capsian. In any event, the first impression seems correct even in the light of the recent thorough investigations. They confirm - even though in an extremely general context (A.E. Close in this Volume) - the Epipaleolithic character of the Ti-n-Torha industry. With a view to defining with a greater accuracy the most likely territories of origin, one may refer to the Nubian Shamarkian facies, through the medium of the 'Early Neolithic' of the Egyptian Sahara. I have recently specified what meaning should be ascribed to the reference to the Iberomaurusian context (7).

The process of change identified by G. Aumassip in the Wadi Mya area, subsequently corroborated at Bordj Mellala (Tixier 1975), could also apply to the interior of the Sahara, excluding the Capsian-type developments detected. Epipaleolithic cultures of that type could have been the means by which the 'neolithisation' process affected the Sahara. The introduction of ceramics might have been adopted with all respect for the stone tool industry, that is symptomatic of the initial prosecution of economic activities already being carried out. We have already seen that an ultra-microlithic, a-ceramic complex, dating from the end of the 10

---

6) Cf. Goetz 1967 (Kristelian); Cadenat 1963,1970 (Columnatian); Roubet 1968 (Elassolithic).

7) Cf. Barich 1985. In this paper, the Nubian Shamarkian is independently reported as one of the facies allowing a more specific reference.

Fig. 5.6 - Wadi Ti-n-Torha. Pottery in the dotted wavy line style from Ti-n-Torha East.

a

b

Fig. 5.7 - Wadi Auis. a: schematical engraving representing a fish; b: schematical engraving from Uan Giriual.

a

b

Fig. 5.8 - Wadi Auis. a: rock-wall with superimposed engravings; b: engraving representing a bovid (?) on the East Shelter vault.

millennium, was found at the base of the Torha East sequence. At that time, the site appeared as a hunter-fisher camp just beginning to make use of ceramics. It was, in substance, a proto-ceramic horizon co-existing with an Epipaleolithic litho-technical tradition.

Ti-n-Torha allows us to reconstruct a model of the cultural process that might have prevailed in a number of other places with a similar background. In fact, no attempt is being made to ascribe to the area a dominating role in respect of other 'wavy line' complexes through diffusion; I am sceptical of the possibility of reconstructing exceedingly circumscribed lines of descent. In my opinion, the affinities among the different groups should be seen in the way suggested by S.M. Puglisi (1976, 1981). Economic inducements determine the search for sites having definite requirements - shelters, rocky areas rich in water, closeness to rivers or catchment basins - thus resulting in a context uniformity. At the same time, the autonomy of the experiences warrants the "individual originality" that is denoted, in particular, by the lithic and bone tools (gouges, celts, arrow-heads and harpoons).

It is within this framework of small, post-Paleolithic cultures having an early ceramic experience and a rather fluid socio-economic pattern open to different specializations, that one may trace the origin of those differentiations that appear quite conspicuously in subsequent phases.

Moving on to more general considerations, there is no doubt that the sequences recently identified in the central Sahara enhance the problems of relating to the earliest ceramic complexes of the Near and Middle East. The discovery in North Africa of ceramics dating from levels that are comparable with the most ancient sites of the Near East (for example, Umm Dabaghiyah) leads to a reconsideration of the long accepted model for the appearance of ceramics and the closely related acquisition of domestication practices. It is well-known that, traditionally, investigators refer to an 'original process' - and therefore, to a real innovation - only in those places where the wild ancestors of the subsequently domesticated species lived, and that this was thought to be Middle and Near East (Ucko, Dimbleby eds. 1969; Higgs ed. 1972). Elsewhere, they suggested a derivation or introduction from the outside.

The presence of ceramics among groups that were otherwise characterized by an ergology and an organization in the Late Paleolithic tradition, is however, symptomatic of autonomous development that was not necessarily moving in the direction of an agricultural, village economy, but rather towards a model of development in harmony with the local environment. Therefore, recourse to classical definition formulas (such as the "Saharan Neolithic" and similar definitions) should be questioned since they suppose a cultural process proceeding through predetermined stages (Barich 1980 a, 1984 a). The recurrence, in a separate form, of similar experiences may only be explained by taking into consideration the system that produced them. Therefore, changes will only be understood through a thorough examination of how a system worked rather than by supposing the migration of human groups.

This peopling of the Tadrart Acacus - the most ancient so far stratigraphically documented - is characterized by a nomadic way of life with sporadic habitation sites to meet hunting and gathering requirements of the type also found in the Tihedine and Sughd wadis. In the 'Two Caves' and 'Torha East' shelters, faunal determinations point to the exclusive presence of wild species. A. Gautier, from the study of the birds, suggests a winter habitation of the sites and their abandonment during the summer rainy period (Gautier, Van Neer 1977-82:101). This alternation quite likely ended with the climatic stability that came about 8500-8000 B.P.

It is at this time that the type of settlement changes, acquiring a character that is more markedly sedentary as evidenced by the stone structures present at Ti-n-Torha East. The inception of more continuous occupation is also recognized in the "developing" features of the lithic and ceramic finds and is warranted by the stability of the environment in relation to a more constant availability of water, as pointed out by B. Marcolongo. This is confirmed by the botanical and zoological sequence, both supporting a continuity of the wadi occupation.

On the basis of both the environmental and the locational aspects, it was suggested that the Ti-n-Torha communities might have played a rôle in the preselection of species intended for domestication (Barich 1984a). The area of Central Sahara appears as a generalized ecosystem (Harris 1969:8 <u>et seq.</u>) of hunter-fisher-gatherers, aiming at an intensive exploitation of the bioma. The groups peopling it, taking advantage of a wide spectrum of resources, were no longer forced into long-distance mobility; a limited nomadism in small mountain valleys or in broader plains may have contributed to practices of incipient domestication, in respect of plant and animal species (8).

In the Ti-n-Torha facies I specifically suggest this initial level of exploitation of the territory that is related, not only to a microlithic industry associated with dotted wavy line patterned ceramics but also, and especially, to a preparatory role in food production.

Hunters armed with bows are a recurrent theme of the local rock art. If the whole pastoral art style is to be related to the culture of the herders of the 6th millennium B.P., the representations that fail to fit into that sphere of interests and pictorial context (hunting scenes related to an "ancient pastoral style", schematic figures (see figs. 5.7,8) may perhaps be traced back to this early peopling, the first such event which is stratigraphically documented, and may be ascribed to <u>H. Sapiens Sapiens</u>.

The definition of the Ti-n-Torha facies ascribes a specific background to the pastoral communities that, shortly afterward it will appear in this same area. A similar pattern is found in such far-away regions as the Egyptian Sahara (Wendorf, Schild 1980) and the Republic of the Sudan (Caneva ed. 1983) which confirm the observations in the Tadrart Acacus. The term "Neolithic Revolution" entails a radical transformation of the arrangement and life-style of groups, while it is clear that such a transformation could not have occurred unless a number of fundamental conditions such as

---

8) Further details as regards the botanical species will result from the study by K. Wasylikowa. Among the faunal remains I refer to the initial recognition by A. Gautier at Torha East of a "large bovid", that could be paralleled in the evidence of Nabta and Bir Kiseiba (Gautier, Van Neer 1977-82: 105). The author, in this Volume reconsiders this possibility but in any event, one may not rule out the possibility of forms of control involving <u>Ammotragus</u> itself (cf.Close 1980: 51).

sedentism had already come about. The area of the Central Sahara appears as a zone where fishing practices had accustomed human groups to a special, short-distance nomadism that involved the return to home-base settlements (9). A decrease of the available water resource may have led the groups to emphasize gathering and the domestication of local species without bringing about a relevant change in their structural context.

---

9) This is the area of expansion of dotted wavy line decoration on pottery that, in addition to the Hoggar and the Acacus, spread to Mali (Smith 1976) Ténéré (Roset 1983) and Borkou (Courtin 1969).

PART II: Wadi Teshuinat and the Central Acacus

CHAPTER 6

RESEARCH IN THE CENTRAL ACACUS MASSIF

By Barbara E. BARICH

1. PAST INVESTIGATIONS

The initial investigation in the Acacus massif, carried out under the direction of F. Mori, focused on the Wadi Teshuinat and the wadis connected with it (fig. 6.1a). The Teshuinat was found to be a large valley bordered by a number of shelters, horizontal recesses, in sandstone cliffs. It is quite likely that, in prehistoric times, this valley had been frequented over and over again by pastoral groups during times of transhumance. As a matter of fact, the type of occupation in the different shelters points to many short stays, at short time intervals, leaving a sequence of hearths and dwelling floors in small and thin levels.

Seven campaigns were organized from 1955 to 1964 for the purpose of exposing and studying the rock art that, made known by ancient travellers, were subsequently uncovered and published by L. Frobenius (1937) and P. Graziosi (1942; 1962). In the light of these investigations, the Tadrart Acacus became known as a centre of both engravings and paintings in which Mori succeeded in identifying a sequence of stylistic phases. The most significant among these, the 'pastoral' phase, consists of an extremely rich pictorial complex spread along the major wadis around the Teshuinat (Ti-n-Lalan; Imha; Anshal; Tanshalt). Also distinguished, within the same area, were scenes depicting the large African fauna, the 'Big Wild Fauna phase' and schematic style (phase of the 'Round-Heads') that was at odds with the realistic style of the pastoral phase. Both were considered to precede the 'Pastoral' phase. Mori pointed to the difficulty of associating those documents with specific cultural groups since no deposit that could be related to phases preceding the 'pastoral' had yet been discovered (Mori 1965: 233). It should be pointed out that the 'Round-Heads' phase was particularly well represented in the area where most pastoral settlements were found; as for example in the Ghrub and In Taharin shelters, or at Uan Muhuggiag. Moreover, within the pastoral phase itself, Mori pointed to an earlier sub phase characterized by the presence of both a more refined style and individuals, or small groups of people in hunting scenes.

Part of Mori's research was addressed to the survey of archaeological deposits, a number of them being within shelters decorated with paintings. Except for Uan Muhuggiag, where more comprehensive investigations were carried out, most of the survey involved small-scale exploratory excavations or the collection of samples for radiocarbon analyses; reference to these sites will be limited to the Fozzigiaren, Uan Telokat and Uan Muhuggiag shelters.

The Fozzigiaren Shelter, sited beside the wadi of that name, had two cultural horizons: the upper one belonging to the 'pastoral' phase and the lower one associated with "dotted wavy line" ceramics of the best tradition (fig. 6.2*b*); the date 8072 ± 100 (1), the earliest obtained by Mori in this preliminary research cycle, must relate to this horizon. It should be noted that Fozzigiaren reproduces the environmental conditions of the Auis (fig. 1.1) since it is sited in the south-eastern area outside the massif (2).

The Uan Telokat Shelter, sited in the Imha area, directly to the south of the Teshuinat (fig. 6.2*a*), is another important pastoral site uncovered during this preliminary work. It is a large shelter, a few metres above the wadi bottom, where the deposit covered a painted figure which, although vague, was definitely not in the 'pastoral' style. A test exploration showed that the lower archaeological layer, dated to 6754 ± 175 (3), had covered the lower section of the figure. It is evident that the formation of this part of the deposit was subsequent to the execution of the painting and that its date provides a *terminus ante quem* for the painting itself. The deposit was thick and the exploration failed to uncover the deepest levels. Therefore, one may not rule out that, just as at Fozzigiaren, the pastoral phase had been preceded by an earlier one to which would belong the figure, quite likely in the 'Round-Heads' style.

The site where the most noteworthy finds were made was Uan Muhuggiag Shelter on the northern side of the Wadi Teshuinat (fig. 6.1*b*). A. Pasa in 1959-1960 opened two trenches here on the western and eastern sides of the deposit, ("Taglio IL" and "Taglio IA"). Past observations (Pasa, Pasa-Durante 1962) and Mori's comments (1965: 221-28), refer for the most part to Taglio IA, which had good organic preservation (fig. 6.3*b*). The presence of a skull of *Bos*, as *ibericus* (*Brachyceros*) (Pasa, Pasa-Durante 1962:254) dated 5952 ± 120 B.P. is one of the earliest finds of cattle in the Sahara. The skull dating was one of three early dates (all by the Institute of Nuclear Geology, Pisa), the other two being 5405 ± 180 B.P.: from the animal skin wrapping of a child burial (Trench IA, 1959), and 7438 ± 120 B.P.: from a hearth in the deepest level (Trench IA, 1959).

The latter date is particularly significant since it is related to the beginning of the Uan Muhuggiag sequence and, if accepted, would move the beginning of the pastoralism at this site back by over a millennium. This date is different from those obtained from the directly superimposed levels which agree with recent dates from the 1982 excavations (see Chapters 7 and 14), but agrees with the date obtained from the nearby site of Uan Tabu.

Already, when the Pasa excavations were in progress, a discordance had been detected in the sequence and, although the author failed, perhaps, to stress it enough, it shows when correlating the stratigraphic series resulting from Trench IA and IL. Trench IL (opened in the western area of the deposit) suggested, towards the end of the sequence, to the author, a

---

1) Obtained by the Institute of Nuclear Geology of the University of Pisa.

2) The investigations performed in 1982 and 1983 have uncovered in the In Araien (Fozzigiaren) and Afa (Tacharchori) area the presence of some very beautiful scenes in the 'Round-Head' style (Lupacciolu 1978 in press).

3) Obtained by the Geochron Laboratories of Cambridge, Mass., U.S.A. In the text, they will be referred to as "Geo".

a

b

Fig. 6.1 - Tadrart Acacus. a: a pointed panoramic view of the Wadi Teshuinat; b: the Uan Muhuggiag rock shelter in the Wadi Teshuinat.

phase of aridity owing to the presence of dune deposits and the rarity of occupation signs. Charcoal samples coming from it, in a level overlying a boulder that, breaking away from the back of the shelter, had fallen into the deposit, were dated to 4730 ± 310 (Geo). The lower surface of the boulder was decorated with two oxen and, quite clearly, were painted prior to the collapse (fig. 6.3a). Besides providing a fixed point for the subsequent occupation to the cave, this date also sets a terminus ante quem for the boulder paintings (4).

The Uan Tabu Shelter faces the Wadi Teshuinat, not far from the Uan Muhuggiag Shelter. It contained an extremely important deposit over thirty metres in length and about two metres deep. A small test exploration (1 x 2 m) revealed a sequence of adjoining layers extremely rich in charcoal and closely comparable to Uan Muhuggiag. The exploration disclosed three horizons (lower, medium, upper). The upper is associated with a late, wooden hut-structure which was also partly excavated. The middle strata gave the date 7045 ± 175 B.P. (Geo) that may confirm the early age of the Uan Muhuggiag base level.

Reconsidering the major results obtained by the initial research in the Tadrart Acacus, one can affirm that:

1) In the light of conditions as those at Fozzigiaren and Uan Telokat, doubts expressed about the presence in this area of contexts preceding the pastoral cultures are now meaningless.

2) This presence provides an indication in respect of the relations to be established with the 'Round-Head' phase of rock art.

3) The presence at Uan Muhuggiag and Uan Tabu of a developed pastoral economy since the 8th millennium B.P., shows that in this area there existed nuclei of early domestication. In view of this fact, an exact chronological setting of the earliest pastoral events is now essential to an understanding of the cultural process that took place in the Tadrart Acacus.

4) The chronology of such rock art styles as the Uan Tabu (which may correspond to the Uan Muhuggiag sequence), and the Ti-n-Lalan might have developed after 4730 ± 310 B.P. It should be pointed out that the Ti-n-Lalan type long-limbed herders are considered by Mori as the last of the groups that frequented the massif.

2. THE NEW RESEARCH CYCLE: MAIN OBJECTIVES

The resumption of the excavations in the Teshuinat area in 1982 was intended to relate the new findings to the investigations in the Wadi Auis. The results obtained in that area allowed us to suggest that domestication processes might have had a local origin, but in order to check this hypothesis, it appeared necessary to confirm the extremely early dates from the Teshuinat and to define the association between the domesticated species and the archaeological evidence. The investigations in the Auis

---

4) Ante quem date for the boulder paintings (in the "Uan Tabu" style) and post quem date for the paintings depicted on the wall section that resulted further to the collapse (in the "Ti-n-Lalan" style) (cf. Mori 1965: 239).

a

b

Fig. 6.2 - Tadrart Acacus. a: the Uan Telokat rock shelter in the Imha area; b: the site of Fozzigiaren in the wadi having the same name. (After Mori 1965: figs. 147 and 145).

had allowed us to define a specific facies (the Ti-n-Torha facies) in which pre-conditions necessary for food production could be recognized. The existence of a relation between that facies and later pastoralists, could only be ascertained by a comparison of the respective finds.

Special relevance was also ascribed to the study of the links between economic transformation and the new features in the rock art and so to explain the origin of a phenomenon as important as that of food production. Moreover to the hypothesis, formulated in Chapter 5, of a continuity from the purely extractive (hunting-gathering) to the pastoral economy and of how the latter might be considered a further adaptation from a broad-spectrum one. Further investigations in the Teshuinat area and the resumption of the Uan Muhuggiag excavations, have allowed a proper formulation of these problems, even though it is quite clear that by themselves they may not illuminate the whole process.

From a geo-morphological point of view, this central part of the Acacus is quite different from the north-eastern area. The altitude is relatively high (around 900 m above sea level), the wadi pattern is much more developed and converges into systems which became major communication routes. B. Marcolongo in Chapter 11, observes that notwithstanding a greater spread of the wadi system with respect to the northern area, there are no passes through the watershed and, therefore, no communications with the western slope, while these are well developed in the northern area. In view of these facts, the Teshuinat area appears as a rather closed ecosystem, protected against the spread of desertification from the north (Sebha region), and was characterized by a greater relative humidity.

It is most likely that the movements of human groups within the massif occurred along the routes marked by the major wadis (Ti-n-Lalan, Kessan, Teshuinat, Anshal, Tanshalt) while the transhumance movements occurred, out of necessity, along the routes leading towards the east (5). The more secluded and protected conditions of this area favoured good preservation of occupation evidence, both rock art and settlements. The notion of a high occupation density that results from it, is matched by the importance of pastoralism, in respect of the frequency of both human groups and herds. The traces they have left appear to be much more common than those of hunters.

The possibility of making a comparison between the technologies of the two economies may be ascertained. The technology of the pastoral groups continues the traditions of the preceding phase of both ceramic and lithic products, even though the latter show a marked impoverishment. The recovery of individual types has also allowed the establishment of more specific links. As far as chronology is concerned, the new dates obtained from Uan Muhuggiag cover the whole 6th millennium B.P. and, therefore, are consistent with those resulting from the medium-lower levels of Pasa. The lowest Uan Muhuggiag date (7438 $\pm$ 120), had not been matched as yet by dates of a comparable age in the new series. It should be borne in mind that only three samples have been analysed and, thus, an effective control will

---

5) A. Gautier (Chapter 12, this Volume) refers to the 'stopover sites' of the Sebha region: the "fireplace/Brandgrube", reported by B. Gabriel (1977:177). Gabriel has recently re-formulated the problem of these formations (Gabriel 1984).

a

b

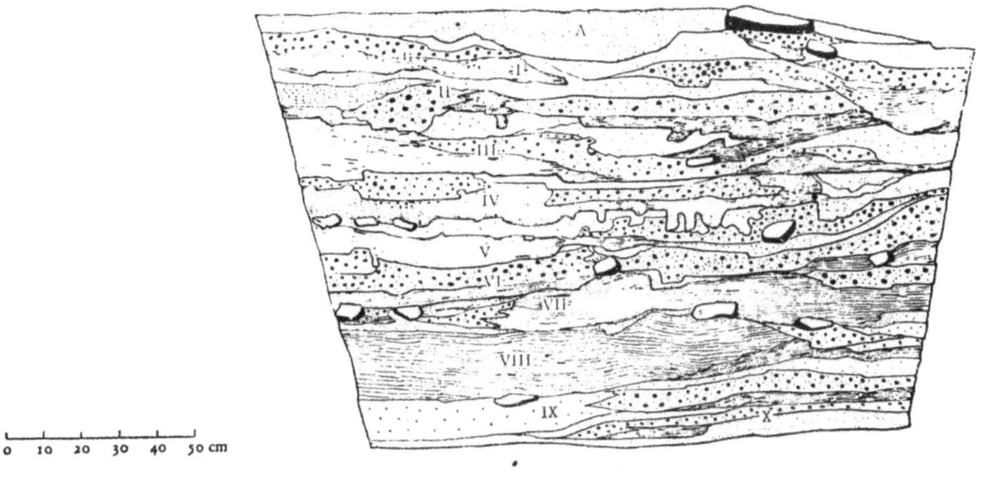

Fig. 6.3 - Tadrart Acacus. a: boulder with paintings recovered from Uan Muhuggiag; b: schematical stratigraphy of Trench IL at Uan Muhuggiag. (After Mori 1965: figs. 156 and 153).

only be feasible when the whole study being conducted by G. Belluomini is complete.

The evidence from the rock art sites in the same area agrees with the changes observed in the sites. Two distinct groups can be defined in the artistic evidence. On the one hand, there are scenes that are linked to the hunting world and might be associated with early domestication experiments (see the Uan Muhuggiag, Uan Tabu and Uan Telokat base levels), and there are scenes of the full pastoral phase, where the herd as a whole (see, for example Uan Tabu) or as a few isolated heads (Uan Amil) represent the dominant theme. The study of the distribution of this evidence with respect to the settlements may contribute to clarifying the various steps involved in developing pastoral-oriented processes. It will also attempt to study the specific correspondence of orientation and the location in preferred settlement sites (cf. Soleilhavoup 1979: 121), and attempt to check and confirm from rock art the archaeological data.

In my opinion, however, the conventional stratigraphic excavation work still plays the main rôle in solving the questions outlined. The complex of information, resulting from the last excavations of Uan Muhuggiag, allowed the reconstruction of an elaborate palaeoeconomic model meeting specific environmental requirements, which was made possible through the combined study of faunal and pollen series. This model may be further refined by a more specific reconstruction of both environment and climate, through the study of macro-botanical remains (6). It is also feasible to ascribe to this general outline specific archaeological traits, in the lithic and ceramic technology, tracing back their development through time. Uan Muhuggiag is only one example of the potential that the Teshuinat area appears to have.

---

6) It was already stated in Chapter 1 that the study, entrusted to K. Wasylikowa, could not be included in this monograph and will be published separately.

CHAPTER 7

THE UAN MUHUGGIAG ROCK SHELTER

By Barbara E. BARICH

1. THE SITE AND THE EXCAVATIONS

The resumption of the excavations at Uan Muhuggiag in 1982 was determined by the information that had resulted from the initial excavations by Pasa. The sequence appeared to shed light on the questions outlined in the preceding chapter.

The shelter is located on the northern side of the Wadi Teshuinat, at approximately 3.50 m above the wadi bottom, and is approximately 130 m long (fig. 7.1a). It had been studied on account of the pictorial complexes spread over a broad section of its back wall. At least five scenes belonged to different pictorial styles, from the so-called "Round-Heads" style to rather late scenes in the "Horse" style (1).

Moving from West to East, the scenes appear in the following sequence:

- Scene quite likely depicting a boat with 'Round-Head' figures inside. There was no archaeological deposit and the ledge in bedrock at this spot is very smooth and 1-1.50 m above the wadi bottom; there was one of the so-called "Giant's Kettles" (2) (fig. 7.1b).

- Groups of figures in the 'bitriangular' style that do not allow an easy reading. There was no deposit and the bedrock was at the same height as before.

- Scene of cattle with herders. The characteristics of the ledge in bedrock remain unchanged.

- Scene with a figure of a dead person, with a matt background filling, attributed to the final phase of the 'Round-Heads'. Beside it, but unrelated to it, there are some running hunters. There is no deposit and, from this point, bedrock begins to slope in a S/E direction.

- Scene foreshortened, of red female figures in a row. The archaeological deposit starts here, in association with some large boulders, resting upon bedrock and resulting from an ancient collapse.

- "Egyptian"-type scene depicting running individuals, attributed to the "Horse" phase.

---

1) The pictorial complex is published (Mori 1965).

2) Forms of natural erosion that were subsequently extended by man to store liquids and grains.

a

b

Fig. 7.1 - Uan Muhuggiag, Wadi Teshuinat. a: outer view of the shelter; b: rock floor in the western side of the shelter.

a

b

Fig. 7.2 - Uan Muhuggiag, Wadi Teshuinat. a: rock-wall in front of the deposit; b: the settlement before the 1982 excavations.

The boulders had aided deposition and subsequently the accumulation had covered the sector of the shelter on the opposite side to the paintings (fig. 7.2 a, b). In front of the paintings the shelter floor is bare, very smooth and has, in several spots, "kettle"-like erosions. This separation between the dwelling and the pictorial areas was ascribed to ritual. It should be pointed out, however, that the ledge, the actual floor of the shelter, has a steep W/E inclination and, thus, at the western end, where the so-called boat scene was observed, the space between floor and ceiling does not provide as convenient dwelling area as that close to the protection offered by the large sandstone blocks.

A. Pasa conducted two test explorations, a western (Trench IL) and an eastern (Trench IA), at each end of the area where the deposit reaches its greatest depth. Chapter 6 summarised the data resulting from these excavations (3).

The untimely death of Pasa did not allow to conduct a systematic study on the archaeological material that, indeed, might have shed light upon the inner variability of the deposit. The relative data have however been concisely presented and commented upon by F. Mori (1961, 1965:221-28) on account of their chronological value. Different matters were left open: the need to specify the course of domestic species; the potential relations with the Early Holocene facies; in addition to, quite clearly, the attempt to establish a relation - still ill-defined - between the inhabitation levels and the pictorial context. Hence, the interest in a resumption of the investigations.

The general layout (fig. 7.3) clearly shows that the two trial excavations had only marginally affected the deposit. This resulted in the selection of an intermediate area between these two, delimiting a perfectly preserved area of approximately fifty square meters. This area was covered by a 1x1 m grid and, inside it, two excavation sectors were defined, identifying them as UM A and UM B. In Sector A, the outer front was set at the large boulders resulting from the ancient collapse, while in Sector B it was drawn back toward the back wall.

As I have already pointed out, the type of sediment shows a closely connected sequence of hearths and compacted levels. It seems that, in both sectors, the identified levels should be attributed only to two cultural events.

Sector A (figs. 7.4, 5)

Level 1 - Incoherent level consisting of sand and loose soil mixed with vegetable elements (seeds, berries, fibres, tiny pieces of wood).

---

3) cf. Pasa, Pasa-Durante 1962. The work emphasized the biological aspects of the data, including a mummified child (Mori, Ascenzi 1959, republished in Ascenzi et Alii 1983) and the examination of sub-fossil insects (Pierre 1963a, 1963b). It was completed by a geo-morphological study focusing on the Tadrart Acacus (Bartolomei 1966). Further information as regards fauna may be obtained from Sorbini et Alii 1974.

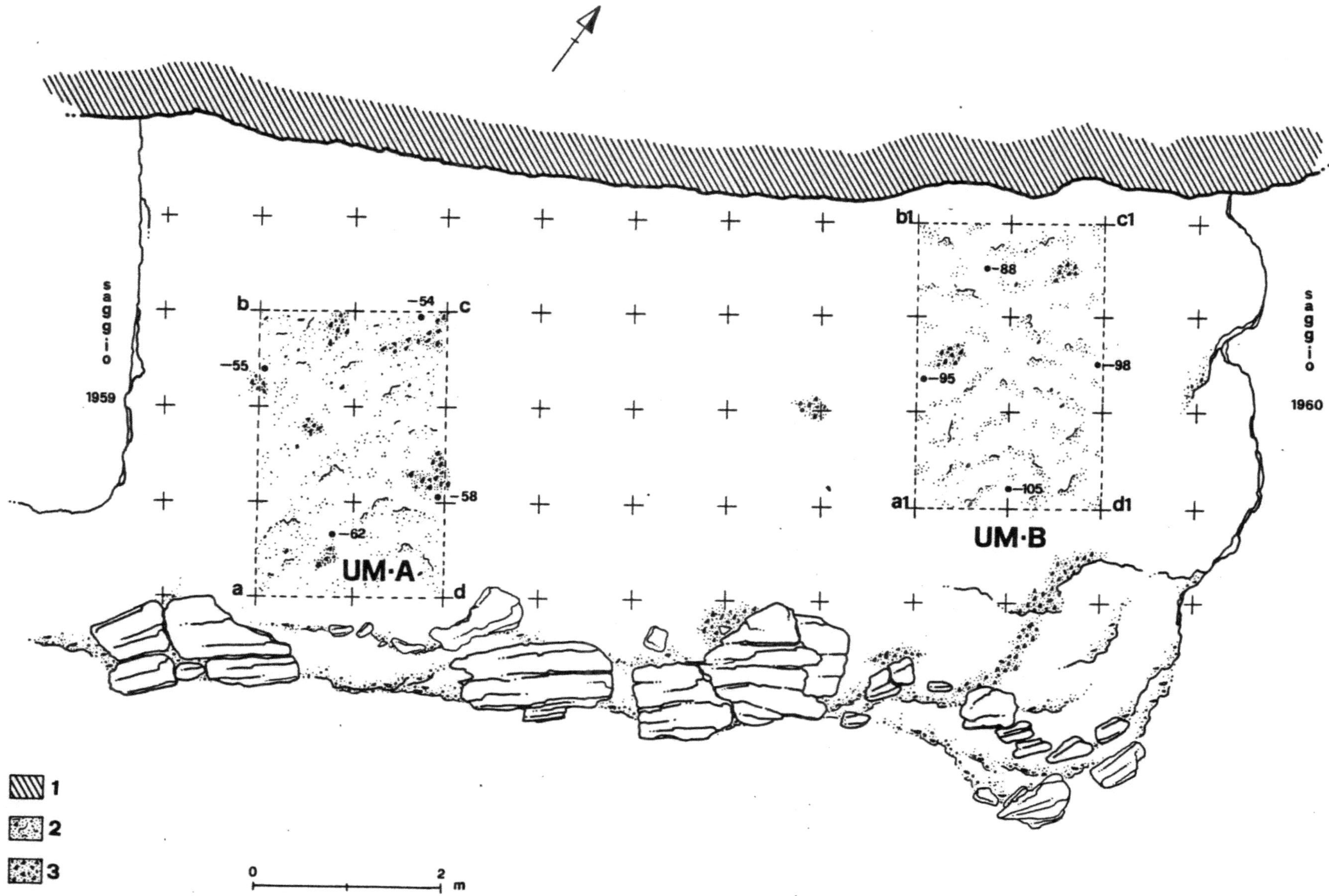

Fig. 7.3 - Uan Muhuggiag, Wadi Teshuinat. General layout of the excavated area. Key: 1) sandstone; 2) sand; 3) charcoals.

It is closed at the top by "crusta" type compacted soil, while on the south-eastern side it rests against a large boulder. A number of hearths are visible and their presence is marked by extremely compacted soil.

1a - The soil, ranging in colour from reddish brown (5YR 4/3 - 5YR 5/3) to brownish yellow (10YR 6/6), has initially a composition similar to that of the preceding level but gradually acquires greater firmness, tending to accumulate into clods. Towards the base it becomes looser due to the presence of rock collapses and secondary weathering of the sandstone, until it acquires a sandy texture. The presence of the collapses restricted the excavation area.

Radiocarbon dating of this level: 3700±200 B.P. (Ud 224) (4).

Level 2 - A blackish soil (5YR 3/1 - 5YR 2.5/1) with dark reddish brown (5YR 3/2) areas close to the hearths. The soil is very loose, mixed with ashes, dust and clods of clay in relation to the hearths where charcoal was collected for C14 analyses. It was separated from the preceding level by an extremely compact crust and represented a different cultural horizon.

Radiocarbon dating of this level: 6035±100 B.P. (Ud 225).

The following subdivisions represent separate phases within a single settlement phase. Each level is separated from the one beneath it by a thin concreted layer that represents the inhabitation surface.

2a - Grey (5YR 5/1) coloured soil with lighter zones where therewere patches of sand and vegetable remains. Towards the bottom it acquired a pink (5YR 7/3) colouring. The sediment base was dusty, like the preceding level, and on the western side had a compact surface with hearths from which charcoal was collected.

2b - The soil is more compact, with a dark reddish brown (2.5YR 2.5/4) colour, separated from the preceding level by a lighter coloured crust, partly concreted. In a few places there are accumulations of blackish ashes.

2c - At its surface were hearths characterized by compact clay mixed with charcoal. The soil beneath, ranging in colour from reddish brown (5YR 4/3) to grey (5YR 5/1), was very loose and, especially on the western side, included bundles of plant fibres.

---

4) These are the first three dates, from the C.R.A.D. of Udine, from the series sampled in 1982. Reference to dates further on in the text are from preceding work. Comments on the dating were communicated by me to G. Belluomini of the Dipartimento di Scienze della Terra, University of Rome "La Sapienza", and included in Chapter 14.

It ends on a surface formed by calcareous salts. The outcrop of small and easily flaking stones, anticipates the basal sandstone ledge. In view of the extreme rarity of archaeological finds, the excavation stopped without having fully uncovered the bedrock.

A general observation of the stratigraphy shows that the whole upper part has the appearance of a very thick crust, sloping in a W/E direction which grew markedly thinner in Sector B. This formation (level 1 and 1a) ended on a tumble of sandstone blocks fallen from the roof which covered much of the lowest occupation surface. The collapse provided an unquestionable stratigraphic indication, definitely sealing level 2 which was characterized by a completely different type of sediment,its archaeological and faunal remains being also markedly different. Due attention should be paid to the high incidence of floral remains.

This sequence agrees with what was observed in Trench IL of the Pasa excavation which adjoins Sector A. It was here that the investigators noted the presence of a boulder, collapsed from the back wall, which depicted two bovids in the Uan Tabu style. This corresponds to the roof collapse observed in the 1982 excavation. The new date resulting from Level 1a is consistent with the date obtained by Pasa (4730±310 B.P.). On these grounds, the formation of the entire upper layer should date from the 4th millennium B.P.

Sector B (figs. 7.4, 5)

Level 1 - Yellowish red (5YR 4/6) clayey soil, with a tendency to form very compact clods; there were no vegetable remains. It was a stratum with regular stratification, comprising a limited number of charcoal lenses, sealed at the top by a compacted crust of cattle dung. Another compacted crust marks the separation from the underlying level. The whole plane has a marked NS/WE inclination.

1a - The soil forms a dark grey (5YR 4/1) clayey crust with frequent hearths on the surface. There was much vegetable matter especially round the hearths. A compacted layer at the base marked the change to Level 2.

Level 2 - A blackish soil like Level 2 in Sector A of which this was a continuation. The stratigraphy was limited to the times of abandonment and periodic re-use by human groups.
The first sublevel showed three hearths with accumulations of grey soil (5YR 5/1) mixed with ashes and charcoal. The almost intact skeleton of a small animal, probably a sheep (fig. 7.6) was recovered by one of the hearths at the centre of the inhabited area. Charcoal for C14 analysis was collected from close to this area.

2a - Loose, grey-coloured (5YR 5/1) soil, mixed with extremely frequent vegetable remains of the Typha type that, in certain areas, took on the appearance of weaving or matting.

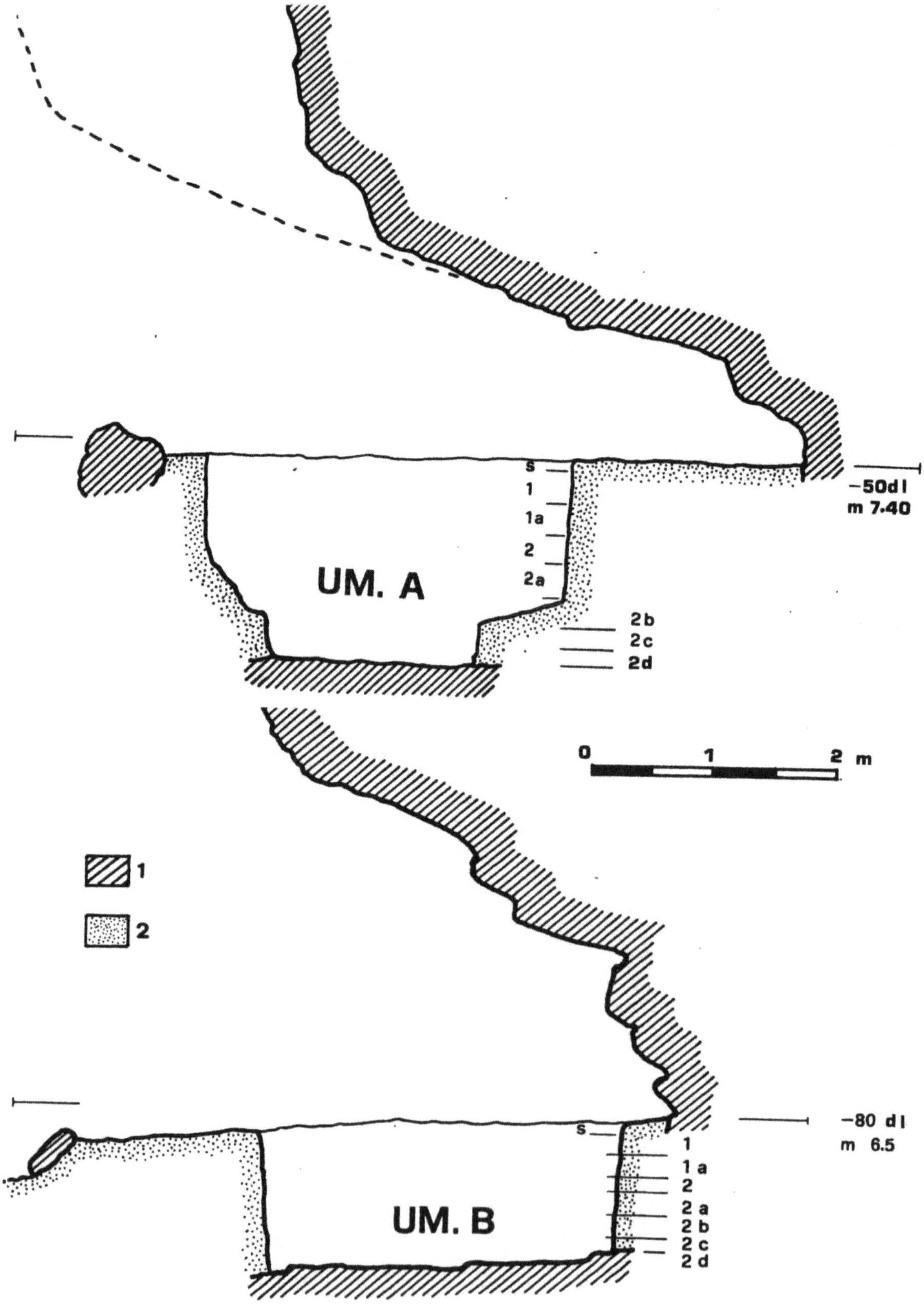

Fig. 7.4 - Uan Muhuggiag, Wadi Teshuinat. Cross-sections. Key: 1) sandstone; 2) sand.

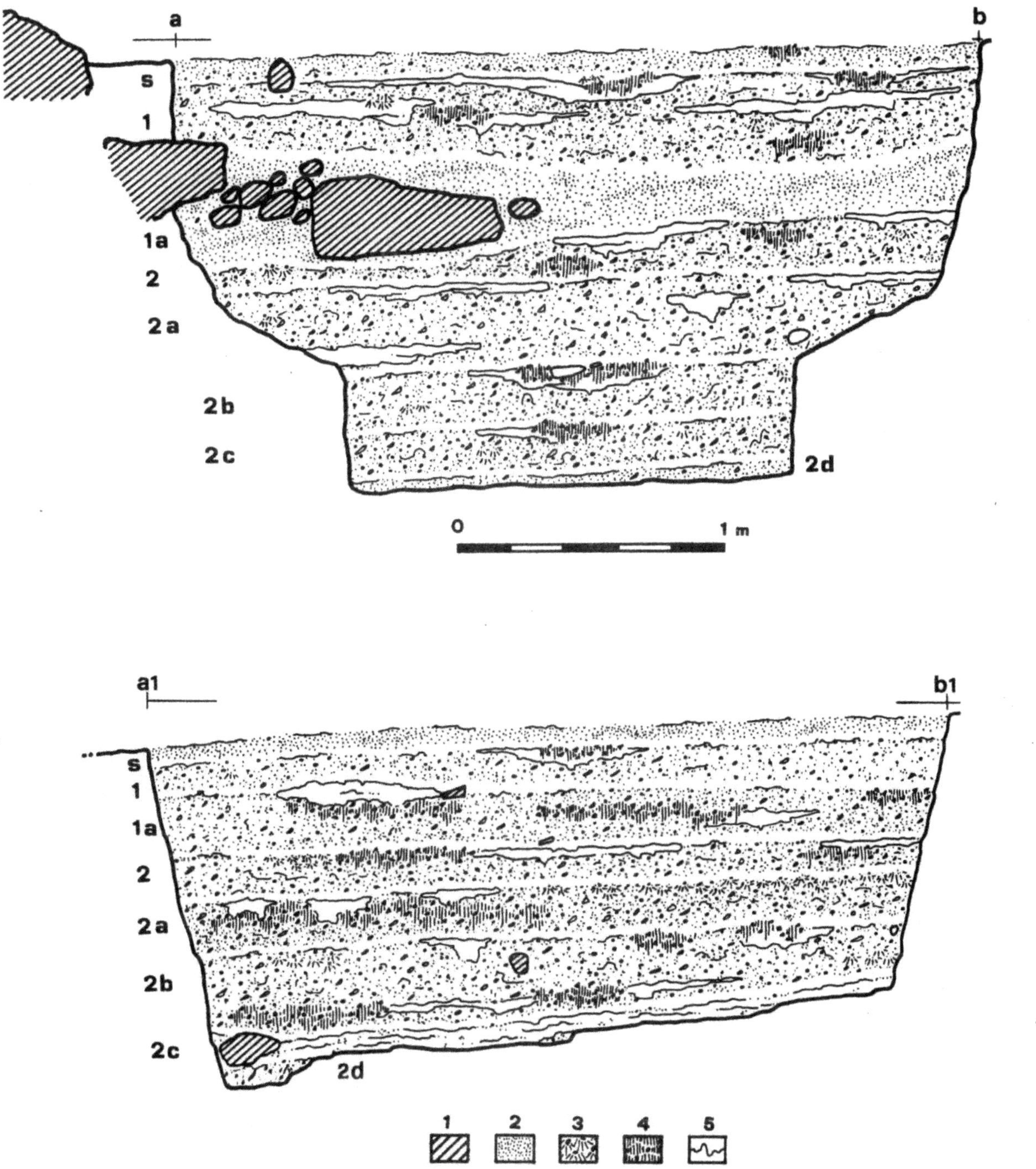

Fig. 7.5 - Uan Muhuggiag, Wadi Teshuinat. Stratigraphies of Sector A (above) and Sector B (below). Key: 1) sandstone; 2) sand; 3) fibres; 4) coals; 5) clay.

2b - Very dark grey (5YR 3/1), dusty soil mixed with ashes and much charcoal including, in the deepest part, macroscopic vegetable remains (berries, tiny pieces of wood, fragments of branches) that rested on an unmistakably inhabited surface. In the south-eastern corner this was broken by stones forming a hearth with a wooden pole with a pointed end stuck inside.

Radiocarbon dating of this level: 5350±200 B.P. (Ud 226).

2c - This sublevel was mixture of ashes and dust, with several carbonate concretions that gave the soil a light grey colour (5YR 6/1). The soil rested on a stone layer that anticipated bedrock.

2d - The initial habitation was marked by a very thin layer of light grey (5YR 6/1) dust that rested directly on the bottom stone layer following its contour and sinking into two pockets on the north-western front and the south-eastern area. Charcoal was collected.

The stratigraphy observed in Sector A served as a guideline during the excavation of Sector B, which was more complex on account of the hearths. The upper part of the deposit failed to show that separation from the underlying part that had been observed in Sector A and in profile the stratigraphy appeared as an uninterrupted sequence of thin levels, rich in ashes and charcoal, suggesting many closely successive spaced abandonments and re-utilisations of the site. Most of the archaeological evidence found came from this sector.

In Chapter 12, Gautier asks what meaning should be ascribed to the recovery of the almost intact skeleton of a juvenile sheep. No special positioning was observed in respect of the skeleton that lay covered by soil in the central hearth, it may merely have been a sick animal buried to avoid scavengers.

2. THE ARTIFACTS

The analysis of the complex followed this order: lithic industry, bone industry and ceramic industry.

Lithic Industry

The lithic industry, unlike the Ti-n-Torha industry, failed to show a specialization on specific blade-types, being characterized for the most part by products made on flakes by a rather poor technique. Compared with the limited number of identifiable tool-classes, exclusively macrolithic, the cores/debitage component was numerous and allowed a study of the production method. It was deemed advisable to use the method already applied to industries from the Egyptian Sahara by A.E. Close (Close ed. 1980: 49-54).

The tool-classes were analysed using the typological model resulting from the Ti-n-Torha study. This showed the disappearance of a few classes, such as geometric microliths, truncations and the microburins, while only

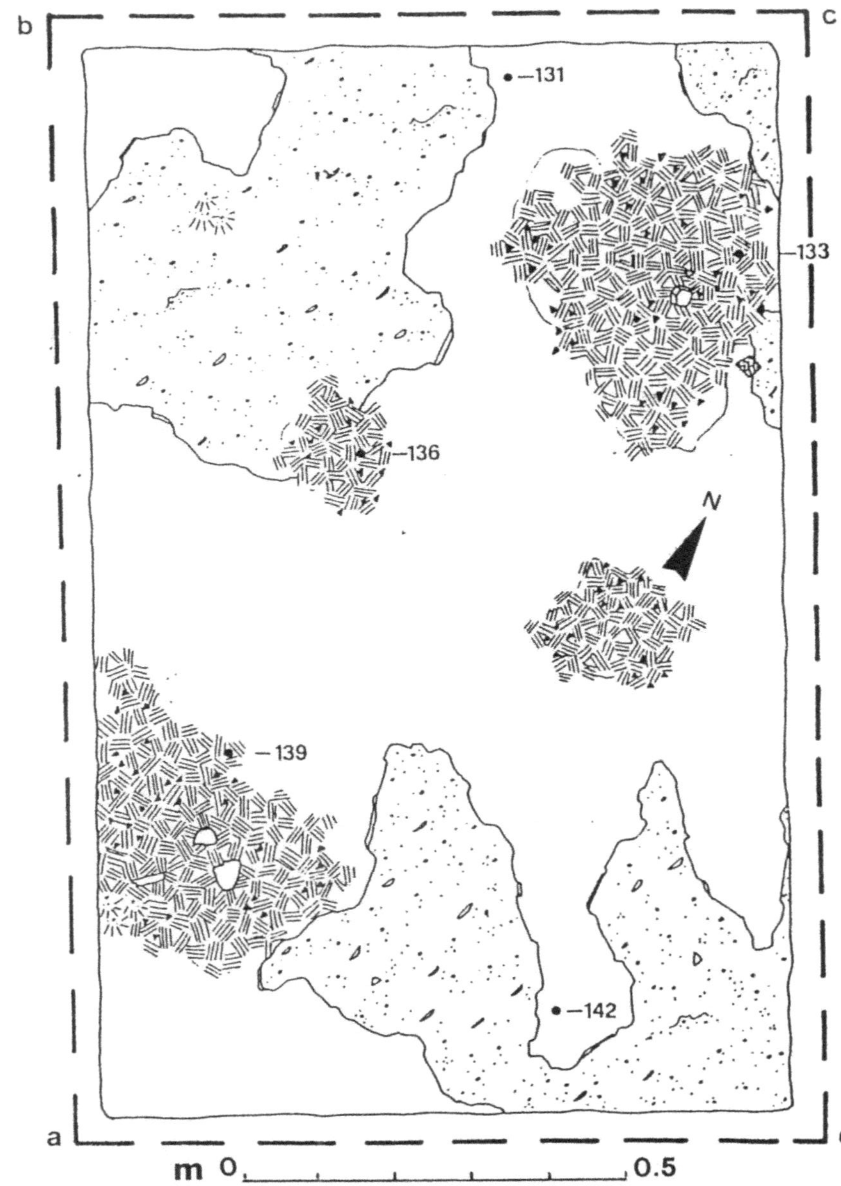

Fig. 7.6 - Uan Muhuggiag, Wadi Teshuinat. Skeleton of a juvenile sheep in Level 2, Sector B and (to the right) plan of the same level.

rare specimens of the variety of backed bladelets observed at Tin-n-Torha were present. The frequency of notched and denticulated pieces was still high and there was an increase in pieces with less specific typology (varia), such as 'continuously-retouched' flakes and side-scrapers. There were also, although limited to a few levels, arrow-heads belonging to types from the later phase of the Ti-n-Torha horizon. Table 7.2 shows the variety of the Uan Muhuggiag assemblage taken as a whole. As was done for the Ti-n-Torha assemblage, blades were separated from the debitage and included among tools to allow a comparison of the blade indexes.

Within the artifacts was a group of partially polished "Large Stone Tools". These lacked specific characterization, perhaps being intended for a variety of uses, with the exception of the small axes, part polished and part flaked and the grinding-implements. The latter, few in number, were almost exclusively spheroid querns.

The analysis has not revealed any appreciable variance in the utilization of raw materials, even in specific classes of lithic artifacts. Coarse-grained, non-Devonian sandstone was the material used with the highest frequency (nearly 85-90%), in the various debitage-types and grinding-implements, followed by the finer-grained quartzitic sandstone and, less frequently still, by siltites. Chert and quartz were also present in small amounts, the latter being used for 'chips' or as microlithic pebbles (especially in the Upper Series of the stratigraphic sequence). The utilization of flint, was exceptional. Specific cases of raw material utilization will be detailed later.

The existence of an unmistakable stratigraphic break in Sector A, led us to make a basic distinction between an Upper Phase (Levels 1a-1) and a Lower Phase (Levels 2d-2) as two different cultural horizons; each horizon, being considered an uninterrupted sequence. Since the stratigraphy in Sector B showed the absence of a real stratigraphic break and general continuity of the sequence, except for Level 1, the materials coming from the two sectors have been analysed and presented separately. This stresses differences in the occupations of the site that will be dealt with in the general comments.

Debitage

The frequencies of the various debitage-types from Sectors A and B and their distribution throughout the stratigraphic sequence are outlined in Table 7.1.

Sector A

Flakes s.s. (unretouched flakes or flakes with partial use-retouch) were fifty-five in the Lower Phase (Level 2c-2), corresponding to 45.1% of the whole cores/debitage group. Fifty-one pieces were recovered from the Upper Phase (Levels 1a-1), corresponding to 34.0%.

In Levels 2c-2 ten pieces were primary flakes of both the unbroken bidirectional (opposite and ninety-degree) and three simple broken ones. Platform-types were lisse or faceted. The three early-stage core-preparation flakes were secondary flakes with unidentifiable, lisse and cortex platforms. In the non-primary group, flakes from single platform were the most frequent type (21.8%). They were all tertiary, with the exception of one secondary, with lisse or unidentifiable platforms; one in Level 2a had inverse use-retouch. Eleven flakes from ninety-degree cores were the second most frequent type (20%). They tended to be rather short and significantly wider than primary flakes and flakes from single platforms: six were tertiary and the rest were secondary, with lisse platform except for two which were faceted and dihedral; two of the pieces with lisse platform showed inverse scaling. Of the five flakes from multiple platform cores, three were tertiary and two secondary flakes. The faceted one was the most recognizable platform-type (fig. 7.15, 2). One showed an additional continuous proximal retouch while another had an obverse proximal notch. Flakes from multiple platform cores represented only 9.1% of the flake-class. They were short and were the thickest among the various non-primary types. The percentage of the unidentifiable flakes was high (25.4%).

No significant variations were observed among the raw material of the different flake-types. The most common raw material was sandstone (85%), while quartzitic sandstone was the second most common. There were five pieces of flint.

The average length, width and thickness of the primary flakes was 36 mm, 32 mm and 9 mm respectively. The average dimensions of flakes from single platform cores were 29 mm in length, 12 mm in width and 6 mm in thickness. The flakes from ninety-degree cores measured on average 25 mm in length, 33 mm in width and 9 mm in thickness. Finally, the average dimension of flakes from multiple platform cores were 43 mm in length, 37 mm in width and 9 mm in thickness.

The artifacts in the Upper Phase were concentrated in Level 1 since only five out of the fifty-one specimens were recovered from Level 1a. Compared with the Lower Phase a significant reduction in the range of types was observed. Nearly half of the flakes (twenty-two pieces) were from single platform cores, either tertiary or secondary, with prevailingly lisse platforms. They ranged from 19 to 48 mm in length, from 18 to 33 mm in width and from 4 to 9 mm in thickness. Fifteen pieces were early-stage core-preparation flakes, all of which were secondary with cortical or unidentifiable platforms. One, made from a chert pebble, showed an inverse scaled retouch, while another had a denticulated sinister retouch. The dimensions of this group ranged from 23 to 45 mm in length, from 30 to 54 mm in width and from 7 to 19 mm in thickness. Flakes from multiple platform cores were the third most common type of those present in this Upper Series. All fourteen specimens came from Level 1 and both the tertiary and secondary varieties were represented. Platform-types were prevailingly lisse, with few (three) faceted examples. They ranged from 12 to 27 mm in length, from 29 to 50 mm in width and from 6 to 13 mm in thickness. Apart from a few quartzitic sandstone and chert, all these flakes were of rough, coarse-grained sandstone.

An increase in the frequency variations of the flakes was noted in the Lower Phase starting in Level 2b (37.5%) moving towards Level 2 (55.3%). In the Upper Phase, while the percentage frequency remains roughly constant, there is an increase from Level 1a to Level 1 (ratio 1:9).

Core-trimming elements amounted to twenty items in the Lower Phase, equal to 16.4% of the entire cores/debitage, while only seven specimens, equal to 4.6%, were recovered from the Upper Phase. Among the rejuvenation products, we have differentiated two specific groups (lames à crête and

Table 7.1 - Uan Muhuggiag, Wadi Teshuinat. Absolute and percentage frequency of cores/debitage-classes.

	SECTOR A															
	1	%	1a	%	Total	%	2	%	2a	%	2b	%	2c	%	Total	%
Flakes	46	34.6	5	29.5	51	34.0	21	55.3	24	46.1	9	37.5	1	12.5	55	45.1
Primary flakes	-		-		-	-	7		2		1		-		10	(18.1)
Early-stage core-prep. flakes	13		2		15	(29.4)	1		1		-		1		3	(5.5)
Flakes from single plat. core	19		3		22	(43.1)	5		7		-		-		12	(21.8)
Flakes from opposed plat. core	-		-		-	-	-		-		-		-		-	-
Flakes from 90° core	-		-		-	-	6		3		2		-		11	(20.0)
Flakes from multiple plat. core	14		-		14	(27.5)	1		2		2		-		5	(9.1)
Unidentifiable flakes	-		-		-	-	1		9		4		-		14	(25.4)
Core-trimming elements	3	2.3	4	23.5	7	4.6	6	15.8	8	15.4	3	12.5	3	37.5	20	16.4
Lames à crête	3		1		4		3		5		-		-		8	
core-tablets	-		-		-		-		-		-		1		1	
core-trimming flakes	-		3		3		3		3		3		2		11	
Chips	62	46.6	8	47.0	70	46.7	7	18.4	17	32.7	9	37.5	1	12.5	34	27.9
Cores	22	16.5	-	-	22	14.7	4	10.5	3	5.8	3	12.5	3	37.5	13	10.6
Total	133	100.0	17	100.0	150	100.0	38	100.0	52	100.0	24	100.0	8	100.0	122	100.0

SECTOR B

	1	%	1a	%	2	%	2a	%	2b	%	2c	%	2d	%	Total	%
	41	42.7	41	32.0	52	54.7	115	41.9	82	38.8	28	45.9	29	46.7	347	41.7
	14	(34.2)	6		8		21		11		3		1		50	(11.4)
	4	(9.8)	7		12		13		6		1		-		39	(11.2)
	1	(2.4)	9		4		23		19		5		6		66	(19.1)
	-	-	5		-		8		4		3		2		22	(6.4)
	-	-	5		3		9		8		-		-		25	(7.2)
	2	(4.8)	3		3		15		12		1		4		38	(10.9)
	20	(48.8)	6		22		26		22		15		16		107	(30.8)
	4	4.1	10	7.8	10	10.5	28	10.2	29	13.8	11	18.1	8	12.9	96	(11.6)
	2		2		1		7		7		2		1		20	
	-		1		4		1		2		-		-		8	
	2		7		5		20		20		9		7		68	
	43	44.9	60	46.9	21	22.1	108	39.5	83	39.3	21	34.4	19	30.7	312	37.6
	8	8.3	17	13.3	12	12.7	23	8.4	17	8.1	1	1.6	6	9.7	76	9.1
Total	96	100.0	128	100.0	95	100.0	274	100.0	211	100.0	61	100.0	62	100.0	831	100.0

core-tablets) and a less specific group which, in addition to rejuvenation flakes, also includes core-sides and overpassed blades. The eight specimens of lames à crête in the Lower Phase were either tertiary or secondary flakes with faceted platforms. All of them were made on sandstone. They ranged from 16 to 58 mm in length, from 6 to 56 mm in width and from 2 to 11 mm in thickness. A single sandstone core-tablet was recovered from Level 2c: having a faceted platform, was obtained by striking the core in a transverse direction (22 x 29 x 11 mm). The remaining eleven pieces included rejuvenation flakes (eight), overpassed blades (two) and a core-side (fig. 7.18, 7). Except for a single item made on flint, all were of sandstone. They ranged from 14 to 50 mm in length, from 14 to 46 mm in width and from 6 to 8 mm in thickness.

Four tertiary lames à crête of sandstone were recovered from Levels 1a-1. Of these, one showed an additional backing retouch (mean dimensions: length 25 mm; width 13 mm; thickness 7mm). As for the rest, one was a core-side (fig. 7.22,8) and two were rejuvenation flakes (29 x 19 x 9 mm; 19 x 7 x 11 mm).

Table 7.1 shows how the core-trimming elements tend to decrease from the bottom to the upper levels. Their percentage frequency was 37.5% in Level 2c and 15.8% in Level 2. In Level 1a, this class of debitage totalled 23.5%, while in Level 1 it was only 2.3%.

Chips and unidentifiable fragments - This class included pieces measuring less than 15 mm (chips), and unidentifiable fragments of all other artifact classes (cf. Close, Wendorf, Schild 1979: 34). In the Lower Phase there was a total of thirty-four pieces having these characteristics, equal to 27.9% of the cores/debitage group. In the Upper Phase seventy pieces represented 46.7% of the total. This component is important throughout the Uan Muhuggiag series. They have their highest incidence in Levels 2b and 2a (37.5% and 32.7% respectively) while their incidence is the same (46-47%) in Levels 1a and 1. The chips are for the most part made of quartzitic sandstone or quartz.

Sector B

Flakes s. s. totalled 347 in Levels 2d-1a which here represent the Lower Phase, corresponding to 41.7% of the whole cores/debitage group. In level 1 there were forty-one pieces, equal to 42.7%.

Primary flakes were one of the most diagnostic types in the whole flake-class. A total of fifty pieces (14.4%) were recovered from Levels 2d-1a and were of the unbroken simple (fig. 7. 8,9), unidirectional, bi-directional and multidirectional varieties. A few specimens of the broken simple or unidirectional varieties were also recovered. Lisse platforms were the most common type, but many cortical platforms were also recovered, together with a few faceted and one dihedral one (Level 2c). A few examples showed some exceptional marginal use-retouches; one specimen (from Level 1a) had an inverse, "single blow", sinister notch. Considering the non-primary flakes, those from single platform cores, were the most common type (sixty-six, that is 19.1% of the whole flake-class). They were tertiary for the most part, with only a small percentage of 3% of secondary pieces, often with an oval (fig. 7.16,3) or elongated triangular shapes. Platforms were generally lisse, but few examples of the faceted type were also present. On

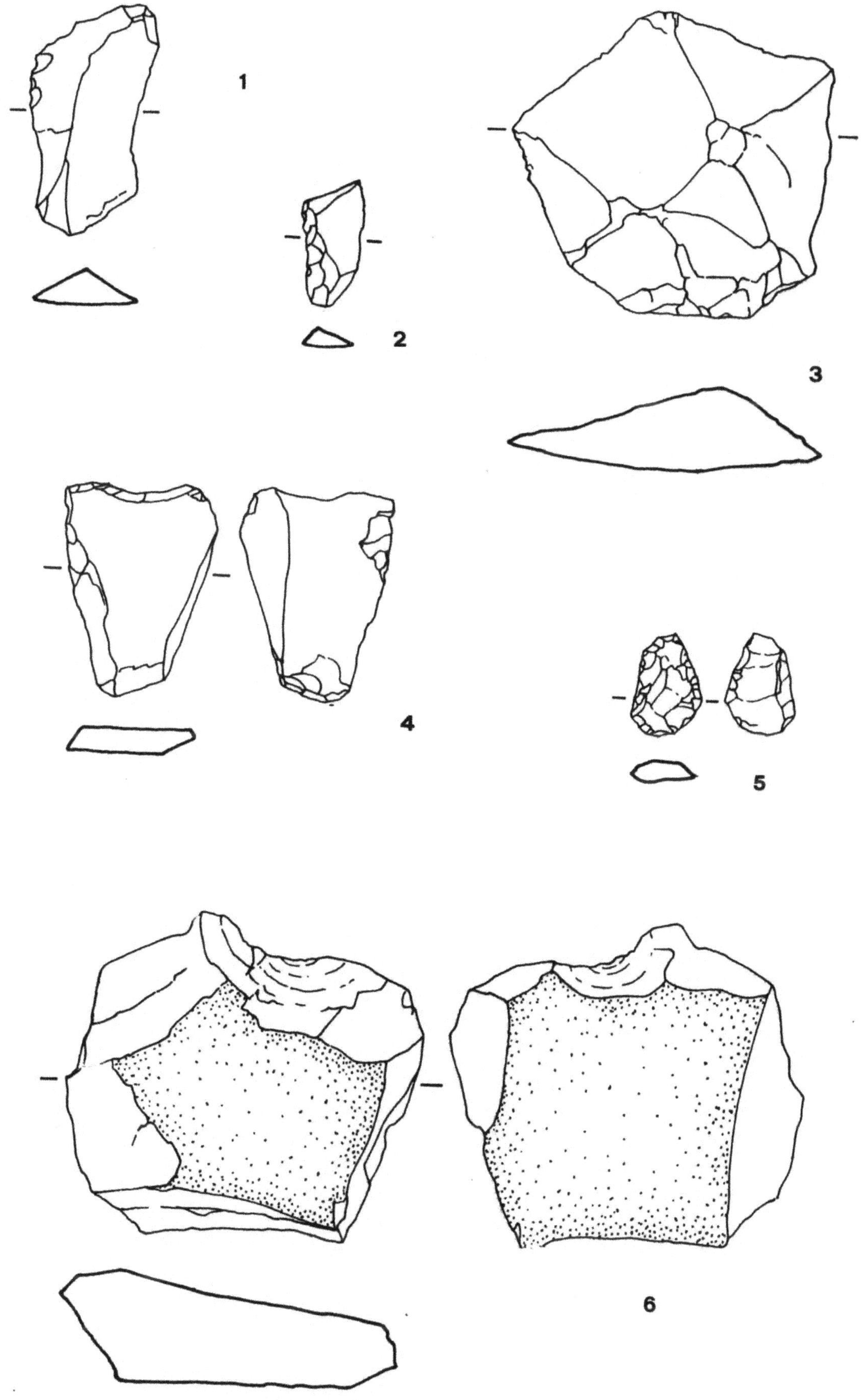

Fig. 7.7 - Uan Muhuggiag, Wadi Teshuinat. Level 2d, Sector B. 3, flake; 6, end-scraper; 2, backed bladelet; 4, notch; 1. denticulates; 5, arrow-head (1:1).

one secondary piece recovered from Level 2a, an obverse dexter proximal notch was observed. Early-stage core-preparation flakes totalled thirty-nine items (11.2%); they were all secondary pieces with lisse or faceted platforms. One specimen from Level 2c showed a very rough dexter denticulated retouch, while another piece had a discoidal shape. There was also a significant number of flakes from multiple platform cores: thirty-eight pieces (10.9%), most of which were circular or sub-circular. Lisse platforms were the only identifiable platform-type. Flakes from ninety-degree cores were less common than in Sector A, twenty-five, being 7.2% of the whole flake-class. They were tertiary for the most part, with only a few secondary examples. Platforms were prevailingly of the lisse or faceted type, but a few examples of either pointed or cortex platforms were also represented. Sector B also yielded a few specimens of flakes from opposed platform cores that had not been recovered from Sector A. Twenty-two specimens (6.4%) were recovered, starting from the base level; with the exception of three secondary they were all tertiary. Platforms were of the lisse type, with only two examples, one of which fragmentary, of the faceted type. Finally, one should mention flakes lacking specific features (unidentifiable flakes), which were frequent as in Sector A. A total of 107 specimens, corresponding to 30.8% of the entire flake-class.

As in Sector A, no variations were observed in the utilisation of raw materials. Sandstone was the most frequently used raw material, with quartzitic sandstone and flint being much less used.

A certain degree of variability was noted in the average sizes of the main types. The mean dimensions of primary flakes were: length 42 mm; width 32 mm; thickness 13 mm. The average length, width and thickness of flakes from single platform cores were 40 mm, 25 mm and 11 mm respectively, and from multiple platform cores: length 35 mm; width 36 mm; thickness 12 mm. Flakes from ninety-degree cores averaged 39.5 mm in length, 31mm in width and 7 mm in thickness. Finally, the group from opposed platform cores had the following average dimensions: length 43 mm; width 30 mm; thickness 11 mm.

The forty-one flakes from Level 1 are divided into a limited number of types; the most representative group are fourteen primary flakes with cortical platforms (except for one which had a lisse platform). Four were early-stage core-preparation flakes and two flakes from multiple platform cores. A single flake from a single platform core was recovered. The incidence of the unidentifiable flakes is high, nearly 50% of the whole flake-class. All are made of sandstone.

Throughout the Lower Phase the frequency index of flakes s.s. remained on average close to 40-45% of the entire cores/debitage, with the highest value, 54.7%, in Level 2 and a minimum, 32.0%, in Level 1a.

Core-trimming elements (96 pieces in Level 2d-1a, 11.6% of the whole cores/debitage group and four pieces in Level 1, 4.1%). This seems to have the same pattern as shown in Sector A.

Lames à crête (twenty items) were found in each one of the levels of the Lower Phase. They were either tertiary or secondary with lisse platforms (fig. 7.20,6). One from Level 2c was a double lame à crête (fig. 7.8,7), and a second, coming from Level 2b, showed a sinister use-retouch to form a perforator. A third one, with a crescent shape, showed a dexter denticulated retouch (fig. 7.17,5). Their dimensions ranged from 25 to 32

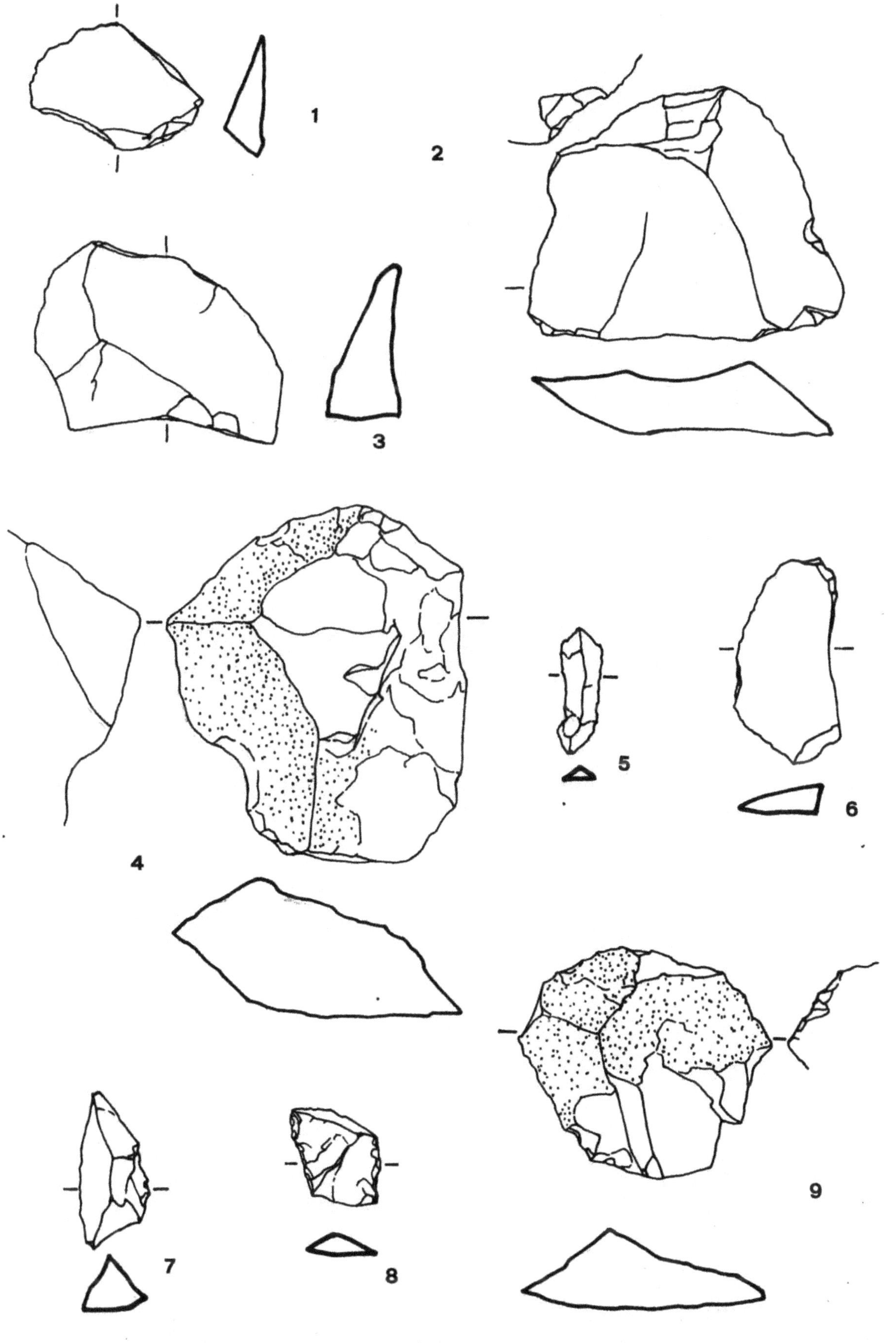

Fig. 7.8 - Uan Muhuggiag, Wadi Teshuinat. Level 2c: Sector A (1-3), Sector B (4-9). 9, flake; 1-3, 7, core-trimming elements; 4, end-scraper; 5,6, blunt unretouched blades; 8, side-scraper (1:1).

mm in length, from 5 to 12 mm in width, and from 3 to 12 mm in thickness. One primary and seven tertiary or secondary sandstone core-tablets were recovered from Levels 2b-1a (fig. 7.16,4). They ranged from 15 to 38 mm in length, from 12 to 35 mm in width and from 4 to 9 mm in thickness. The sixty-eight core-trimming elements *sensu lato* were mostly simple rejuvenation flakes. A few of them had a denticulated retouch (fig.7.8,6) while one had a sinister notch. Seven overpassed blades were recovered. Sandstone was the only raw material used for all these pieces. The entire group ranged in length from 11 to 52 mm; in width from 12 to 44 mm and in thickness from 4 to 12 mm.

Of the four items collected from Level 1, two were tertiary *lames à crête* with faceted and unidentifiable platforms, one of quartzitic sandstone and the other of flint (22 x 9 x 7 mm; 12 x 2 x 3 mm). The remaining two items were rejuvenation flakes.

The percentage frequency values of this debitage class throughout the stratigraphic sequence shows a constant decrease from the bottom towards the top, as was observed in Sector A. The indexes are 12.9%, 7.8% and 4.1% in Level 2d, Level 1a and Level 1, respectively.

Chips and unidentifiable fragments - In the Lower Phase, a total of 312, 37.6% of the cores/debitage group, were found. These products increased in number throughout the sequence being 30.7% in Level 2d and 46.9% in Level 1a. This tendency continued in Level 1, where forty-three pieces, 44.9% of the whole cores/debitage group, were recovered.

Cores

Sector A

Thirteen cores were recovered from the Lower Phase corresponding to 10.6% of the whole cores/debitage group. Twenty-two pieces, 14.7%, came from the Upper Phase. Table 7.1 shows the percentage frequency of cores throughout the sequence.

In the Lower Phase there were three single platform cores and two patterned multiple platform cores with flaking-surfaces on adjacent and opposite sides, from Levels 2c and 2b (fig. 7.11,4). Unpatterned multiple platform cores - again with flaking-surfaces on adjacent and opposite sides - were also found (fig. 7.15,8); one showed a denticulated retouch along one of the sides. An initially struck core was recovered from Level 2. Finally, three examples were unclassified fragments. All the cores had irregular globular shapes. Platform-angles were roughly similar (always less than 90°) and the platform-types were usually *lisse*. Three were cortex, of which two were for initially struck cores and one for a fragment. No sign of preparation was observed on the core sides or back. All, except for one of quartzitic sandstone, were on sandstone.

The twenty-two coming from the Upper Phase may be ascribed to two types; twelve were multiple unpatterned platform cores and the rest were unclassifiable. Platform-types were *lisse* or unidentifiable and platform-angles were less than 90°. None of the cores shows platform preparation and

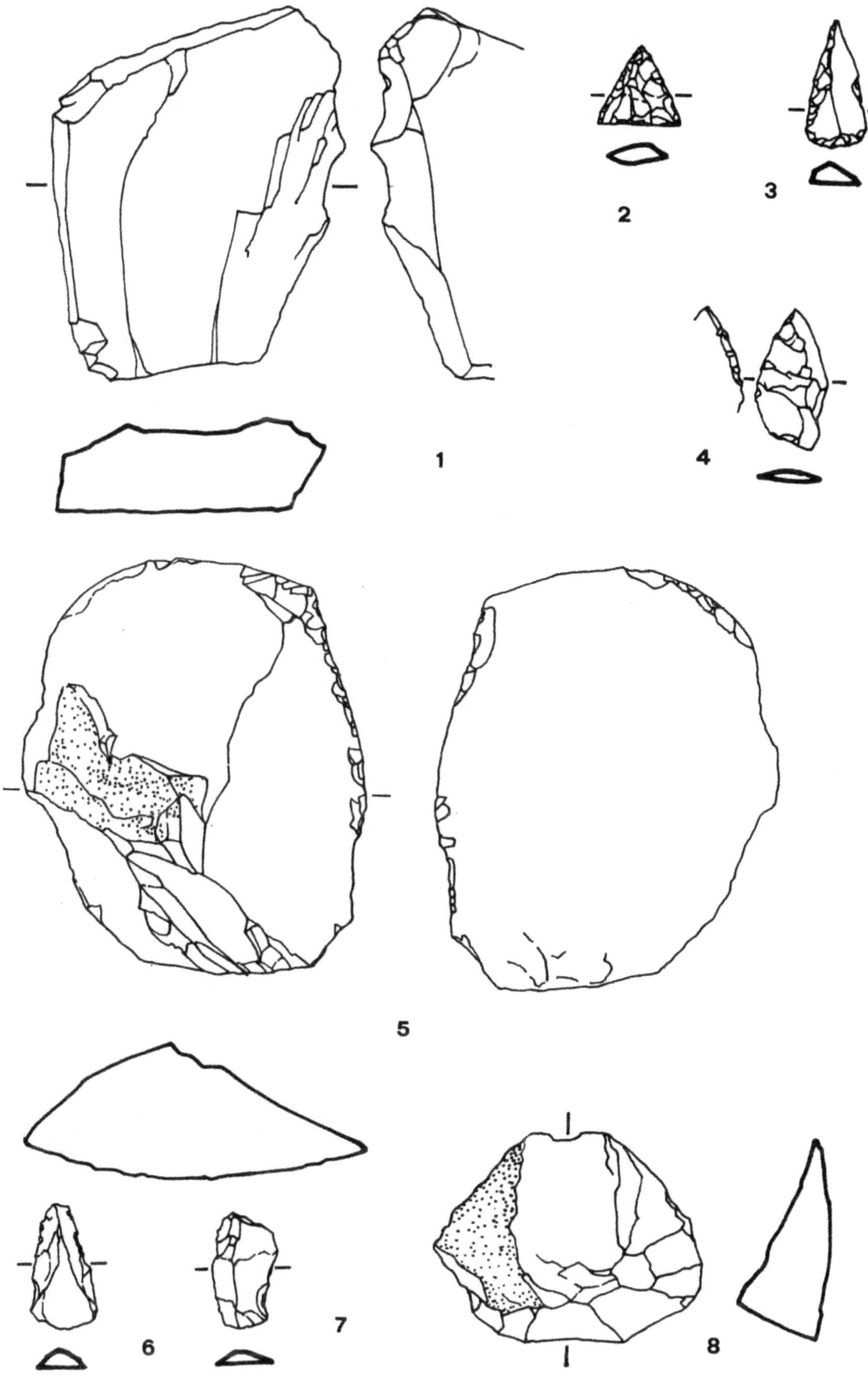

Fig. 7.9 - Uan Muhuggiag, Wadi Teshuinat. Level 2c, Sector B. 8, end-scraper; 7, blunt unretouched blade; 3,6, backed bladelets; 5, scaled piece; 1, denticulate; 2,4, arrow-heads (1:1).

all were made on sandstone (5).

There were some variations in size among the different core-types. Single platform cores ranged from 18 to 52 mm in length, from 10 to 42 mm in width and from 8 to 18 mm in thickness. The two multiple patterned platform cores had the following dimensions: 42 and 80 mm in length, 40 and 55 mm in width, 30 and 40 mm in thickness, respectively. The multiple unpatterned group ranged from 18 to 67 mm in length, from 10 to 41 in width and from 17 to 42 mm in thickness. The single, initially struck core, measured 18 mm in length, 10 mm in width and 8 mm in thickness. Fragments were on average 25 mm long, 30 mm wide and 15 mm thick.

Sector B

A total of 76 cores was collected from Levels 2d-1a (9.1% of the whole cores/debitage group). Table 7.1 shows their distribution throughout the sequence and clearly points to the significant increase in the class frequency starting from Level 2b. Only eight specimens, equal to 8.3%, come from Level 1.

There were fourteen single platform cores, six of which showed a simple, continuous retouch on one of the sides. One from Level 1a was an opposed platform core with flaking surfaces on the same side. Two ninety-degree cores were recovered from Level 2d, and these too had flaking-surfaces on the same side. The most common category was represented by multiple platform cores, of both the patterned (eighteen) and the unpatterned (twelve) types. All of them showed flaking-surfaces on adjacent or opposite sides (fig. 7.12,10). A specimen recovered from Level 2a had a sub-pyramidal shape (fig. 7.17,8), while another one coming from the same level was globular. Two discoidal cores came from Levels 2b and 2: one had flaking-surfaces on opposite, and the other one (Level 2) on adjacent sides showing a denticulated retouch (fig. 7.19,4). Nine specimens (Levels 2b, 2a, 2 and 1a) were initially struck cores. One of these (Level 2) presented a "single blow", proximal notch. Sixteen unclassifiable examples were recovered; two, from Levels 2b and 2, were whole pebbles.

As in Sector A, all the cores had irregular shapes. The mean value of platform-angles on ninety-degree and multiple platform cores, be they patterned or unpatterned, was about 85°. While it was difficult to measure it on the other types, platform-angles were always less than 90°. The most frequent type of platform was lisse (thirty-eight examples) but the cortex type was also common (twenty examples) especially in the multiple platform cores. Five were faceted while the rest were unidentifiable. There were none with a platform-preparation technique. The most common raw material was sandstone, followed by quartzitic sandstone. A multiple unpatterned microlithic core from Level 2d was made on flint and two (whole pebble and single platform) were of quartz. Four small microlithic pebbles of quartz coming from Level 2a were very similar to those recovered from Levels 1a-1 of Sector A; they were not included in the Table.

---

5. We should also refer to a number of small (microlithic) quartz pebbles coming also from the Upper Phase. They may not be actual nuclei and are not shown in the table but their freqeuncy is significant; eleven in Level 1a; forty-four in Level 1. They have rounded profiles and diameters ranging between 0.5 and 2mm.

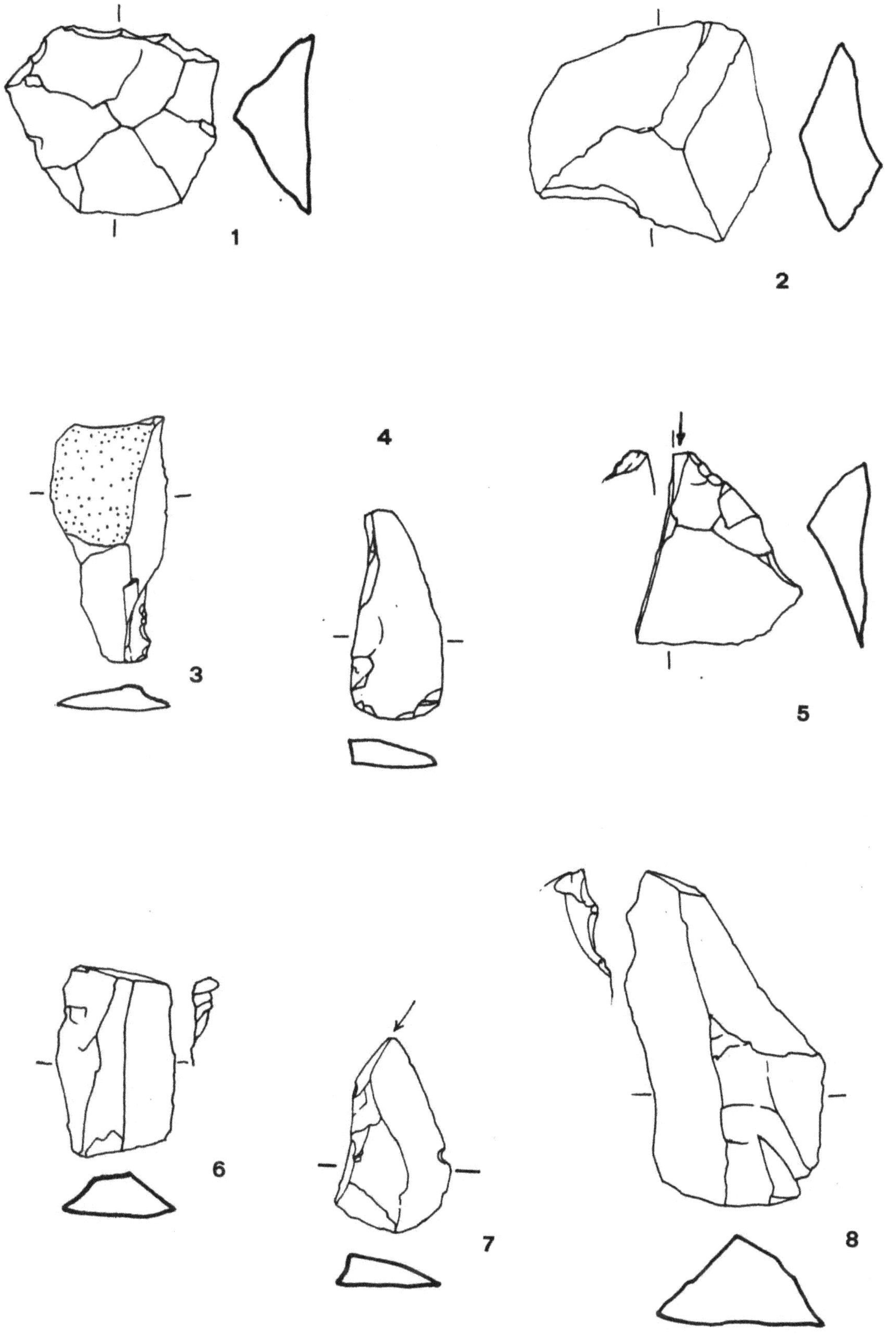

Fig. 7.10 - Uan Muhuggiag, Wadi Teshuinat. Level 2b, Sector A. 1, end-scraper; 5,7, burins; 3, blunt unretouched blade; 6, piece with use-retouch; 2,4,8, notches (1:1).

Eight cores came from Level 1; three were single platform cores, one of which had use-retouch on one side. Two were multiple platform unpatterned cores with flaking-surfaces on opposite sides. The remainder were unclassifiable or fragments. No platform preparation was observed and all of the samples were roughly flaked. Except for two faceted examples, platforms were <u>lisse</u>. While sandstone was the most commonly used raw material, two examples were of quartz.

Dimensional variations were observed in respect of the various core-types. Single platform cores ranged from 12 to 29 mm in length, from 20 to 32 mm in width and from 12 to 24 mm in thickness. Multiple patterned platforms ranged from 22 to 52 mm in length, from 33 to 51 mm in width and from 17 to 30 mm in thickness. Practically, the same values were reported in respect of unpatterned platforms that ranged from 19 to 49 mm in length, from 5 to 38 mm in thickness, while width was about 40 mm. The single, ninety-degree core was 37 mm long, 33 mm wide and 12 mm thick. The two opposed platform and discoidal cores were 54 and 53 mm long, 43 and 50 mm wide and 21 and 24 mm thick, respectively. The initially struck cores ranged from 30 to 58 mm in length, from 53 to 66 mm in width and from 21 to 30 mm in thickness. The unclassifiable specimens and the fragments were, on the average, 25 mm long, 26 mm wide and 12 mm thick.

<u>Retouched Tools</u>

Retouched tools are arranged in a limited number of classes (Table 7.2) and were analysed according to the following criteria.

1 - The presence of each class within the Lower and Upper Series is considered separately.

2 - An attempt has been made to show as many illustrations as possible of each class, especially in respect of the most significant examples.

3 - Dimensional data (length, width, thickness) are provided as average dimensions measured over the total number of pieces for each class. Because the illustrations are full-size, no further details are provided for single items included in the illustrations.

4 - Due consideration was given to the percentage trend, throughout the sequence, of each class.

5 - As for the raw material being used, reference should be made to what was mentioned above, in the general remarks. In the description, only special instances will be considered.

Sector A

<u>End-Scrapers</u> - Levels 2c-2: a total of 11; general index 13.6%

Levels 1a-1: a total of 2; general index 5.9%.

In Levels 2c-2, seven specimens were simple end-scrapers with distal or proximal scraping-end (fig. 7.8,<u>1</u>). (In one of them, the scraping-end was unidentifiable). Two specimens were on primary and tertiary flakes while a third was on a fragment of tertiary blade. As far as the other specimens

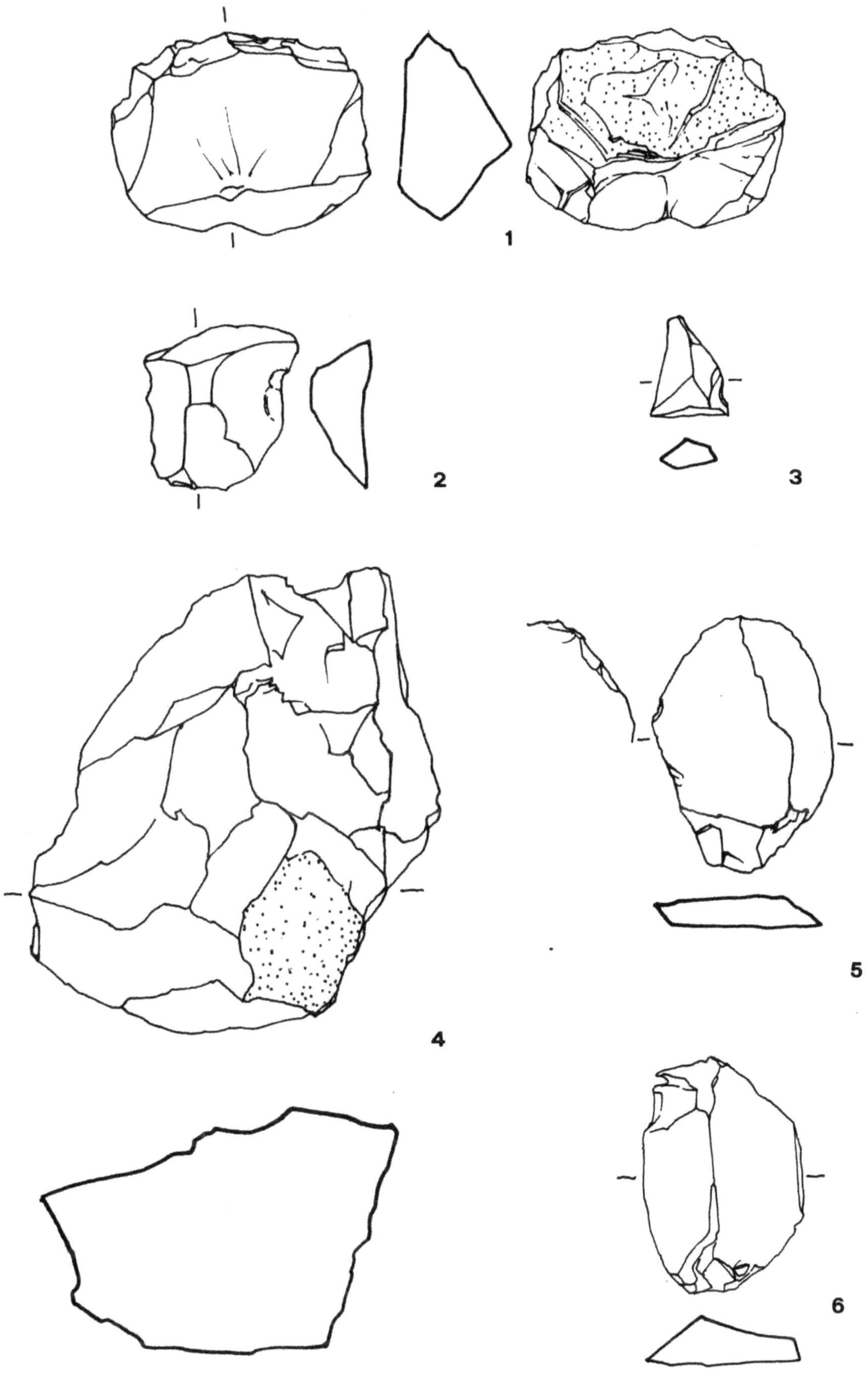

Fig. 7.11 - Uan Muhuggiag, Wadi Teshuinat. Level 2b, Sector A. 4, core; 1, end-scraper; 6, blunt unretouched blade; 3, backed bladelet; 2,5, pieces with use-retouch (1:1).

Table 7.2 - Uan Muhuggiag, Wadi Teshuinat. Absolute and percentage frequency of tool-classes.

	SECTOR A															
	1	%	1a	%	Total	%	2	%	2a	%	2b	%	2c	%	Total	%
End-scrapers	2	7.7	-	-	2	5.9	1	4.7	5	20.8	3	10.7	2	25.0	11	13.6
Burins	1	3.8	-	-	1	2.9	1	4.7	-	-	2	7.1	-	-	3	3.7
Perforators	-	-	1	12.5	1	2.9	-	-	-	-	-	-	-	-	-	-
Blades	17	65.4	4	50.0	21	61.9	7	33.5	9	37.5	11	39.3	1	12.5	28	34.6
<u>Varia</u>	<u>2</u>	<u>7.7</u>	<u>1</u>	<u>12.5</u>	<u>3</u>	<u>8.8</u>	<u>6</u>	<u>28.6</u>	<u>1</u>	<u>4.2</u>	<u>3</u>	<u>10.7</u>	<u>1</u>	<u>12.5</u>	<u>11</u>	<u>13.5</u>
Scaled pieces	1	3.85	1	12.5	2	5.9	3	14.3	-	-	1	3.6	-	-	4	4.9
Continuously r.pieces	-	-	-	-	-	-	-	-	-	-	-	-	-	-	-	-
Side-scrapers	1	3.85	-	-	1	2.9	3	14.3	1	4.2	2	7.1	1	12.5	7	8.6
Pieces with use-retouch	2	7.7	-	-	2	5.9	2	9.5	5	20.8	2	7.1	2	25.0	11	13.6
Notches and denticulates	1	3.85	1	12.5	2	5.9	2	9.5	3	12.5	7	25.1	2	25.0	14	17.3
Arrow-heads and points	1	3.85	1	12.5	2	5.9	2	9.5	1	4.2	-	-	-	-	3	3.7
Total	26	100.0	8	100.0	34	100.0	21	100.0	24	100.0	28	100.0	8	100.0	81	100.0

SECTOR B

	1	%	1a	%	2	%	2a	%	2b	%	2c	%	2d	%	Total	%
	5	31.3	4	13.3	2	6.1	5	11.4	3	6.2	2	9.1	3	25.0	19	10.0
	-	-	-	-	-	-	-	-	1	2.1	-	-	-	-	1	0.5
	1	6.2	-	-	-	-	2	4.5	-	-	-	-	-	-	2	2.1
	7	43.8	15	50.0	20	60.6	19	43.2	24	50.0	7	31.8	2	16.7	87	46.0
	-	-	3	10.0	5	15.2	7	15.9	8	16.7	5	22.7	1	8.3	29	15.4
	-	-	1	3.3	2	6.1	4	9.1	5	10.4	3	13.7	1	8.3	16	8.5
	-	-	-	-	-	-	1	2.3	1	2.1	1	4.5	-	-	3	1.6
	-	-	2	6.7	3	9.1	2	4.5	2	4.2	1	4.5	-	-	10	5.3
	1	6.2	3	10.0	3	9.1	3	6.8	1	2.1	-	-	-	-	10	5.3
	-	-	5	16.7	3	9.1	7	15.9	2	4.2	5	22.7	4	33.3	26	13.8
	2	12.5	-	-	-	-	1	2.3	9	18.7	3	13.7	2	16.7	15	7.9
Total	16	100.0	30	100.0	33	100.0	44	100.0	48	100.0	22	100.0	12	100.0	189	100.0

are concerned, one was on a flake from a scaled piece; two were on simple core-trimming elements and another one was on a lame à crête (fig. 7.14,4). As for the rest, three were denticulated on a core-trimming element with distal scraping-end (fig. 7.11,1). A further specimen coming from Level 2b was a denticulated end-scraper on a circular tertiary flake (fig. 7.10,1). From Level 2a came an end-scraper on a secondary flake with a denticulated inverse sinister proximal plus central retouch (fig. 7.14,5). On the average, these pieces were 28 mm long, 33 mm wide and 13 mm thick.

With reference to the two specimens recovered from Level 1, one was a simple end-scraper on a tertiary blade with proximal scraping-end (36x12x6 mm), while the other one was a distal end-scraper on a secondary flake with inverse invasive sinister plus proximal retouch (fig. 7.23,6).

Table 7.2 shows the absolute and percentage frequency of the class. In the Lower Phase frequency is 25.0% in Level 2c and, after a sharp drop in Level 2b (10.7%) it rises once again in Level 2a and finally becomes stable around 4.7%. A similar figure is reported from the Upper Phase: 5.9%.

Burins - Levels 2a-2: a total of 3; general index 3.7%

Levels 1a-1: a total of 1; general index 2.9%.

Two examples were recovered from Level 2b. One was a distal angle burin (fig. 7.10,7) made on a tertiary flake with additional retouch forming a sinister central scraping zone. The other was a composite tool, an end-scraper burin: end-scraper on a core-trimming element with a distal burin on an abrupt retouch (fig. 7.10,5). A third burin came from Level 2 and consisted in a sinister angle burin on a straight, normal truncation. The blank was a tertiary flake (fig. 7.18,3).

The single specimen from Level 1 was a multiple angle burin on an inverse retouch, made on a small pebble of quartz (fig. 7.23,5).

The incidence of the class was low in both the Lower and the Upper Phases. The scarcity of burins corresponds to a very poor standard of manufacture. Sandstone and quartzitic sandstone were generally used; the item from Level 1 was made on quartz.

Perforators - Levels 2c-2: absent

Levels 1a-1: a total of 1; general index 2.9%.

The type was not recovered from the Lower Phase and only one item was found in Level 1a. This was a simple, distal perforator on a primary, blade-like flake formed by alternate retouch (fig. 7.22,4).

Blades - Levels 2c-2: a total of 28; general index 34.6%

Levels 1a-1: a total of 21; general index 61.8%.

As pointed out in the introduction, blades have been considered separately from the debitage and have been included among tools to stress their incidence in the assemblage and to allow a comparison with the Ti-n-

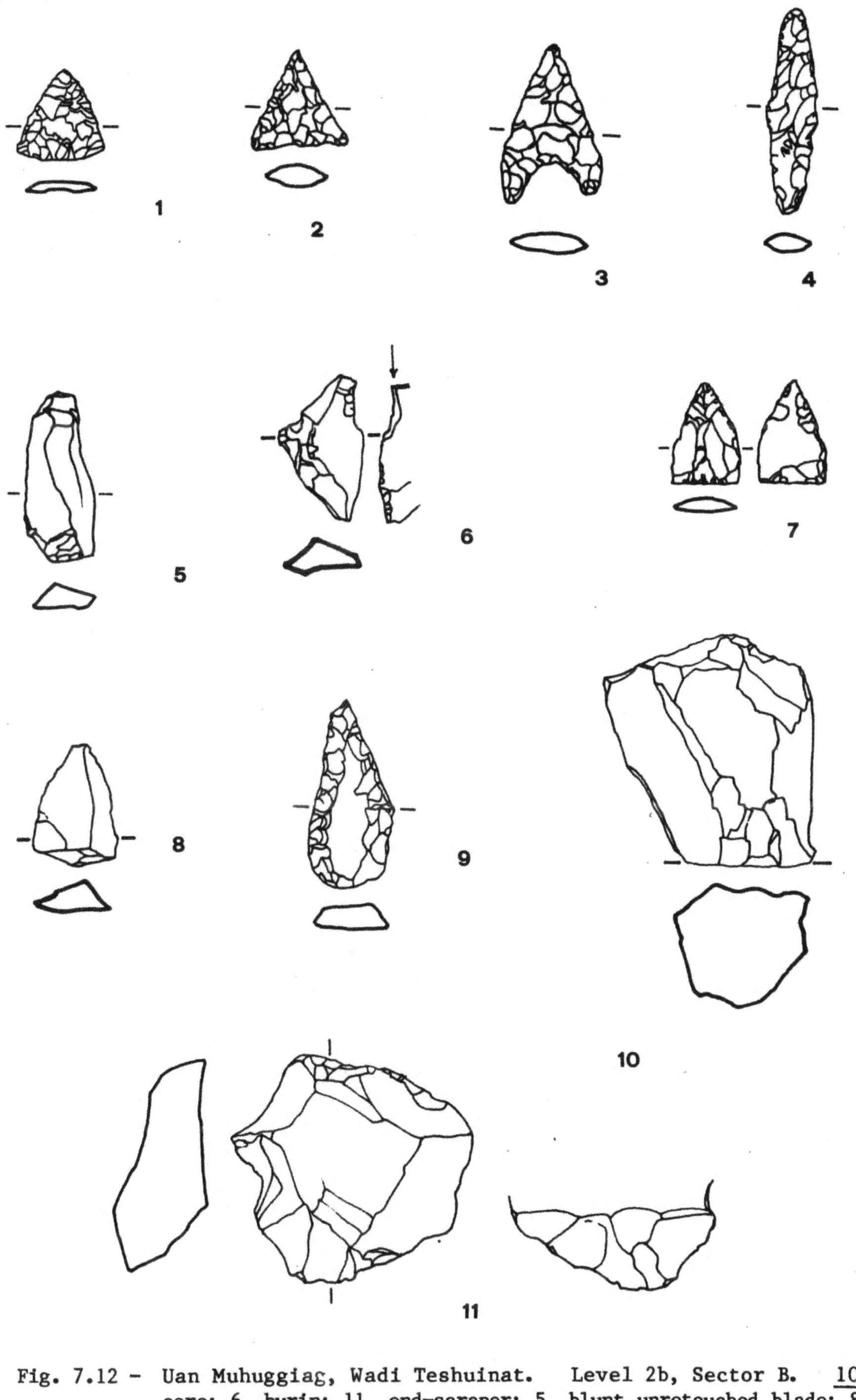

Fig. 7.12 - Uan Muhuggiag, Wadi Teshuinat. Level 2b, Sector B. 10, core; 6, burin; 11, end-scraper; 5, blunt unretouched blade; 8, pointed blade-like flake; 1-4, 7, 9, arrow-heads (1:1).

Torha industry. This involved, as it had already been done for the latter, the computation of those pieces that, although inconsistent with the conventional blade definition provided by Tixier (1963), showed marked, blade-like features. These pieces are thus considered as "blade-like flakes". Notwithstanding this expedient, the incidence of the blade component is rather low in both sectors and quite different from the quantities observed in the Ti-n-Torha industry. Compared to the latter, not even the diversification into different specimens could be detected and the various specimens may be grouped under the heading of blunt unretouched blades, comprising for the most part blade-like flakes. Only a few examples, even including fragments of backed bladelets, may be accounted for separately.

Six specimens collected from Levels 2b and 2a are really blunt blades from single platform cores, three of which, quite fragmentary, were secondary (figs. 7.10,3; 14,3) while one showed a simple retouch at the proximal end (fig. 7.11,6). Two whole pieces were tertiary (fig. 7.15,4,6). Another proximal fragment of an unretouched tertiary blade from single platform core, showing the original bulb, was recovered from Level 2. Blade-like flakes were much more numerous, with twenty-one specimens. While only one of them was recovered from Level 2c, they increased rapidly in the following levels and fourteen examples were recovered from Levels 2b and 2a. All except for a secondary one were blade-like tertiary flakes from single platform cores and nine of them showed a pointed distal end. As for their dimensions, they ranged from 16 to 40 mm in length, from 10 to 35 mm in width and from 3 to 12 mm in thickness.

A single backed bladelet was recovered from Level 2b. It was a distal fragment with an obverse dexter backing (fig. 7.11,3).

Blade-like flakes represented the most common class in the Upper Phase; sixteen tertiary specimens from single platform cores. Out of these pieces, eleven showed distal pointed ends and one had an inverse sinister distal retouch. Only three examples from Level 1 - one of which fragmentary - were true blunt unretouched tertiary blades from single platform cores (fig. 7.22,2,7).

Finally, two backed bladelets on core-trimmings (lames à crête) were recovered from Levels 1a and 1 respectively. One had obverse central and distal backing, the other one obverse backing along the entire edge and additional inverse invasive retouch (fig. 7.23,2). The dimensions of the pieces coming from the Upper Phase ranged from 22 to 45 mm in length, from 9 to 28 mm in width, from 3 to 9 mm in thickness.

Almost all the blades from Sector A were made on quartzitic sandstone and sandstone; a few quartz or flint specimens were noted.

The incidence of blades throughout the sequence is far smaller than those in the Ti-n-Torha industry. As far as the Lower Phase is concerned, the index is 12.5% in Level 2c with a 33.5% peak in Level 2. The incidence is significantly higher in the Upper Phase with a general index of 61.8%.

Varia

This heading groups classes of tools that lack definite typological and manufacturing features. The fact that what falls within varia represents

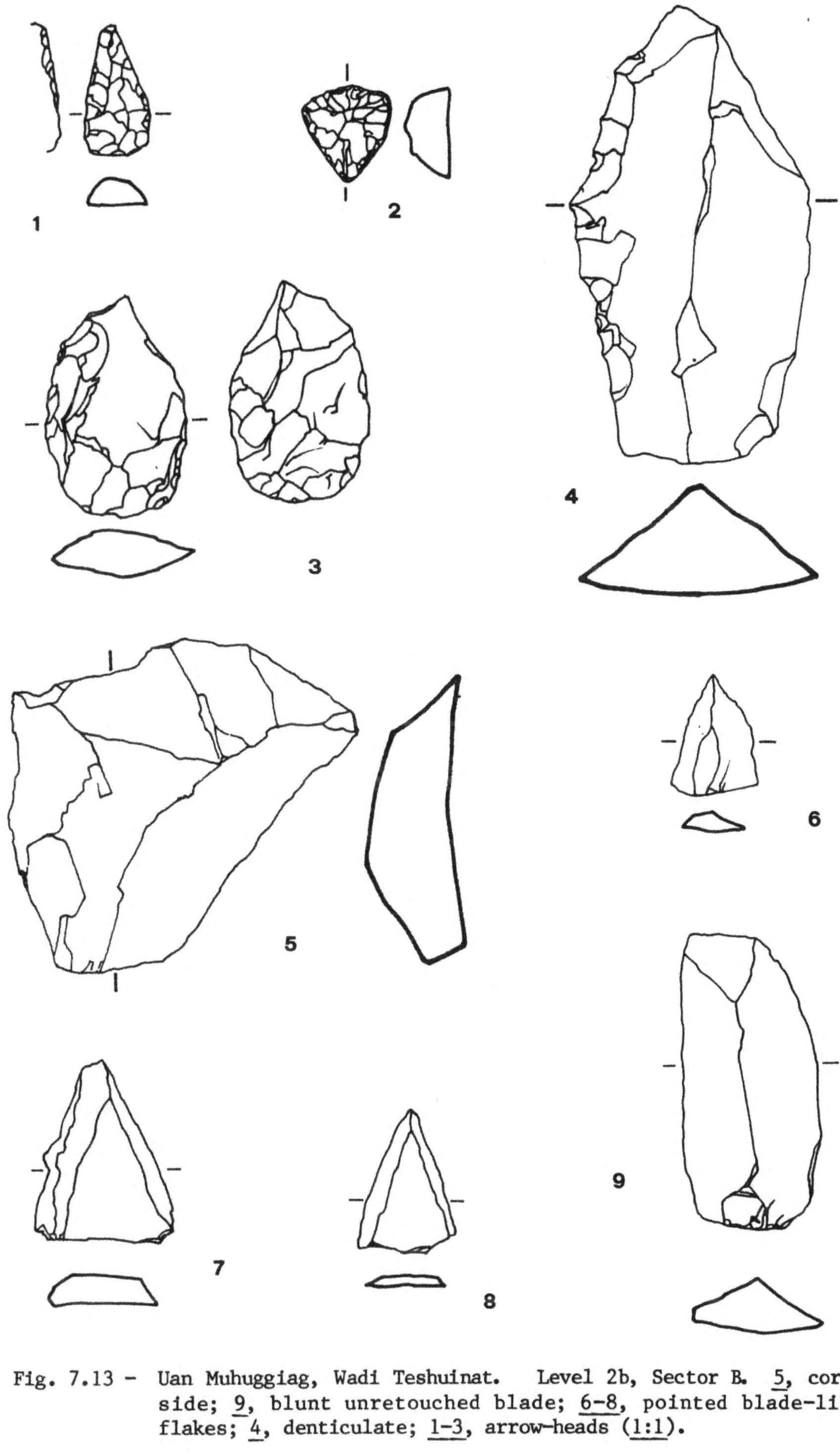

Fig. 7.13 - Uan Muhuggiag, Wadi Teshuinat. Level 2b, Sector B. 5, core-side; 9, blunt unretouched blade; 6-8, pointed blade-like flakes; 4, denticulate; 1-3, arrow-heads (1:1).

13.5% and 15.4% of the total of retouched tools in Sectors A and B (Lower Phase) respectively, stresses the poor quality of this industry compared with that of the Wadi Ti-n-Torha.

Scaled pieces - Levels 2c-2: a total of 4; general index 4.9%

Levels 1a-1: a total of 2; general index 5.9%.

The four examples coming from the Lower Phase included three on tertiary flakes, all with obverse scaling: obverse distal in one piece and proximal plus central in the others (average size: 22 mm long, 24 mm wide, 7 mm thick). The fourth specimen, coming from Level 2, was a scaled piece on a core-trimming and showed a bifacial alternate scaling: obverse, proximal and inverse distal (fig. 7.18,1).

Two pieces were recovered from the Upper Phase; both were on tertiary flakes with obverse proximal scaling. One of them had an additional inverse dexter proximal notch ("single blow"). Their dimensions were as follows: 16 and 55 mm in length, 10 and 24 mm in width, 4 mm in thickness.

In view of their extremely limited number, very little may be said about variations throughout the series, but an increase may be noted in the upper levels.

Side-scrapers - Levels 2c-2: a total of 7; general index 8.6%

Levels 1a-1: a total of 1; general index 2.9%.

Three side-scrapers were on tertiary flakes with obverse or inverse (fig. 7.11,5) convex scraping-edges. Two nucleiform side-scrapers were recovered in Levels 2b and 2a: one of them was on a spheroidal core and showed an inverse, bilateral, convex scraping-edge; the other piece, made on a nucleiform secondary flake, was a double, obverse dexter and inverse sinister straight side-scraper (fig. 7.14,2). Level 2 yielded a side-scraper obtained from a scaled piece fragment by bifacial proximal scraper-retouch (fig. 7.18,2). A single denticulated side-scraper was recovered from Level 2; made on a tertiary flake it had denticulated obverse sinister straight scraping-edge. The pieces had the following average dimensions: length 38.5 mm; width 35.5 mm; thickness 13 mm.

The single specimen recovered from Level 1 was a bifacial sinister straight side-scraper on a blade-like secondary flake, pointed at the distal end (fig. 7.23,1).

This tool-class was reported from the base Level 2c, with an index of 12.5%. While this value decreases in Levels 2b-2c (7.1% and 4.2%, respectively), it rises again in Level 2 (14.3%). The incidence of this class in the Upper Phase was very low.

Pieces with use-retouch - Levels 2a-2: a total of 11; general index 13.6%

Levels 1a-1: a total of 2; general index 5.9%.

This class is composed for the most part of flakes or blades in which the use-retouch produced similarities with traditionally defined tool-types. They may be roughly divided into two groups: denticulates (six) and side-

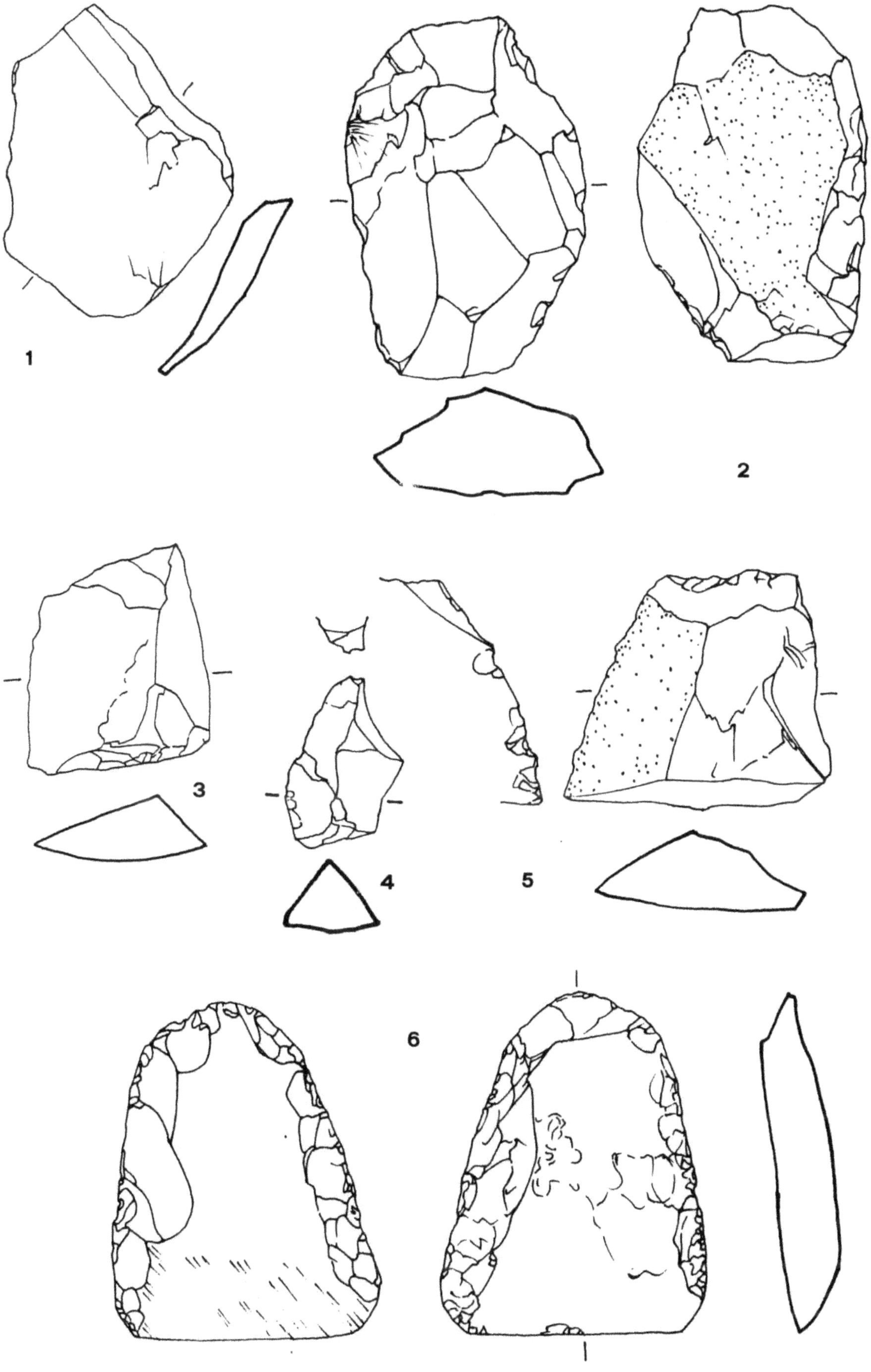

Fig. 7.14 - Uan Muhuggiag, Wadi Teshuinat. Level 2a, Sector A. 4,5, end-scrapers; 3, blunt unretouched blade; 2, side-scraper; 1, piece with use-retouch; 6, axe (1:1).

scrapers (five). The first group included three examples on tertiary flakes that showed an inverse sinister entire edge (figs. 7.11,2,5; 14,1) or an obverse sinister distal use-retouch (fig. 7.18,8). Another specimen was on a secondary flake with inverse sinister entire edge retouch (fig. 7.18,4). Two tertiary flakes with a notch-like use-retouch at the distal end, one of which obverse (fig. 7.8,3) and the other one inverse (fig. 7.8,2) were recovered from Level 2c. Out of the pieces used as side-scrapers, three (Level 2a) were obtained from scaled pieces and showed an obverse scraper retouch. One was a tertiary blade with an inverse dexter distal scraping-edge (fig. 7.10,6). The average dimensions for the class were: length 37 mm; width 32 mm; thickness 9.5 mm.

The two examples of the Upper Phase came from Level 1. One was a denticulate-like piece on a tertiary flake with obverse dexter distal use-retouch (fig. 7.23,4) and the other, on a quartz pebble, showed a straight normal truncation (fig. 7.22,3).

Pieces with use-retouch were present from the Lower Phase and their incidence remained roughly stable, except for Level 2a where the frequency was 20.8%.

Notches and denticulates - Levels 2c-2: a total of 14; general index 17.3%

Levels 1a-1: a total of 2; general index 5.9%.

The eleven notched flakes or blades were the most diagnostic group in Levels 2c-2. Almost all of them were on tertiary flakes or blades, except for four which were on core-trimmings. With the exception of a "single blow" sinister distal piece, all were retouched (seven with obverse and three with inverse retouch). One tertiary bladelet (Level 2b) had a pointed distal end and an additional retouch at the opposite proximal end (fig. 7.10,4). A multiple notch on a tertiary flake coming from the same level was also pointed (fig. 7.10,8). Another multiple, obverse dexter proximal plus central notch on a nucleiform flake was recovered from Level 2a (fig. 7.15,3) together with an obverse distal notch with additional flaking at the proximal end (fig. 7.15,7). Of the remainder three denticulated pieces, two were on blade-like tertiary flakes with obverse denticulations at the distal end, one, on a core-trimming element, had proximal denticulations (fig. 7.15,5). The specimen sizes were very homogeneous in the various levels, ranging on the average from 19 to 59 mm in length, from 11 to 48 mm in width and from 4 to 15 mm in thickness.

A single specimen of a notch on a core-trimming element was recovered from Level 1a. It had an obverse, "single blow" dexter proximal retouch (22 x 32 x 13 mm). The specimen from Level 1 consisted of a denticulated piece with inverse sinister distal retouch (fig. 7.23,3).

The frequency of the class was quite high in the Lower Phase: notched and denticulated tools were 25.0 and 25.1%, in Levels 2c and 2b, dropping to 9.5% in Level 2. In the Upper Phase the class was only 5.9% of the lithic artifacts.

Arrow-heads and points - Levels 2c-2: a total of 3; general index 3.7%

Levels 1a-1: a total of 2; general index 5.9%.

Of considerable significance is the presence of bifacially retouched

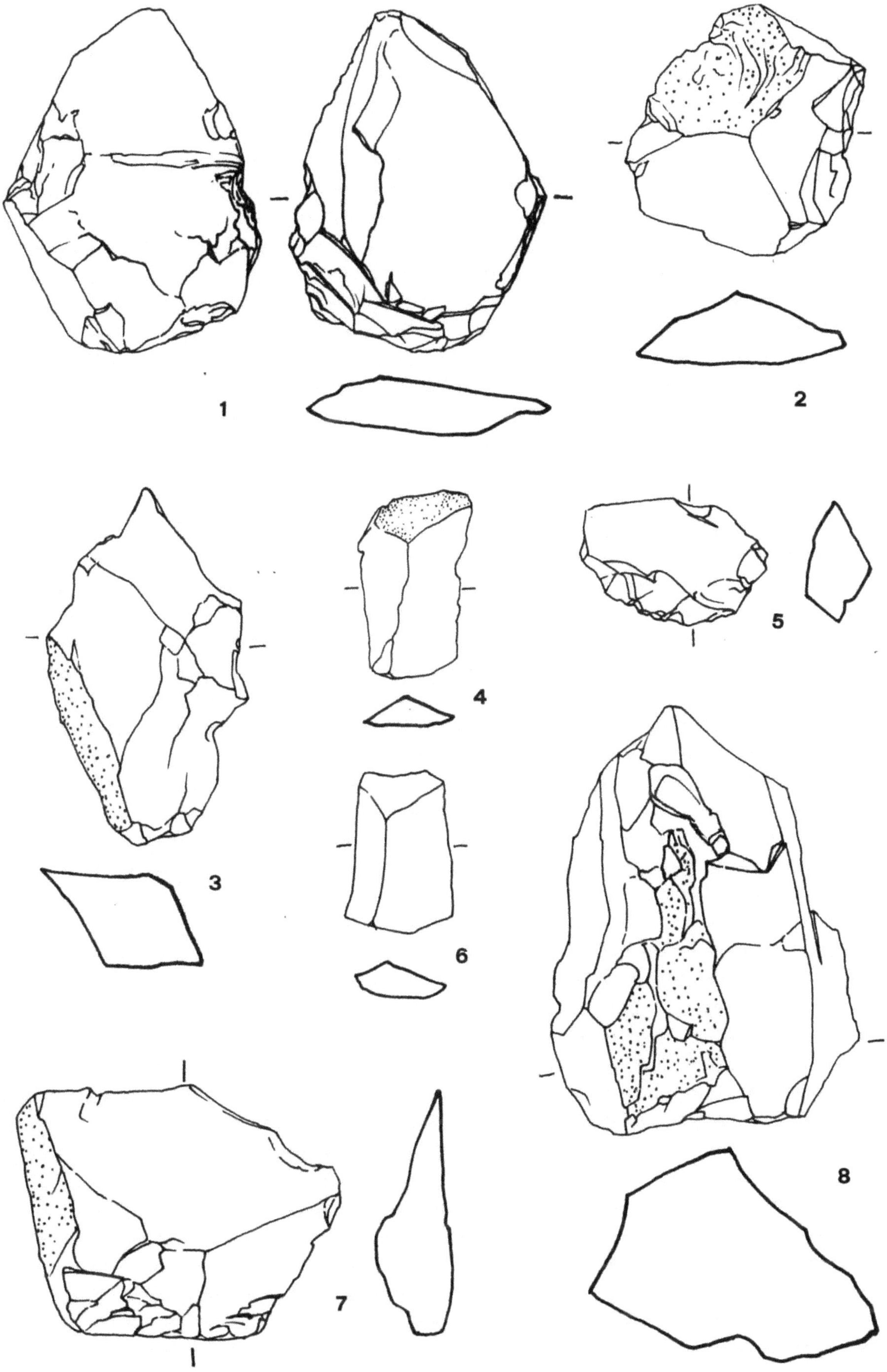

Fig. 7.15 - Uan Muhuggiag, Wadi Teshuinat. Level 2a, Sector A. 2, flake; 8, core; 4,6, blunt unretouched blades; 3,7, notches; 5, denticulate; 1, silt tablet (1:1).

pieces. A few of these are actual arrow-heads with fully bifacial retouch. Other ones, where the retouch is localized may be more properly defined as points with bifacial partial retouch. Their presence is one of the major elements of continuity with the Ti-n-Torha facies, since the examples, those of Sector B in particular, match the types already observed at Torha East (CI) and Torha North (V-IV). Together with the latter, they fit in quite easily in the general Saharan typology.

The few examples recovered from Sector A do not correspond to the classical definition. Two ogival-shaped points with flat retouched bases were recovered from Levels 2a and 2. One of sandstone, showed obverse proximal retouch and denticulations along the dexter edge. The other one, smaller (18x12x4 mm) of quartzitic sandstone, was bifacially retouched and showed a plano-convex cross-section (fig. 7.18,6). A specimen from Level 2, made on a tertiary blank of silt, was an elongated leaf-arrow-head with round base. It had obverse invasive and inverse sinister abrupt retouch. The example approximates to type C4 of Hugot's classification (Hugot 1957) (fig. 7.18,9).

The Upper Phase yielded two bifacially retouched pieces. One, from Level 1a, was a triangular point of flint with a slightly convex base. The retouch was unifacially invasive with an inverse microburin scar (fig. 7.22,1). A point on a blade-like tertiary flake of sandstone, with use-retouch on the sinister side (fig. 7.22,6), was recovered from Level 1.

In sector A, this class is rare throughout the stratigraphic sequence. While arrow-heads were not recovered from the base levels, they appeared from Level 2a onwards and their frequency is quite stable in all the subsequent levels.

Sector B

End-Scrapers - Levels 2d-1a: a total of 19; general index 10.0%

Level 1: a total of 1; general index 31.3%.

Twelve simple end-scrapers from Levels 2d-1a had distal (figs. 7.16,2; 21,7) or proximal (fig. 7.21,6) scraping-ends. They were all except for one on a flake from a scaled piece, tertiary and secondary flakes or blades. A nucleiform end-scraper with a distal scraping-end and an additional "single blow" obverse sinister proximal notch (fig. 7.8,4) came from Level 2c. Another nucleiform end-scraper with a proximal scraping-end was recovered from the same level (fig. 7.9,8). A third end-scraper (Level 2b) shows additional denticulated obverse sinister (entire edge) and inverse proximal invasive retouch (fig. 7.12,11). An additional denticulated obverse dexter central use-retouch was present in a piece coming from Level 2a (fig. 7.16,2). The remainder were denticulated end-scrapers; one from Level 2d was a real notched distal end-scraper showing an additional continuous obverse dexter retouch (fig. 7.7,4). Two were on cores (fig. 7.19,3) and one on a tabular piece (fig. 7.19,8). The average dimensions ranged as follows: from 16 to 61 mm in length; from 17 to 52 mm in width; from 4 to 29 mm in thickness.

Three of the five items from Level 1 were simple end-scrapers with distal scraping-ends. One was on a secondary blank and the others on tertiary blanks (fig. 7.14,6,9,10). The same level yielded also two

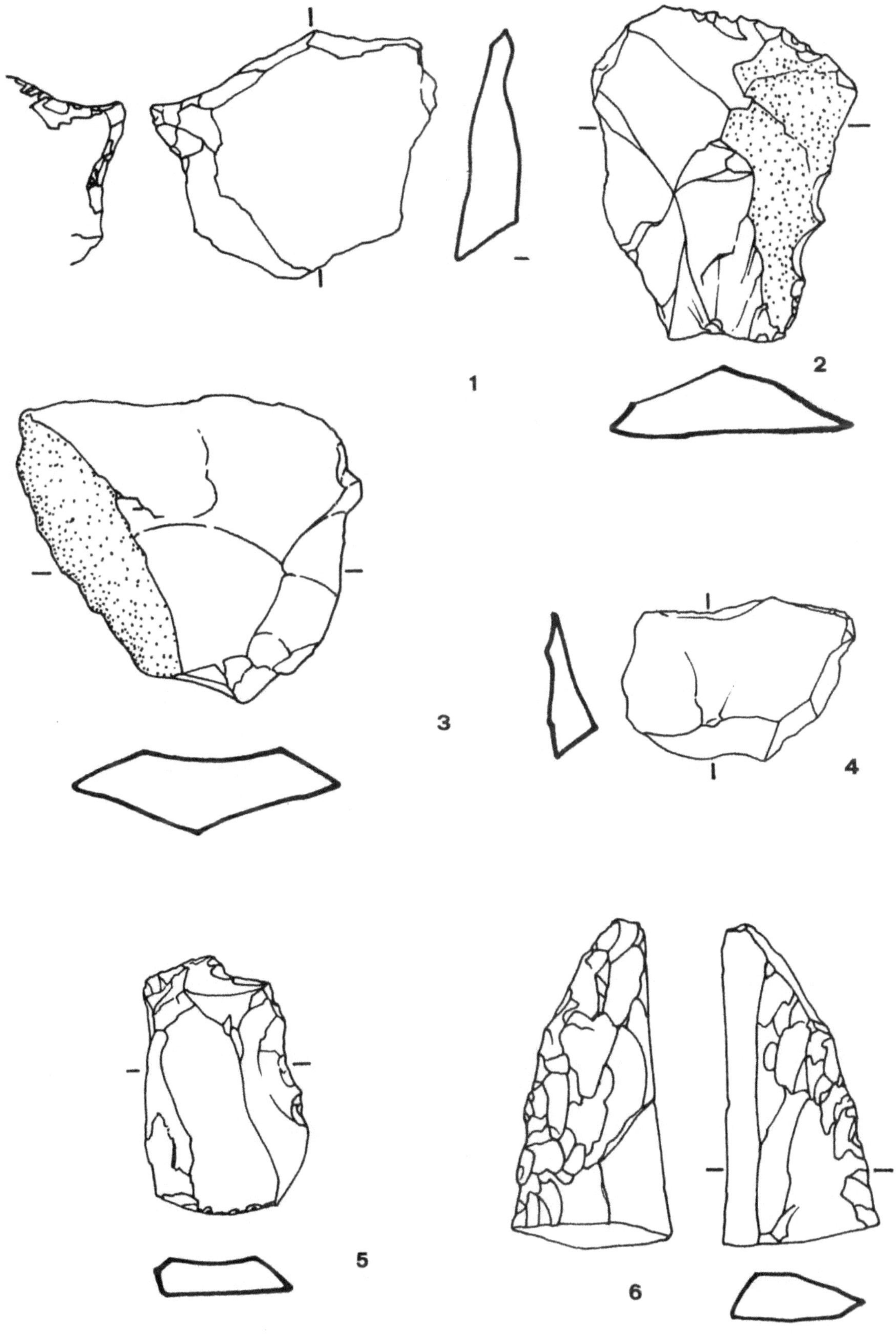

Fig. 7.16 - Uan Muhuggiag, Wadi Teshuinat. Level 2a, Sector B.3, flake; 4, core-tablet; 2, end-scraper; 1, perforator; 6, side-scraper (from Level 2); 5, denticulate (1:1).

denticulated pieces, both with distal scraping-ends; one on a circular nucleiform blank (fig. 7.24,8) and the other on a tertiary flake (fig. 7.24,5). The full-size illustrations show the actual dimensions.

Table 7.2 shows the absolute and percentage frequency of the class in the Lower Phase. It shows a certain degree of fluctuation from Level 2d (25.0%) to Level 2a, where the index is nearly halved (11.4%). A further decrease is observed in Level 2 (6.1%) while the index rises once again in Level 1a (13.3%). In Level 1 this class is among those with the highest frequency (31.3%).

Burins - Level 2d-1a: a total of 1; general index 0.5%

Level 1: absent.

The only specimen recovered was an angle distal (dexter) burin with an additional inverse dexter proximal retouch. It was made on a core-trimming of quartzitic sandstone (fig. 7.12,6).

The absolute frequency is even lower than in Sector A. This confirms the fact that this tool-class represents a very marginal element in the lithic production.

Perforators - Levels 2d-1a: a total of 2; general index 1.1%

Level 1: a total of 1; general index 6.2%.

The two examples from the Lower Phase are from Level 2a. One is a distal déjeté perforator, formed by bifacial sinister retouch and made on a tertiary blade (fig. 7.16,1). The second item was also a déjeté perforator, made on a tertiary blade with distal pointed end. It was formed by alternate retouch, and showed an additional obverse invasive retouch (fig. 7.17,6).

The single piece coming from Level 1 was a simple distal perforator on a core-trimming, formed by an obverse sinister simple and a dexter notched ("single blow") retouch (fig. 7.24,2).

As in Sector A perforators are very rare. They are not reported lower than Level 2. All the recovered items were of sandstone.

Blades - Levels 2d-1a: a total of 87; general index 46.0%

Level 1: a total of 7; general index 43.8%.

The remarks made in the description of the Sector A blades apply here but the greater wealth of material allowed more detailed comment. The types are the same as those in Sector A, with a better documentation of the backed bladelets.

In the Lower Phase, seventeen pieces were blunt unretouched blades (fig. 7.13,9). They were all tertiary from single platform cores, except for one (from Level 2b) which was from an opposed platform core (fig. 7.12,5). One piece (Level 2c) showed sinister use-retouch (fig. 7.9,7) while another had a dexter central notch formed by use-retouch (fig. 7.19,7). Five

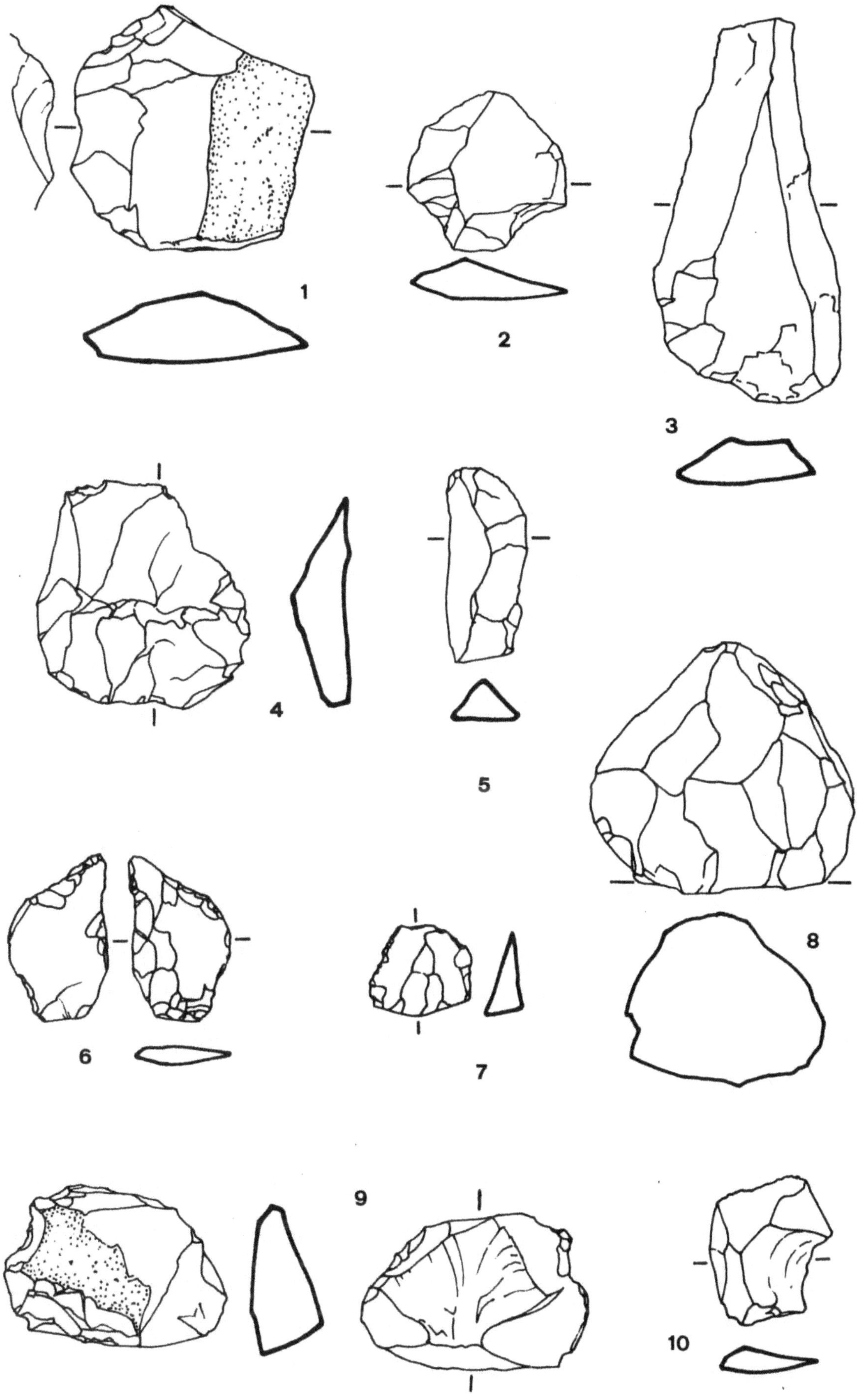

Fig. 7.17 - Uan Muhuggiag, Wadi Teshuinat. Level 2a, Sector B. 5, lame à crête; 8, core; 6, perforator; 3, blunt unretouched blade; 7,10, scaled pieces; 1,4,9, notches and denticulate; 2, point (1:1).

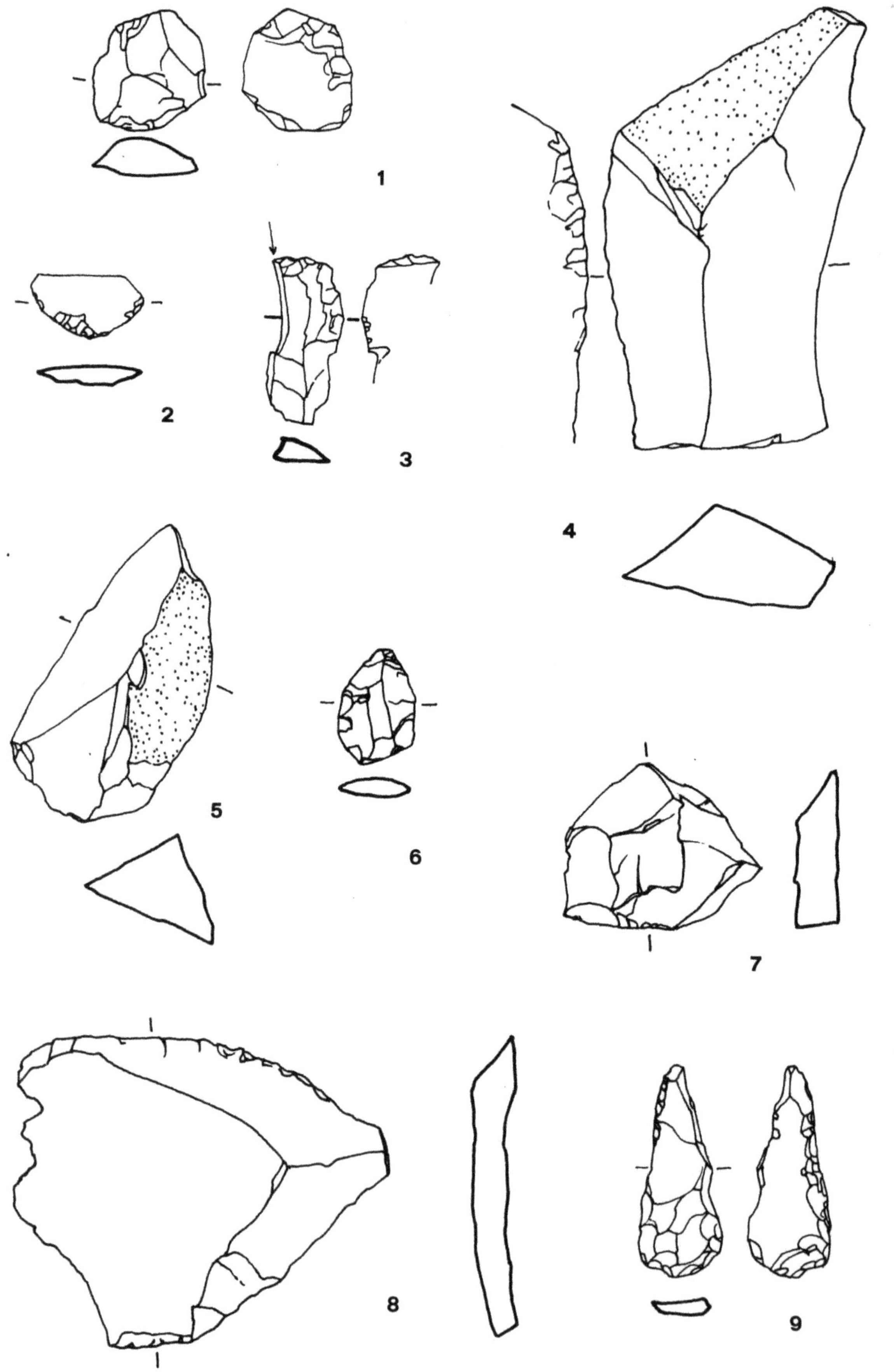

Fig. 7.18 – Uan Muhuggiag, Wadi Teshuinat. Level 2, Sector A. 5, flake; 7, core-side; 3, burin; 1, scaled piece; 2, side-scraper; 4,8, pieces with use-retouch; 6,9, points (1:1).

pieces (Levels 2d, 2b, 2a, 2) may be considered bladelets (an average of 26 mm _ca_ in length and 9 mm _ca_ in width); six were microbladelets: three from Level 1a (fig. 7.21,_2,3_) and the rest, from Level 2a, showed pointed distal ends. They were of quartzitic sandstone and had the following mean dimensions: length 20 mm; width 7 mm; thickness 2 mm. In addition to these unbroken pieces, thirteen fragments of blades from single platform cores were also recovered. They were mainly proximal fragments (figs. 7.8,_5_; 17,_3_; 19,_1_; 21,_10_). An inverse sinister simple continuous retouch was observed in one of them.

The group of fifty-four blade-like flakes were mostly tertiary, and from single platform cores with one example from a ninety-degree core. About 50% had distal pointed ends and the platform-types were identifiable as mainly faceted and _lisse_ (fig. 7.19,_6_). An example of secondary blade-like flake showed a _déjeté_ perforator (fig. 7.20,_5_).

Ten of the pointed pieces might be likened to the Levallois points; they came, for the most part, from one of the lower levels, 2b: (figs. 7.12,_8_; 13,_6-8_) but were also recovered from Level 2 (fig. 7.19,_2_). The item shown in fig. 7.13,_7_ had a sinister polished side, while that shown in fig. 7.12,_8_ had an obverse abrupt proximal retouch. The size ranged from 20 to 41 mm in length, from 10 to 30 mm in width and from 2 to 12 mm in thickness.

The base levels of the deposit (2d and 2c) yielded three examples of backed and pointed bladelets showing a very high standard of manufacture with the classical obverse or inverse backing abrupt retouch. A specimen from Level 2d was the proximal section of a backed bladelet with inverse sinister backing (fig. 7.7,_2_). The best example was a straight backed and pointed bladelet with rounded base (Tixier's type 46) coming from Level 2c. It was a quartzitic sandstone tertiary bladelet with sinister entire edge _sur enclume_ backing (fig. 7.9,_3_). A third also came from Level 2c and was a partially backed and pointed bladelet (Tixier's type 63) on a tertiary blank of quartzitic sandstone with obverse sinister distal plus central backing (fig. 7.9,_6_). These may be compared to those from the Two Caves and the Ti-n-Torha East assemblages (Levels RInf-CII).

From Level 1 four tertiary blades from single platform cores were recovered. Three were proximal fragments (fig. 7.24,_4_) and the other a microbladelet (18x6x4 mm). Three blade-like flakes from single platform cores with pointed distal ends were also found in the same level; one of these had an oval shape. Their dimensions ranged from 19 to 36 mm in length, from 12 to 21 mm in width and from 3 to 9 mm in thickness.

The incidence of this class increased gradually from the base to the upper part of the sequence, the percentage being similar in Levels 1 and 1a.

Varia

Scaled pieces - Levels 2d-1a: a total of 16; general index 8.5%

Level 1: absent.

Scaled pieces were recovered only from the Lower Phase in Sector B. Ten out of the sixteen were obtained from tertiary flakes through generally

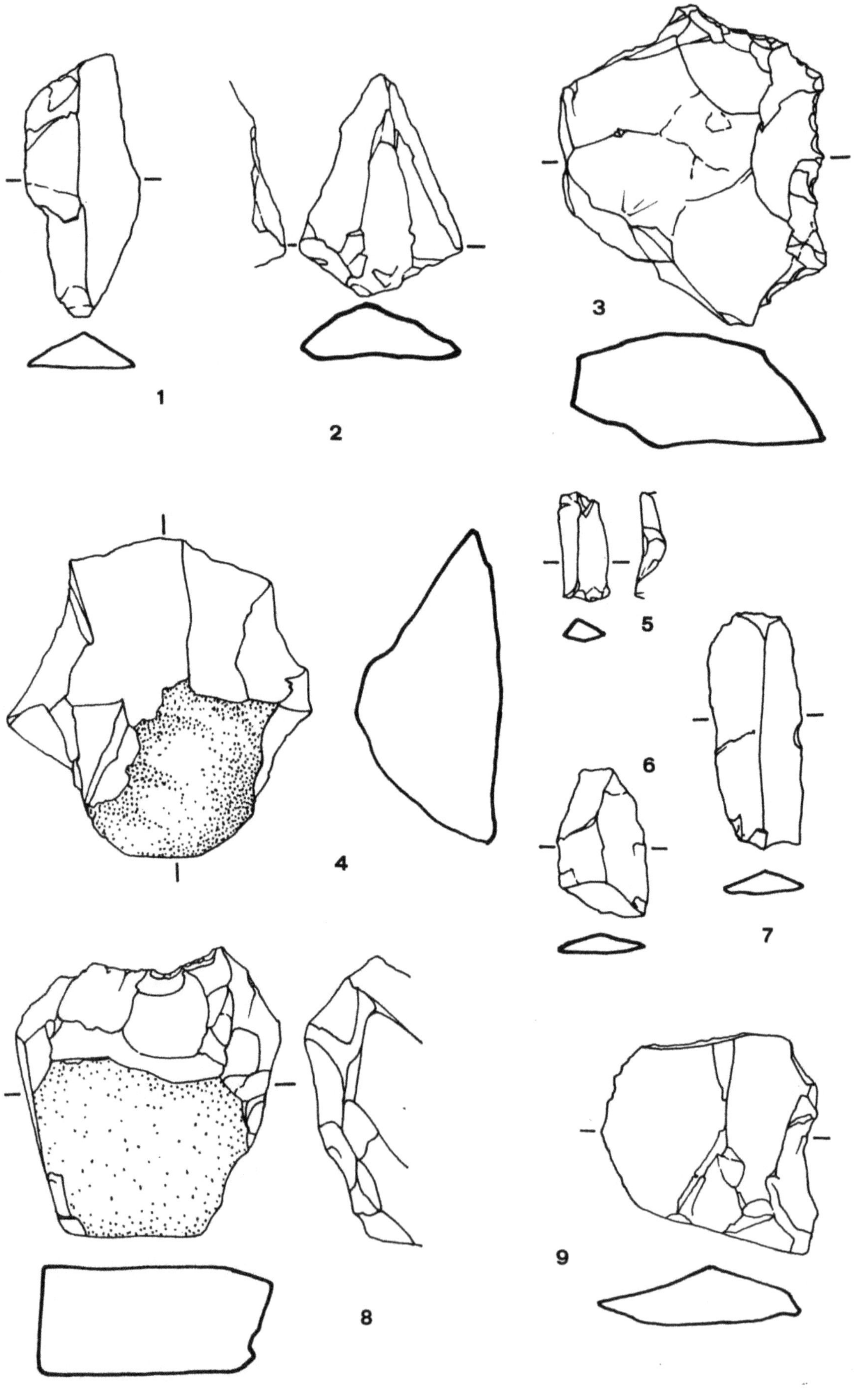

Fig. 7.19 - Uan Muhuggiag, Wadi Teshuinat. Level 2, Sector B. 4, core; 3,8, end-scrapers; 1,6,7, blunt unretouched blades; 2, pointed blade-like flake; 5,9, notches (1:1).

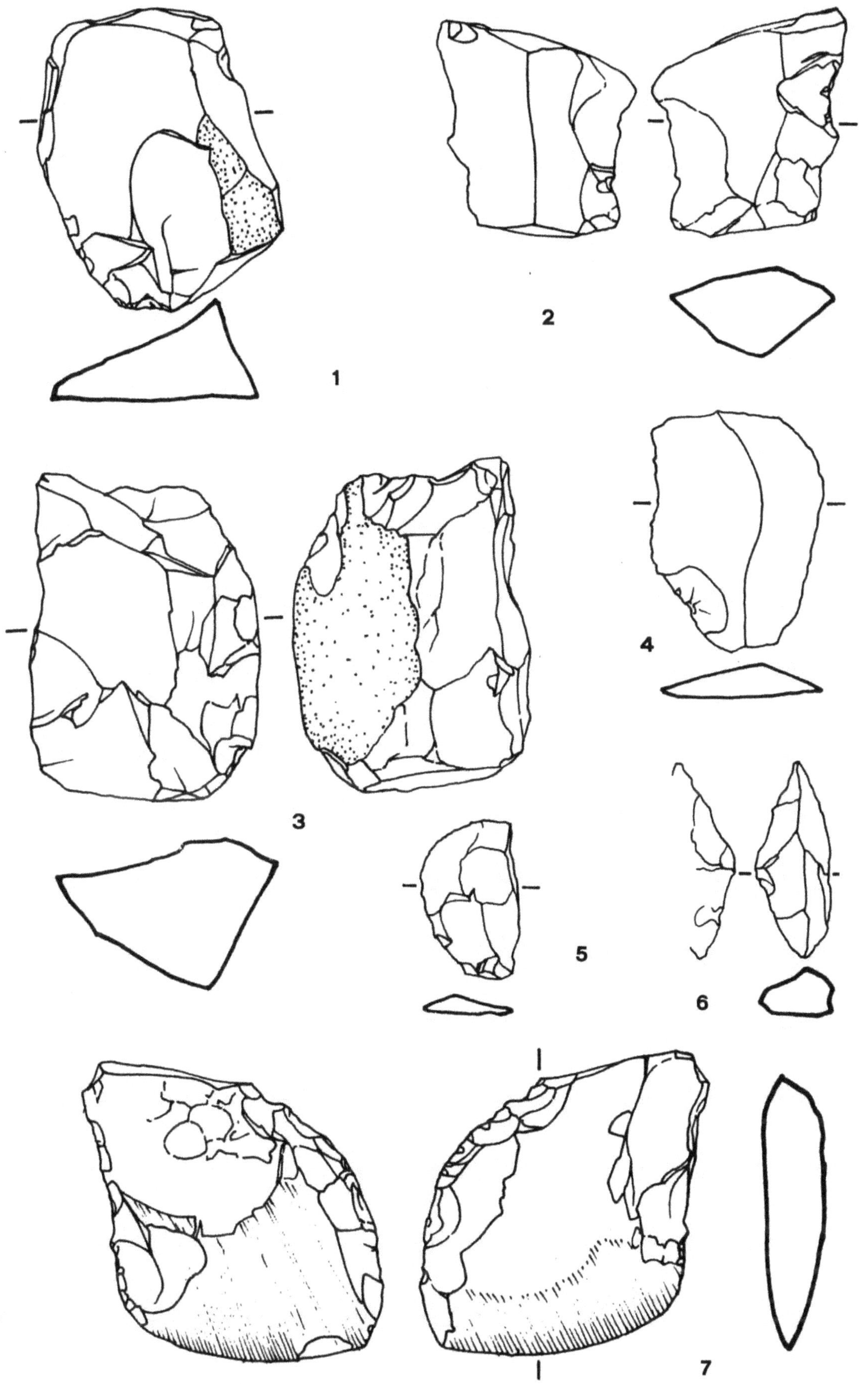

Fig. 7.20 - Uan Muhuggiag, Wadi Teshuinat. Level 2, Sector B. 3, primary flake; 6, lame à crête; 5, blade-like flake; 2, side-scraper; 1, piece with use-retouch; 4, notch; 7, axe (1:1).

central plus proximal obverse scaling (fig. 7.17,7). One example (Level 2b) had a bifacial scaling, while another (Level 2a) had an additional obverse "single blow" notch (fig. 7.17,10). Of the remaining six pieces, two (Levels 2d and 2a) were on core-fragments with unifacial, proximal scaling; two (Level 2c) on core-trimmings and showed obverse central scaling, while another specimen (same level), on a secondary flake, presented an obverse central scaling together with dexter central retouch (fig. 7.9,5). Finally, Level 1a yielded a scaled piece on a quartz chip with obverse proximal scaling.

Most of the pieces were of sandstone but quartzitic sandstone and quartz ones were also recovered. The dimensions ranged from 15 to 55 mm in length; from 7 to 24 mm in width and from 4 to 10 mm in thickness.

The peak incidence (13.7%) of this class was in Level 2c and gradually decreased to 3.3% in Level 1a.

Continuously retouched pieces - Levels 2d-1a: a total of 3; general index 1.6%

Level 1: absent.

This class was only reported in the lowest section of the stratigraphy; the three specimens were all recovered from Levels 2c, 2b and 2a. Two pieces were on secondary sandstone flakes; one of them was characterized by alternate retouch obverse sinister, distal plus central, and inverse dexter distal, the other had only an obverse dexter distal retouch. The third was on a tertiary flake with obverse retouch at the distal end. The pieces were, on average, 34 mm long, 32 mm wide and 11 mm thick.

Side-Scrapers - Levels 2d-1a: a total of 10; general index 5.3%

Level 1: absent.

Side-scrapers were only recovered from the Lower Phase. Four (Levels 2b, 2a, 2) were on tertiary flakes (one on a secondary flake) with convex, dexter or sinister, scraping-edges. One had an invasive bifacial flaking (fig. 7.16,6); two were on scaled pieces, one (Level 2c) was a double side-scraper with concave scraping-edges (fig. 7.8,8), while the other had a dexter convex scraping-edge. Level 1a yielded a sinister convex side-scraper on a core-trimming with an additional, inverse "single blow" sinister proximal notch. Finally, three denticulated pieces came from Levels 2a, 2 and 1a. One of them was on a core-trimming with a dexter straight scraping-edge; another, on a tertiary flake, showed denticulations on both the dexter and the sinister edge (fig. 7.20,2). The third, from Level 1a, was on a primary flake with a sinister straight scraping-edge and an additional inverse "single blow" proximal notch (fig. 7.21,4).

All were of sandstone and their dimensions ranged from 18 to 55 mm in length, from 14 to 45 mm in width and from 7 to 15 mm in thickness.

The class was not represented in Level 2d, but from Level 2c upwards its incidence increased reaching its highest value in Level 2 (9.1%).

Pieces with use-retouch - Levels 2d-1a: a total of 10; general index 5.3%

Level 1: a total of 1; general index 6.2%.

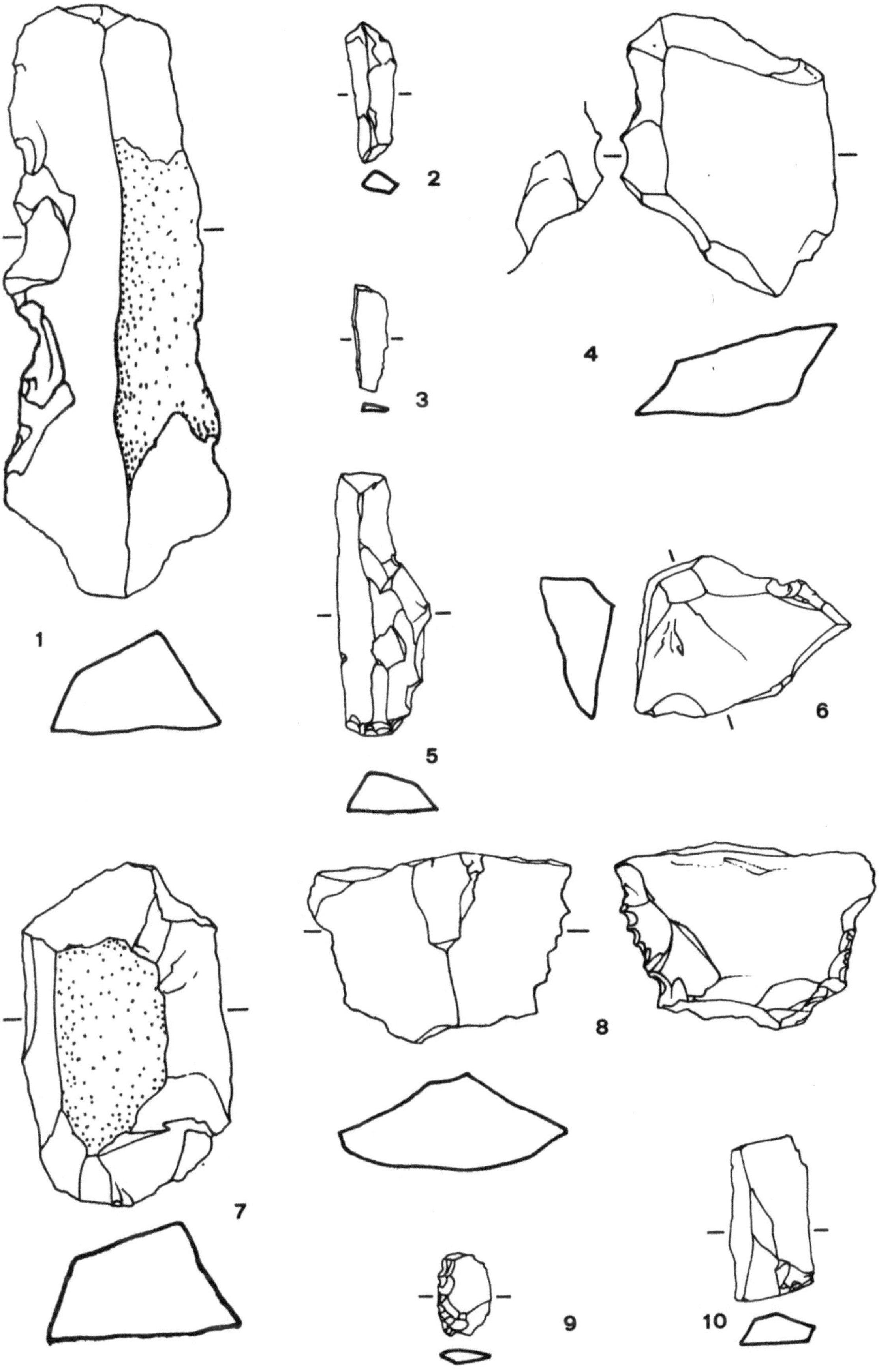

Fig. 7.21 - Uan Muhuggiag, Wadi Teshuinat. Level 1a, Sector B. 5, lame à crête; 6,7, end-scrapers; 2,10, unretouched blades; 3, microbladelet; 4, side-scraper; 1,8,9, denticulates (1:1).

Level 2 yielded a thick secondary flake with a distal use-retouch forming an end-scraper (fig. 7.20,1). Another piece, coming from Level 2b, was a primary flake utilized as a perforator. It had a pointed distal end formed by obverse sinister entire edge retouch and inverse scaling. Four pieces showed denticulated use-retouches. One, coming from Level 2a, was on a blade-like tertiary flake and had obverse dexter central plus proximal use-retouch. Three additional specimens recovered from Level 1a were tertiary or secondary flakes: two had obverse dexter distal and the third inverse sinister entire edge use-retouch. Four more items used as side-scrapers came from Levels 2a and 2; they were all tertiary flakes with convex, dexter, sinister or bilateral scraping-edges. Their dimensions ranged from 19 to 78 mm in length, from 11 to 60 mm in width and from 4 to 18 mm in thickness.

The single item coming from Level 1 was a tertiary flake with denticulated inverse sinister entire edge use-retouch. It had also an inverse notch at the proximal end (fig. 7.24,3).

While pieces with use-retouch did not appear in the basal levels (2d, 2c), they were recovered from Level 2b upwards and their incidence tends to increase gradually from Level 2b (2.1%) towards the top of the Lower Phase (Level 1a: 10.0%).

Notches and Denticulates - Levels 2d-1a: a total of 26; general index 13.8%

Level 1: absent.

The sixteen notched flakes or blades were generally of poor quality and the most frequent notch-forming retouch was the "single blow" type, which could be obverse or inverse, distal (fig. 7.7,6), proximal (figs. 7.19,5; 20,4) or central (fig. 7.19,9). One multiple notched piece was recovered with two obverse notches at the distal end and another, still obverse, at the proximal end. A second item (Level 1a) had two notches on the dexter side, one of which was formed by an abrupt retouch and the other by a simple retouch. Although prevailingly tertiary, blanks were commonly secondary and primary were quite frequent (fig. 7.17,1). One piece (fig. 7.17,9) was on a core fragment; another one was on a core-preparation flake. The size of the notches did not change in the various levels. On the average, the pieces were 32.5 mm long, 30.3 mm wide and 9.5 mm thick.

There were ten denticulated pieces made on tertiary, secondary (figs. 7.13,4; 21,1) or primary (fig. 7.9,1) flakes or blades which were often broken. The obverse or inverse denticulated retouch was on the sinister or the dexter side, but was bilateral on one specimen (fig. 7.21,8). As for the position, the most common was the distal plus central or the distal plus proximal retouch, but that involving the entire edge was also observed (as in fig. 7.13,4). An obverse sinister distal denticulated blade showed also an additional end-scraper at the distal end (fig. 7.7,1). On the average, these pieces were 49.5 mm long, 32.7 mm wide and 14.5 mm thick.

The percentage frequency of the class varied through the Lower Phase. The index, 33.3% in the basal level, dropped rapidly in Level 2b (4.2%) and increased again in the subsequent levels, reaching 16.7% in Level 1a.

Arrow-heads and Points - Levels 2d-1a: a total of 15; general index 7.9%

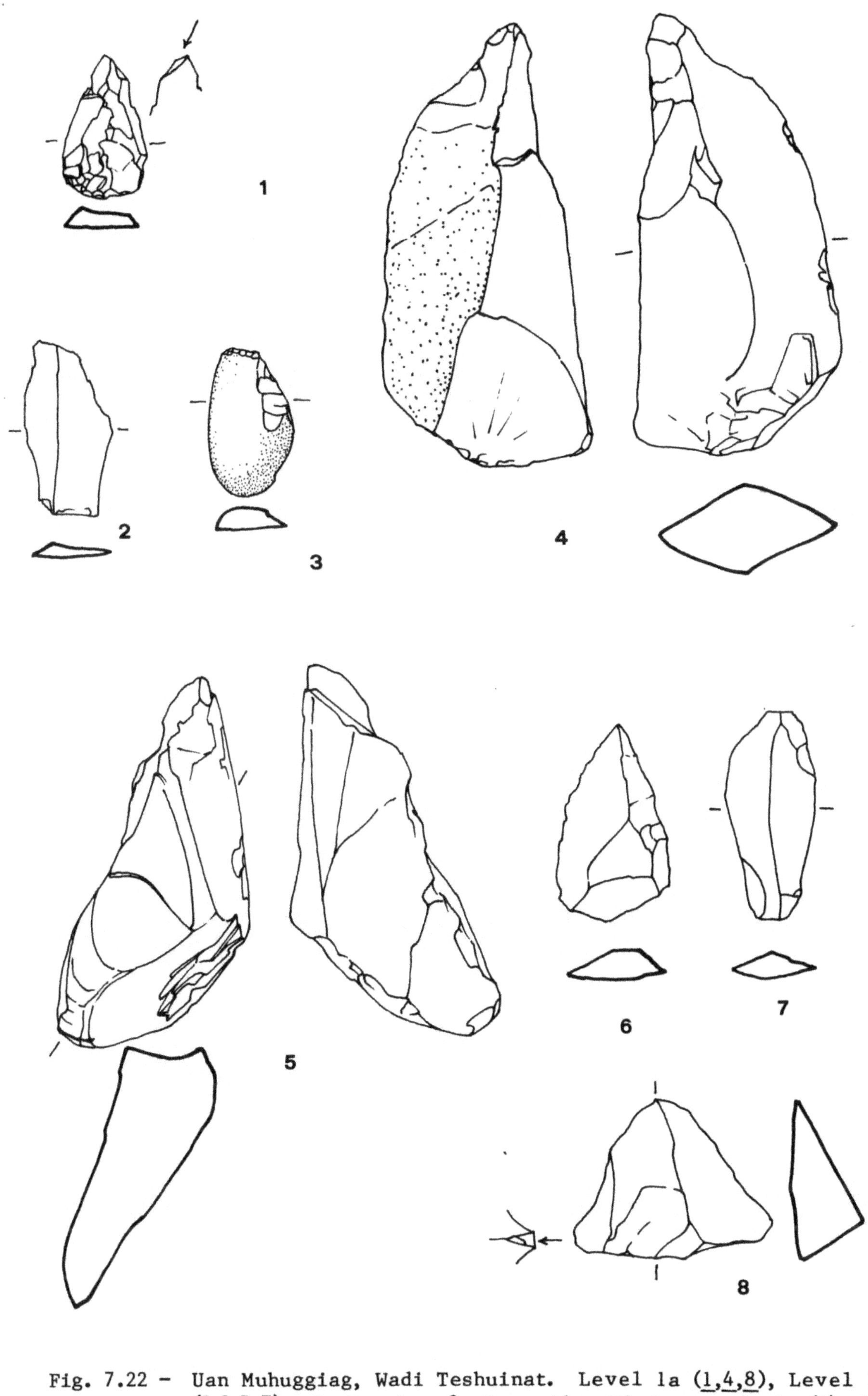

Fig. 7.22 - Uan Muhuggiag, Wadi Teshuinat. Level 1a (1,4,8), Level 1 (2,3,5-7). Sector A. 8, Core-side with inverse burin (?); 4, perforator; 2,7, blunt unretouched blades; 3, piece with use-retouch; 1,6, points; 5, siltite sharpener (?) (1:1).

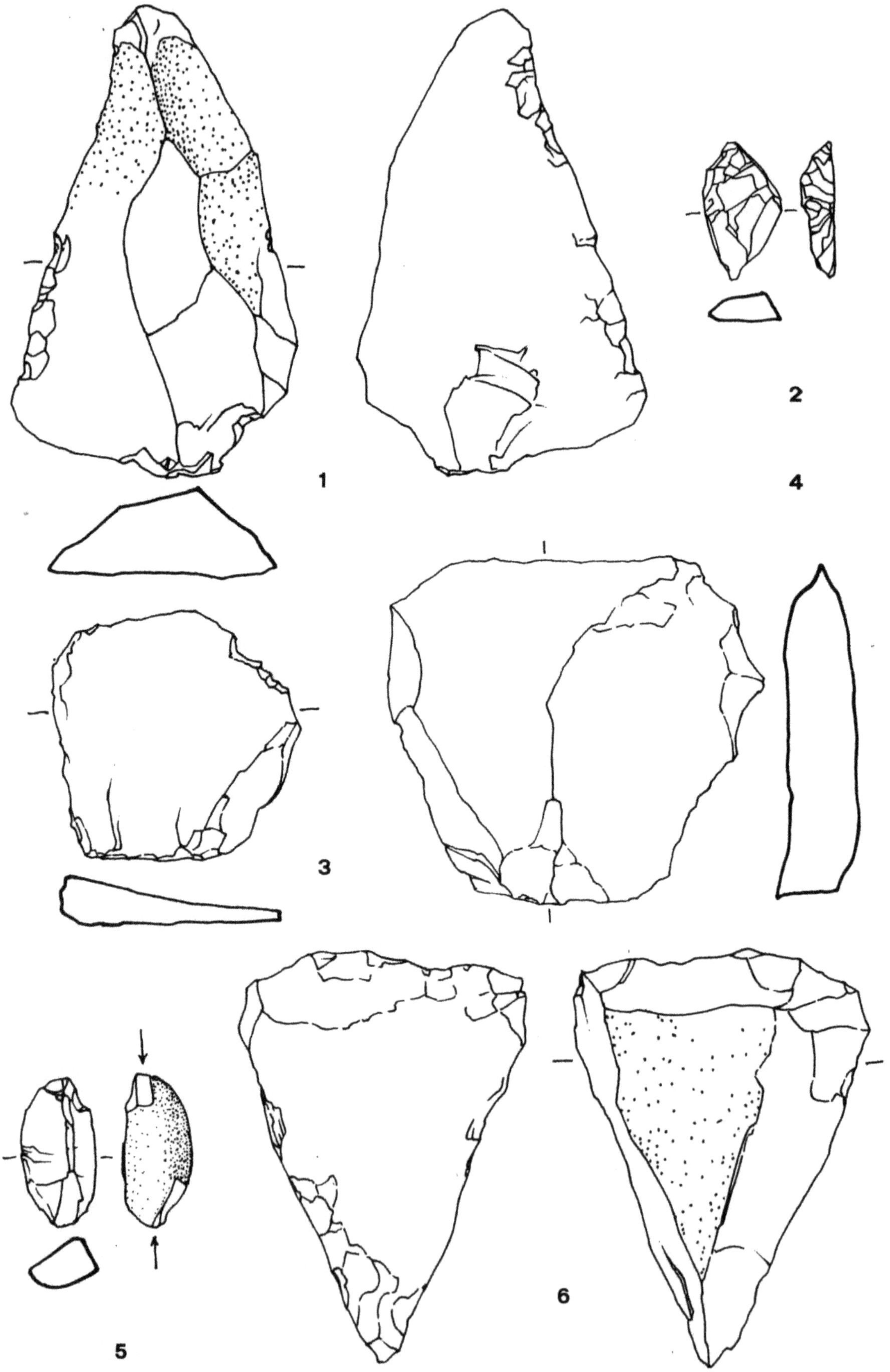

Fig. 7.23 - Uan Muhuggiag, Wadi Teshuinat. Level 1, Sector A. 6, end-scraper; 5, burin; 2, backed blade/flake; 1, side-scraper; 4, p. with use-retouch; 3, denticulate (1:1).

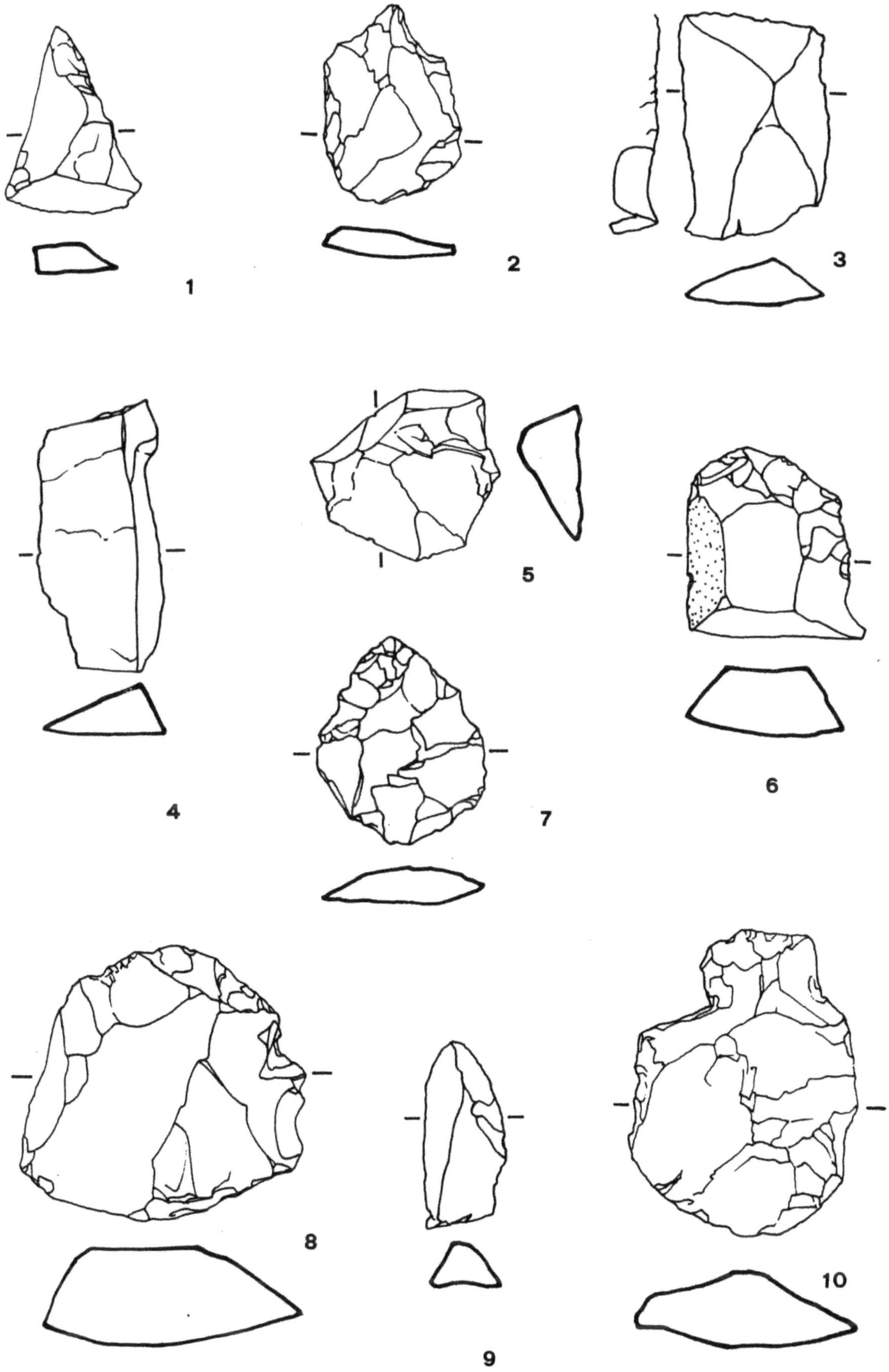

Fig. 7.24 - Uan Muḫuggiag, Wadi Teshuinat. Level 1, Sector B. 5,6,8-10, end-scrapers; 2, perforator; 4, unretouched blade; 3, with use-retouch; 7, arrow-head; 1, point (1:1).

Level 1: a total of 2; general index 12.5%.

This was a larger group than from Sector A; samples recovered in Sector B, and particularly in Levels 2d-2a, allowed a more complete analysis.

Two fragments were recovered from Level 2d: one was a proximal fragment of an arrow-head of quartzitic sandstone with obverse continuous invasive and inverse bilateral retouch (fig. 7.7,5) while the other was a proximal fragment of a sandstone arrow-head with obverse continuous invasive retouch.

Level 2c yielded a triangular flat-based (straight sided) arrow-head on a tertiary quartz blank with bifacial invasive retouch, which measured 12 mm in length, 12 mm in width and 3 mm in thickness (fig. 7.9,2). It corresponds to Hugot's A1 type (Hugot 1957) and to the "flat-based" type of Smith's classification (1976: 186). The same level also yielded a proximal fragment of the same type, together with an initially retouched, crescent-shaped piece on a tertiary blank of flint with unifacial invasive retouch (fig. 7.9,4).

The most numerous group came from Level 2b and may be divided as follows:

Hugot's Type A1

Triangular flat-based (straight sided) arrow-head on a tertiary blank of quartzitic sandstone with bifacial invasive retouch, measuring 16 x 15 x 4 mm (fig. 7.12,2).

Triangular flat-based (convex sided) arrow-head on a tertiary blank of quartzitic sandstone with bifacial retouch, measuring 17 x 11 x 3 mm (fig. 7.12,7).

Hugot's Type A18

Triangular hollow-based arrow-head on a tertiary quartzitic sandstone blank with bifacial invasive retouch, measuring 26 x 16 x 4 mm (fig. 7.12,3). It corresponds to Smith's "hollow-based" type (1976: fig. 10/3).

Hugot's Type B3

Triangular convex-based arrow-head on a tertiary blank of quartzitic sandstone with bifacial invasive retouch, measuring 15 x 13 x 2 mm (fig.7.12,1).

Hugot's Type C4

Leaf-arrow-head with a slightly rounded base on a tertiary blank of quartzitic sandstone with unifacial invasive retouch, measuring 22x10x5 mm (fig. 7.13,1). It corresponds to Smith's "round-based" type (1976: fig. 10/5).

An unfinished specimen. Leaf-round-based arrow-head on a tertiary sandstone blank with bifacial partial retouch, measuring 37x22x9 mm (fig. 7.13,3).

Hugot's Type C1

Leaf-arrow-head, pointed at both ends, on a tertiary blank of quartzitic sandstone with bifacial invasive retouch, measuring 35x7x4 mm (fig. 7.12,4). It corresponds to Smith's "pointed-based" type (1976: 186).

Hugot's Type H5?

The only specimen coming from Level 2a was an atypical tanged arrow-head on a tertiary blade of siltite. Its shape brings to mind examples of the Aterian tradition (perhaps, a re-utilized implement?). It measures 22 x 7 x 7 mm (fig. 7.17,2) and, with reference to Hugot's typology, the only possible comparison is with type H5 "armature sur fragment de lamelle brute".

Two specimens that may not be arrow-heads were recovered from Level 2b. One is a triangular, convex-based point on a tertiary flint blank with unifacial invasive abrupt retouch and a plano-convex cross-section (fig. 7.13,2). The second one is a point on a tertiary blade of quartzitic sandstone showing straight-convex sides and bifacial invasive abrupt retouch with a plano-convex cross-section (fig. 7.12,9).

Two examples were recovered from Level 1, fig. 7.24,7 shows a leaf-arrow-head with round base on a tertiary blank of quartzitic sandstone with bifacial invasive retouch, measuring 35 x 25 x 7 mm. It corresponds to type C1 of Hugot's classification and to Smith's "round-base" type. The other item was a point on a tertiary sandstone blank with unifacial (obverse) dexter retouch and a plano-convex cross-section (fig. 7.24,1).

In conclusion, the lowest levels in Sector B (Levels 2d, 2c and 2b) contained the bifacially retouched arrow-heads also found in other parts of the Central Sahara and whose characteristics fit well within the classification by Hugot. Their presence represents, as will be pointed out later on, continuity with the Ti-n-Torha horizon.

They were located at the bottom of the series and their general index is not high (7.9%); their highest frequency is reported in Level 2b (18.7%).

Large Stone Tools

This definition was preferred to that of "polished stone tools", already used in the classification of the products of the Two Caves assemblage, since only some of the implements belonging to this class show complete polishing. They were a poorly representative class, grinding-stones forming the largest group (Table 7.3).

Sector A

The single smoothing implement from Level 2a had a roughly trapezoidal shape with an approximately oval cross-section. It was made on coarse-grained sandstone and measured 23 mm in length, 16 mm in width and 9 mm in thickness. Two tablets were also recovered from Level 2a. One of these, of siltite, was a fragment (31 x 19 x 3 mm) while the other was an almost circular piece of steatite polished on both faces (fig. 7.15,1). The most interesting object, coming from Level 2a, was a small axe made on siltite. It had a roughly trapezoidal shape and polished surfaces with flaking along the edges (fig. 7.14,6).

In the group of twelve grinding-stones, five (Level 2b) were small fragments belonging to a single lower grinding-stone with a roughly oval surface. The seven remaining specimens were handstones, two of which were fragmentary (fig. 7.25,5); the intact ones had a biconvex or plano-convex grinding-surfaces and a circular shape. They were all made on fine-grained sandstone. Their mean diameter was 100 mm *ca*; while their thickness ranged from 35 to 56 mm (fig. 7.25,2,3).

Level 2 yielded a problematic piece - perhaps a sharpener - consisting of an irregular red siltite fragment (37 x 19 x 5 mm) with only one polished edge. Another came from Level 1a: it was an irregularly shaped siltite piece (65 x 25 x 2 mm) and was the only polished specimen recovered from the Upper Series (fig. 7.22,5).

Sector B

Two smoothing implements were recovered from Level 2b. One was a cylinder made of sandstone (38 x 15 x 7 mm) with only one polished face; the other, made of siltite, had a triangular cross-section and only one of its two ends was polished (54 x 32 x 7 mm). A third, from Level 2, was a siltite pebble of roughly trapezoidal shape which was polished at the base; it measured 60 x 40 x 15 mm. Two tablets were recovered from Level 2b. One had a circular shape (diameter 70 mm *ca*; thickness 7 mm), while the other was shaped as a small trapezoidal axe with flaking on the dexter edge and at the base (fig. 7.27,16). Both sides of the implement show clear traces of a yellow colouring substance suggesting the use of this implement in painting. Two real axes were recovered from Levels 2a and 2: both of them were made on siltite. One was small, flaked and polished on both sides (gouge-like type) (fig. 7.25,6). The other one was a fragment with a very sharp edge although the rest of the two faces was flaked (fig. 7.20,7). A single grooved stone made on limestone and measuring 40 x 21 x 15 mm came from Level 2d.

There were five pieces of sandstone (grinding-stones). Two came from the bottom level (2d) and are fragments of a handstone and a lower grinding-stone. Level 2b yielded an almost intact lower grinding-stone with a slightly concave grinding-surface and a sub-oval shape (diameter 160 mm *ca*) (fig. 7.25,1). Two handstones, with biconvex faces, were also recovered from Level 2a (fig. 7.25,7). Level 1 failed to provide any further specimen.

It is worthwhile mentioning that Level 2a yielded a piece of red ochre.

Bone and Shell Industry

Worked and unworked ostrich eggshell fragments were numerous. As a general rule, bone implements showed little diversification, notwithstanding the high standard of manufacture in some products (spatulae, smoothing tools, punches, etc.).

Sector A

Five items, concentrated in Levels 2b-2a, was recovered from the Lower Phase; no evidence came from Level 2c and a single fragment from Level 2.

Table 7.3 - Uan Muhuggiag, Wadi Teshuinat. Large stone tools.

	SECTOR A							SECTOR B							
	1	1a	2	2a	2b	2c	Total	1	1a	2	2a	2b	2c	2d	Total
Smoothing implements				1			1			1		2			3
Tablets				2			2					2			2
Axes				1			1			1	1				2
Grooved stones														1	1
Grinding-stones															
- handstones			3	1	2	1	7				2			1	3
- lower grinding-stones					1		1				1			1	2
Varia															
- red ochre											1				1
- resharpeners		1	1				2								

Punch - The only specimen comes from Level 2b and was made from a fragment of burnt bovid bone (Gautier's analysis). It was well polished on the outer face (the inner face is untreated) and sharpened at the end (60 x 8 x 4 mm).

Spatulae - One coming from Level 2a was made on the scapula of a large animal, quite likely bovid. The bone is well polished on both faces and presents serrations along one of the two edges (8 mm thick) (fig. 7.26,15).

Smoothing implements - Two specimens were probably used to smooth the surfaces of ceramic vessels. One of these (Level 2b) is quite likely an Alcelaphus cannonbone showing a notch and a polishing on the proximal epiphysis. The other specimen (Level 2a) is less clearly identifiable and indeed is only a very worn fragment of the original implement (25 x 14 x 2 mm; 29 x 19 x 2 mm).

Ostrich Eggshells - The Lower Phase yielded thirty-nine pieces, for the most part unworked and frequently burnt (fig. 7.26,8,13,14). Six finished beads were recovered (Levels 2b, 2a).

In the Upper Phase the bone industry acquires greater significance particularly in Level 1, which yielded five bone implements and twenty-three ostrich eggshell products. The recovered types of implements were largely the same as those described for the Lower Phase. Two punches were recovered from Levels 1a and 1 respectively. Level 1 yielded a long item (fig. 7.26,12) made on a thick bovid splinter that was polished on the outer surface and roughly sharpened (4 mm thick). The punch from Level 1a, the only bone product of this Level, was characterized by a high standard of manufacture and is a beautiful (unfortunately, blunt) sample, flattened by accurate polishing on both sides (3 mm thick). The base is rounded with a flattened and oblique lower margin that might have allowed its use as a spatula in the working of ceramics (fig. 7.26,11). Two awls, both with cylindrical shape and sharp ends, were recovered from Level 1. The best specimen is extremely well preserved, thoroughly polished over its whole area. It has a circular cross-section with slightly rounded ends (5 mm thick) (fig.7.26,9). The second is flat (3 mm thick) and thinned at both ends, of which one is sharpened and the other rounded (fig. 7.26,7). The same level yielded an unfinished spatula. The bone structure (rib from a caprid), is still plainly visible and the two ends are fragmented (6 mm thick) (fig. 7.27,1). A denticulated piece, made of a roughly polished splinter, with both ends still unworked was found. The serrations are very close together and fail to be clearly differentiated suggesting that the piece was used as a small saw rather than as a ceramics comb (5 mm thick) (fig. 7.26,10).

Twenty-three ostrich eggshell products had been recovered from Level 1. Eighteen of these are unworked pieces while five present different stages of bead manufacture, including a small disc (fig. 7.26,1-5). Level 1a yielded a single, small bead and a slightly burnt, unworked fragment (fig. 7.26,6).

Wood

Three wands of wood (6) come from Level 2b. Two have cylindrical shape

---

6) The nature of the wood is being determined by K. Wasylikowa.

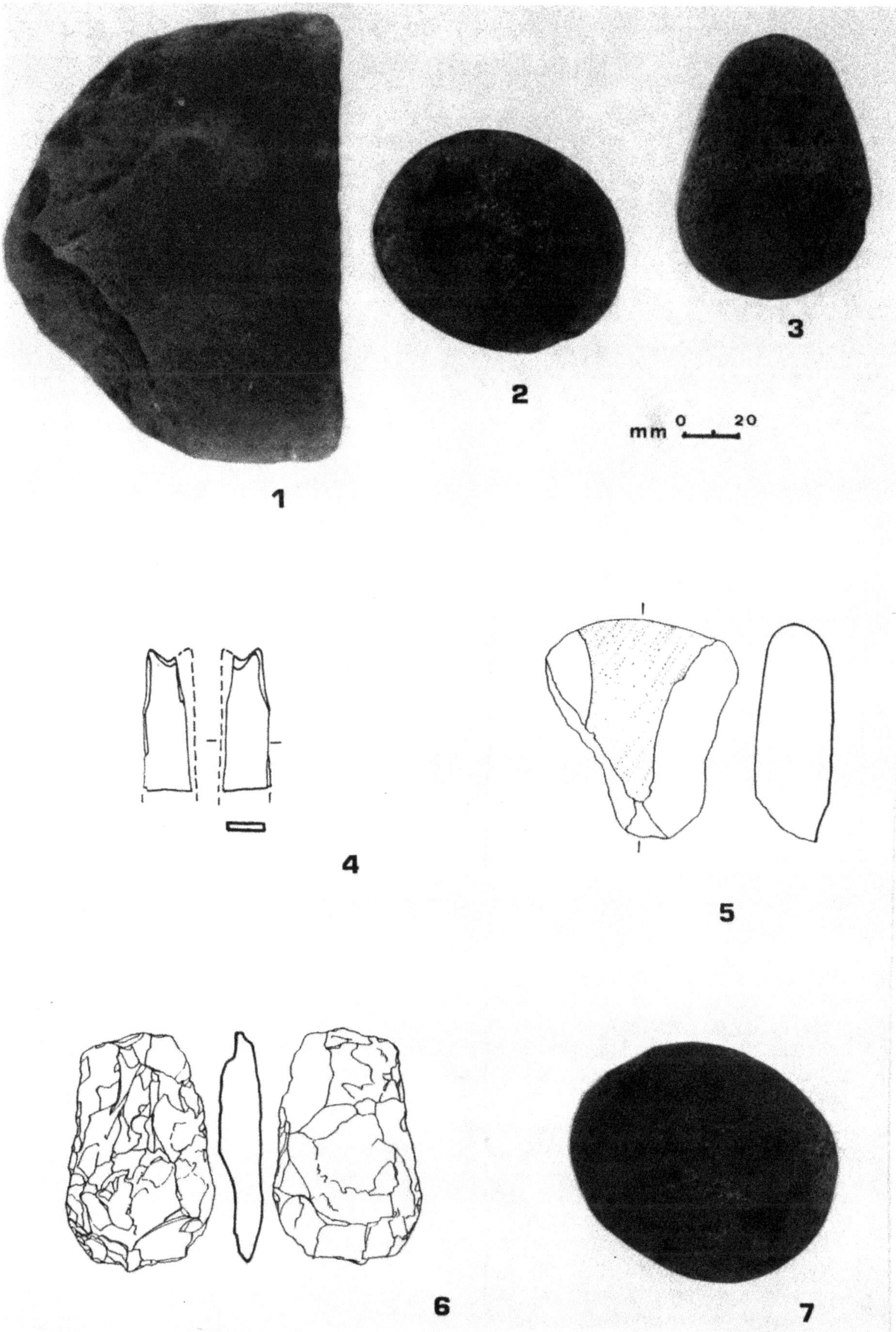

Fig. 7.25 - Uan Muhuggiag, Wadi Teshuinat. Sectors A and B. 2,3,5,7, rubbers; 1, lower grinding-stone; 6, axe; 4, bone implement (1:2. No. 4 is 1:1).

with thinned and blunted ends (mean dimensions: 50 x 10 x 5 mm), while the third specimen is a small stick, broken at one end and sharpened at the other (84 x 11 x 9 mm).

Sector B

All the bone implements from Sector B come from the Lower Phase, Level 1 only yielding five unworked pieces of ostrich eggshell. The nine specimens were concentrated in Levels 2b, 2a, 2, while the basal section and Level 1a only yielded a few ostrich eggshell fragments.

Punches were recovered from Levels 2a and 2. One was made on a small bone that fully preserved its structure, except for the distal end that presented a sharp and oblique point obtained by an accurate polishing (3 mm thick) (fig. 7.27,12). Another, made on a splinter polished on both faces, was long with a flat cross-section (3 mm thick) sharpened at one end and rounded at the other (fig. 7.27,6). Finally, a double-pointed punch was recovered from Level 2a. It is a fragmentary, made on a thin burnt bone splinter, accurately polished on both surfaces so as to obtain two clearly spaced teeth. Even the two implement sides have been flattened through an accurate polishing (3 mm thick) (fig. 7.25,4).

Levels 2a and 2 yielded two specimens of spatulae made with the utmost accuracy. The first had a sub-oval shape, curvilinear at both ends, one of which was broken. It had been accurately polished on the outer side, while the inner side still showed the bone structure (2 mm thick) (fig. 7.27,8). The second and larger specimen had a very similar shape and one of the curvilinear ends was broken. This specimen is thicker (4 mm thick) and polished on both sides (fig. 7.27,14). Three additional examples of spatula at an initial stage of manufacture have been recovered from Level 2b. These are fragments of ribs, ascribable to bovids and caprids, with both sides accurately smoothed without further polishing and without a specific shape (they ranged from 88 to 95 mm in length and from 10 to 21 mm in width, with a mean thickness of 3 mm). A single smoothing implement was recovered from Level 2a. It was made on a burnt bone and was sharply pointed (25 x 15 x 4 mm).

Ostrich eggshell was well represented in Sector B with a total of eighty-three pieces, fifty-three of which were unworked (fig. 7.27,2,5,7,13,15) or at an initial working stage, having been perforated (fig. 7.27,4). There were thirty disc beads having a mean diameter of 8 mm *ca* (fig. 7.27,9,10). Although present in the Lower Phase, as a general rule, they were commoner towards the top of the sequence; eighteen beads being recovered from Level 2 alone (fig. 7.27,11), while Level 1a yielded thirty-eight unworked fragments. Level 1 only yielded five unworked ostrich eggshell pieces, together with two fragments of wooden implements.

Wood, Fibre and Leather

A few fragments falling within *Varia* were collected from Sector B. Level 2b yielded a wooden wand with a circular cross-section (diameter 10 mm *ca*), which was accurately polished throughout. It was cone-shaped at one end, while the other end was broken (fig. 7.27,17). A wooden smoothing implement with a flattened end, was recovered from Level 2 (78 mm *ca* long). Tow braided fibre ropes, measuring 100 and 45 mm respectively, were

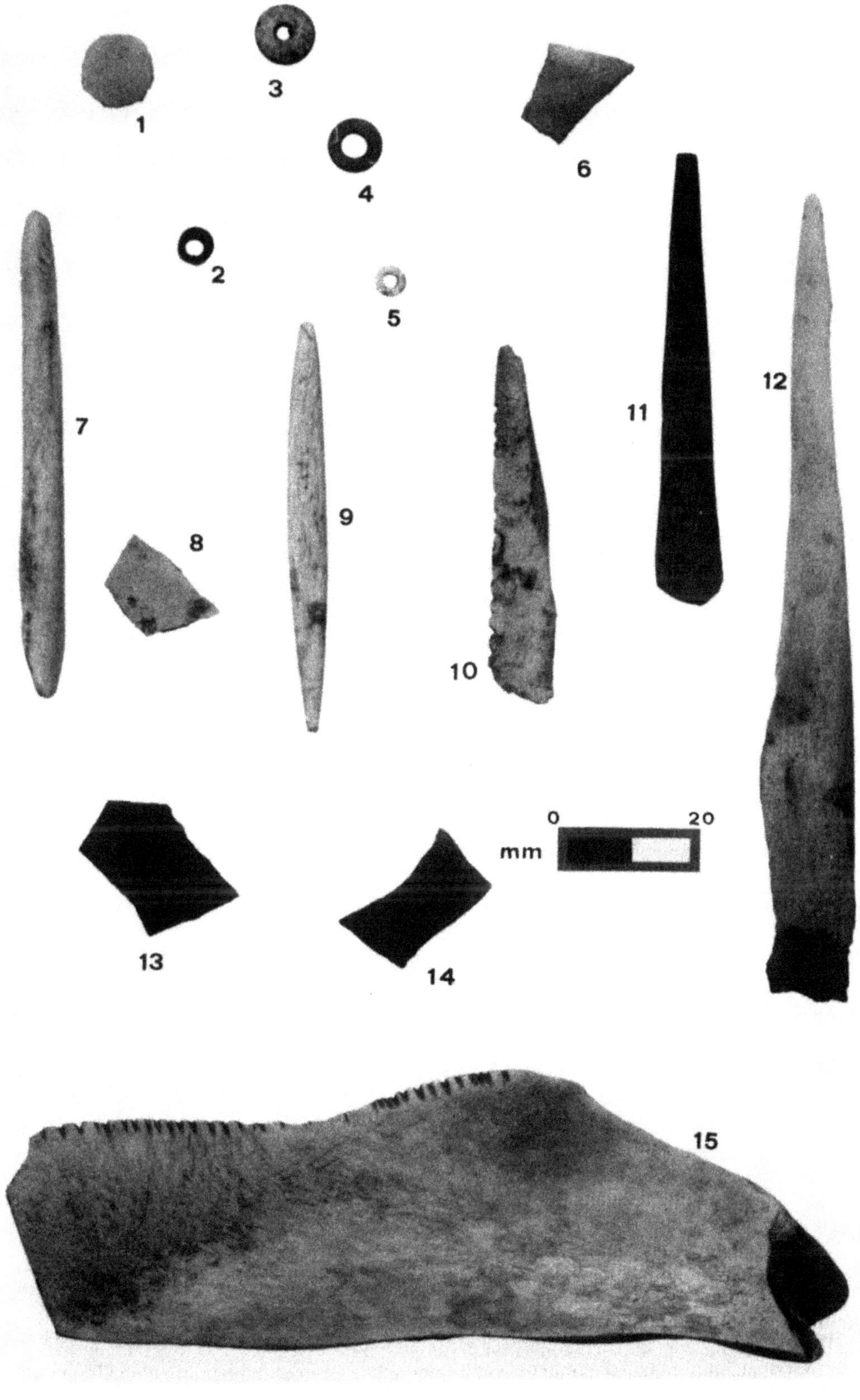

Fig. 7.26 - Uan Muhuggiag, Wadi Teshuinat. Levels 2a, 1a, 1. Sector A. Bone industry. 15, spatula; 7,9,11,12, awls; 10, denticulated; 1-6,13,14, ostrich eggshell (1:1).

recovered from the basal part (Level 2c). Finally, special significance should be ascribed to the recovery from Level 2 of five leather strips which had, perhaps, been used for fastening (ranging in length from 40 to 60 mm).

Ceramics

Compared with the indeterminate character of many of the lithic products, ceramics had greater technological variation. There was a relatively high number of decorated samples, with a good standard of manufacture. The chemical analysis of the components (Palmieri in Chapter 8, this Volume), showed both the major and the trace elements in the tempers. Our classification of potsherds referred to the following diagnostic elements: a paste; b surface treatment; c thickness; d colour; e vessel and rim decoration.

Consideration of points a to d distinguished three different groupings of potsherds (see below). Point e allowed a richer and more varied diversification within the framework of the three groups. Just as in the Ti-n-Torha facies, the only decorative technique to be employed was impression and styles were analysed according to the tool or implement used in the impression. As shown in Table 7.4, the whole ceramic complex of Uan Muhuggiag was decorated with two-toothed tools, spatulae, combs, cords and plant impressions.

The first group had impressed ornaments carried out with punches and quoins (Camps-Fabrer 1966; Barich 1974). Although recognizing the difference between the circular impression made by a punch or the geometric one made with a quoin, the use of single-pointed implements producing single impressions may be ruled out. The most common impressions result from a double pronged implement used like a pair of compasses. I. Caneva (Chapter 9, this Volume) observes that if the two-pronged implement is allowed to rotate through 180°, the result is an extremely uniform pattern with accurately spaced out dots that look like a combing. As a matter of fact, it would be extremely difficult, if not utterly impossible, to achieve so precise a pattern so speedily with comb. This observation led to a reappraisal of all the sherds large enough to be classified with certainty. A variant technique was recognized, the "returned type", in which the decoration was obtained by pivoting the tool in the dots of the preceding row. This may explain the deeper impressions, at times found very close one to the other in a "honeycomb" effect. By accepting this explanation, the ceramic decoration acquires a uniform character and homogeneous pattern that suggests a single origin.

Very little may be said about the vessel shapes. The presence of carenated bowls was definite in upper levels (la - 1) while small globular bowls were found in the underlying layers. The short necks, and the upper part of vessels, show a predominance of convex rims. The size of vessels is unknown although the high frequency of tiny sherds makes small-sized vessels likely.

The diversity between the two Phases found in the stratigraphy was repeated in the analysis of the ceramics. The divergence was particularly evident in Sector A, while in Sector B it would seem that the later occupation only affected superficially this part of the settlement.

The ceramics of the Upper Phase differed in colour, which was always

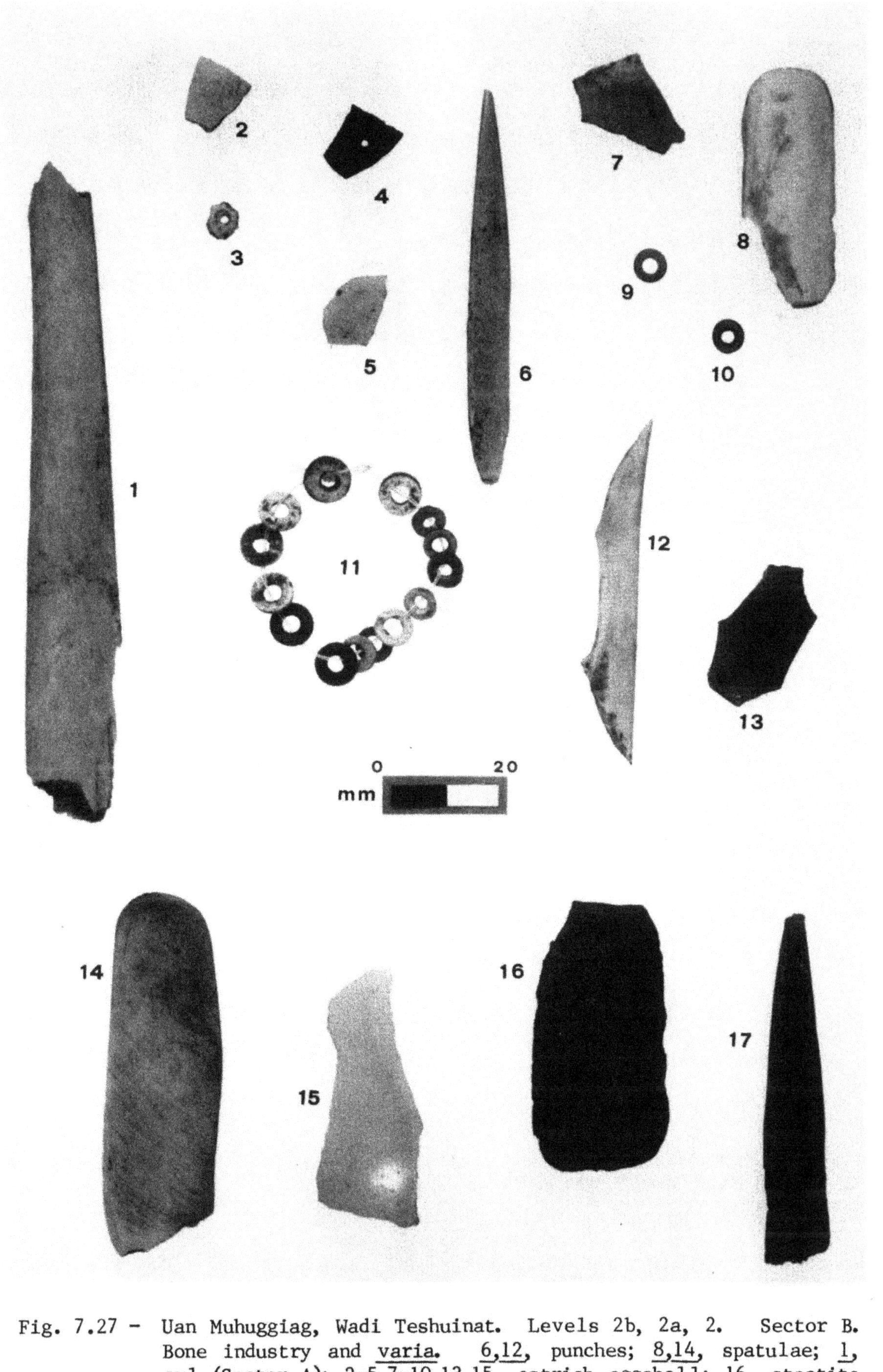

Fig. 7.27 - Uan Muhuggiag, Wadi Teshuinat. Levels 2b, 2a, 2. Sector B. Bone industry and varia. 6,12, punches; 8,14, spatulae; 1, awl (Sector A); 2-5,7-10,13,15, ostrich eggshell; 16, steatite tablet; 17, wooden wand (1:1).

red, and had generally smooth and undecorated surfaces. The space decoration was obtained for the most part by cord impressions. Conversely, the ceramics of the Lower Phase was usually decorated and brown coloured. The following refers to the diagnostic elements:

a) the paste is always compact, with sand the major tempering element;

b) the thickness of the sherds ranged from about 3 to about 10 mm; this may be related to special vessel shapes, but it may also correspond to specific parts of the pot walls;

c) the surface treatment of the inner and outer walls are seldom rough and uneven and indeed they were usually smoothed and, in a few cases, even burnished. The burnishing was particularly evident on the sherds with deep impressions or undecorated;

d) there was a certain variability of colour within the framework of the range of reds and browns, from mat red (2.5 YR 4/6) to brilliant red (2.5 YR 5/8); from dark reddish brown (5 YR 2.5/2) to very dark grey (10 YR 3/1); from reddish grey (5 YR 5/2) to mat grey (5 YR 5/1) (cf. Munsell 1973). As a general rule, both inner and outer surfaces had the same colour, but there are a few examples where the inner and outer walls of the vessels were coloured in a different way. Reddish sherds tend to have dark-coloured cores.

All the potsherds may be placed in three groups.

1 - Those with an average thickness of ca 10 mm, rather dense paste, rough surfaces both inside and outside which, at times, show signs of smoothing. For the most part, they have a reddish-grey to black (10 YR 2/1) colour.

2 - Those ranging in thickness from 4 to 7 mm, with a more dense paste and homogeneous and smoothed surfaces, at times burnished. Their colour is prevailingly from very dark grey to black but there are a few brilliant and mat red sherds.

3 - Those extremely thin (2 mm ca) with very compact paste and smoothed surfaces. Colour is always black in the Lower Phase, red in the Upper.

It is impossible to point to classes of vessels intended for differentiate functions since all the sherds are small (average size 30 x 40.5 mm). While elements as handles, grips and bottoms are missing, many sherds have rims. Simple rims undifferentiated from the walls, straight (fig.7.28,1-6), slightly swelling (figs.7.28,7; 29,5) or convex in relation to closed mouths (fig.7.29,4,8). There are also rims fully differentiated from the walls, either simply bulging and rounded (figs.7.28,5;29,3,6) and markedly everted if related to a very short neck (fig.7.29,1,7).

In conclusion, the prevailing type seems to be a small bowl, narrowed at the mouth with a straight or convex rim or with a short neck and an everted rim.

As already said, the difference between the Upper and Lower Phases showed particularly in the colour and the surface decoration. No special difference was noted in the pastes, thickness and surface treatments. This agrees with the chemical analysis, which pointed to a general homogeneity of

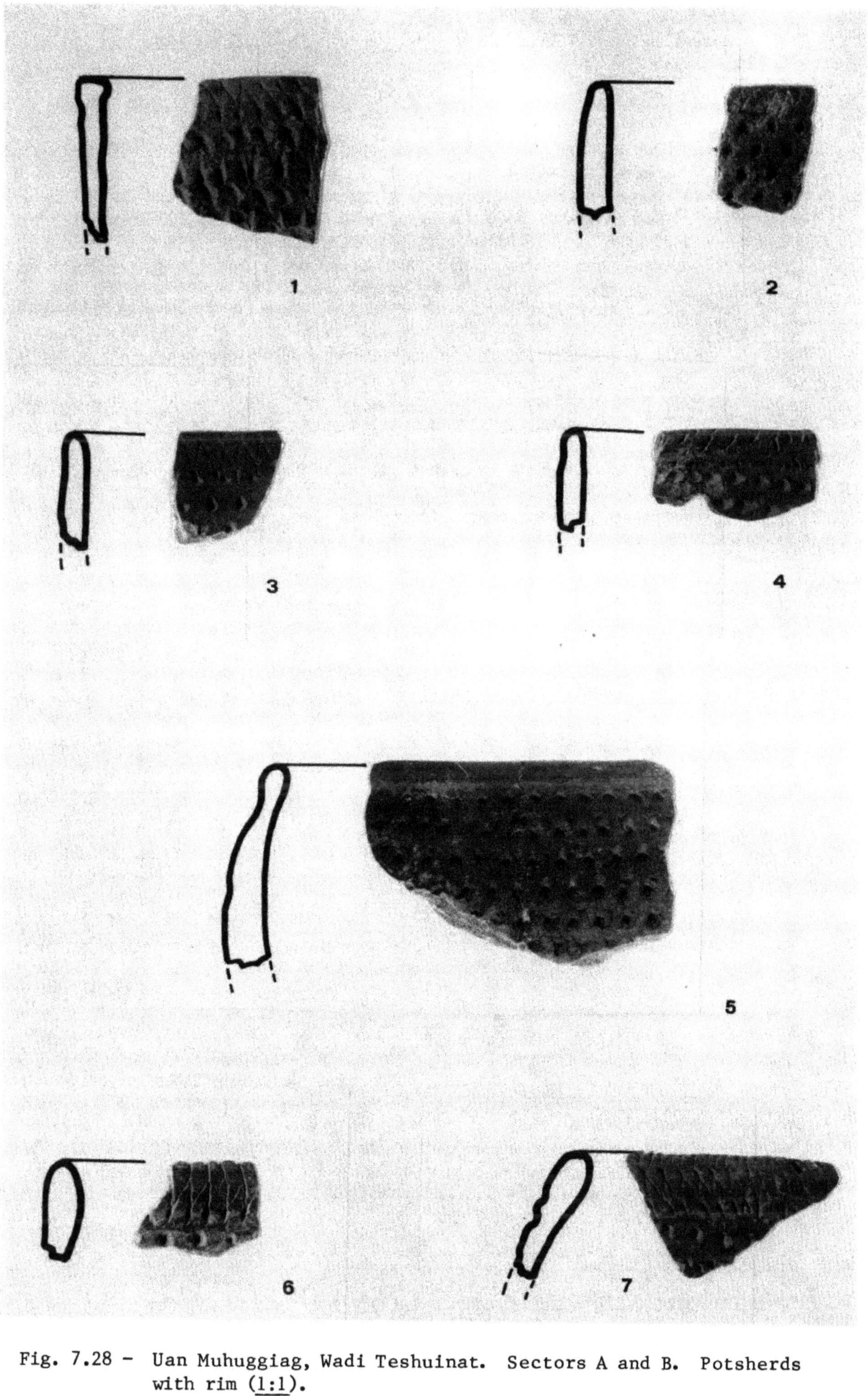

Fig. 7.28 - Uan Muhuggiag, Wadi Teshuinat. Sectors A and B. Potsherds with rim (1:1).

the elements used as tempering materials throughout the different phases.

Classification and study was made on the basis of decoration or its lack. The undecorated sherds caused a problem especially in Sector A where it was observed that undecorated fragments represented the majority of all the fictile products recovered but were parts of only three types of vessel. In Table 7.4, their actual quantity is shown but it is neither included in the grand totals nor calculated as a percentage. The same treatment was adopted for Sector B, Level 1a. The different aspects of each individual impression implement were considered in the following order:

1 - Nature, description of the implement used to decorate.

2 - How the implement was employed and what was the resulting decoration (7). Assignment of the sherds by sectors and levels.

3 - Special patterns (illustrated in the figures).

4 - Percentage frequencies of the use of each implement, with reference to each level and the context as a whole (Table 7.4).

Two-Toothed Implements with Circular Cross-Section: Poinçon

We are dealing here with what was previously defined as a Poinçon, an implement with one or two cone-shaped points that can be applied to produce dots (fig. 7.25,4). The way in which the implement was applied, deeply or superficially, and the angle of application determined the variety of patterns; deep regular dots or rare dots in the case of vertical impression and dots continued as strokes ("stab and drag"), triangles, close zigzags with a *en nid d'abeille* motif ("honeycomb") in cases of oblique impression (8). This applies to both simple impressions and to the use of the implement as pair of compasses. In this way the scheme allows acknowledgement of its homogeneity, showing that the wide variety and wealth of decorations, is related to a specific movement of the implements.

Sector A - In the base level (2c), the sherds were very small. The implement was used obliquely as a pair of compasses to produce regular rows of dots. One example had impressions obtained with returned technique (shallow "stab and drag" pattern) (fig.7.30,5). Another rim sherd shows how the impressions obtained with a double point produce a two-rowed pattern of small triangles (fig.7.30,1). All the subsequent levels are characterized by alternate simple vertical impressions and oblique, regularly spaced ones. The small, triangular-shaped grooves of Level 2b were made by vertical application of the implement (fig.7.33,6,10). Four specimens had dots on very close rows (fig. 7.33,3). A clear, double-pointed effect owing to the returned technique is visible in fig. 9. 3,3. The same type of application produced in different sherds some very characteristic "comma" grooves (fig.7.33,1,8). This effect may also be observed in a sherd coming from Level 2 (fig. 7.40,5).

---

7) This was the preliminary approach and was subsequently enriched by I. Caneva (Chapter 9, this Volume).

8) This decoration is similar to that obtained with rocker impressions of spatula on closely spaced rows (in French "guilloché").

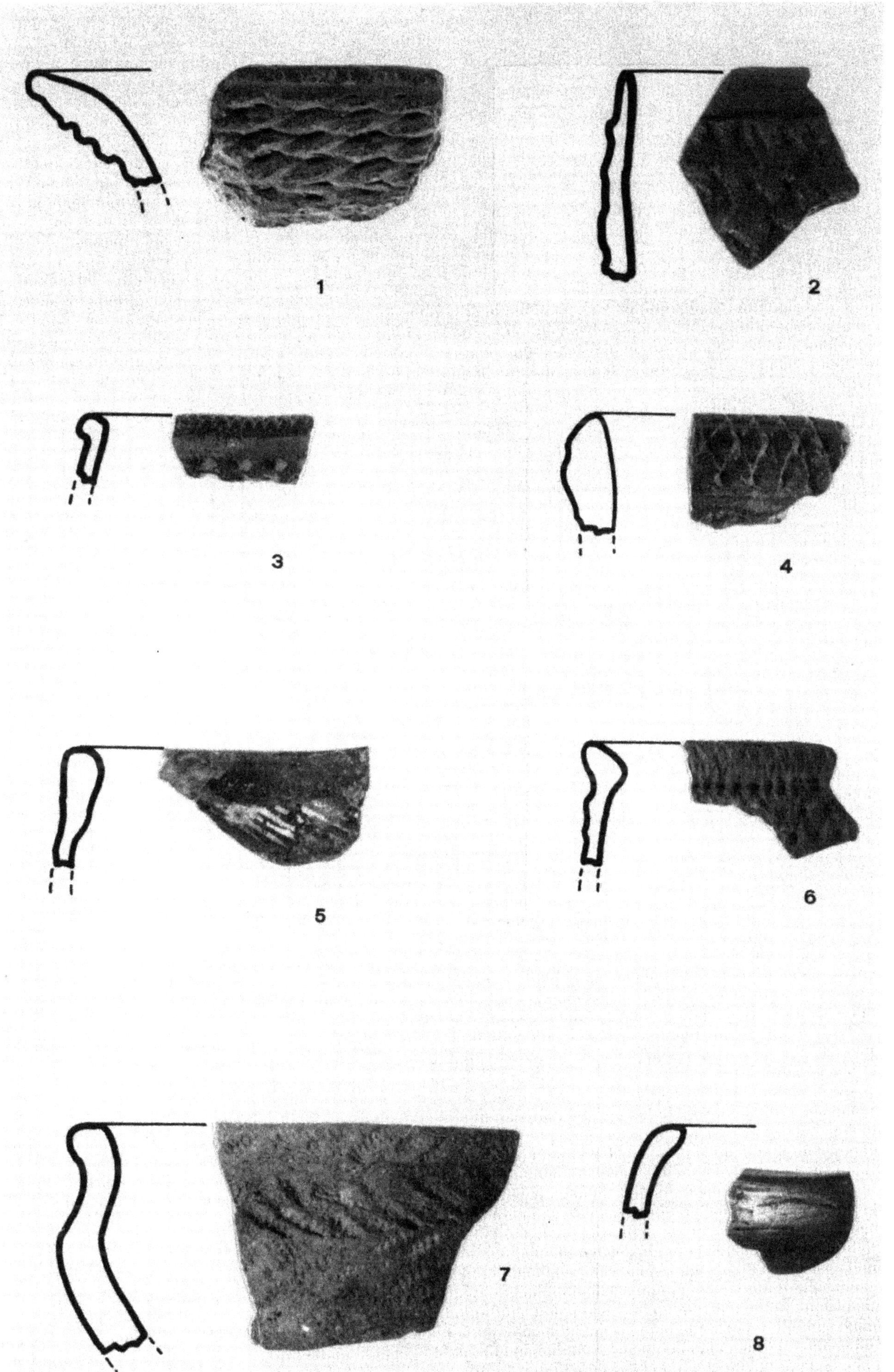

Fig. 7.29 - Uan Muhuggiag, Wadi Teshuinat. Sectors A and B. Potsherds with rim (1:1).

In the Upper Phase (Levels 1a-1) the use of this implement was only noted in a sherd coming from Level 1 (fig. 7.43,3).

The general frequency index was calculated for Levels 2 as 40.7%. The highest incidence was in Level 2c (66.6%) and although it was still high in Level 2b, it dropped sharply in Level 2a (25.0%) to rise once again in Level 2. As previously pointed out, it was negligible in Levels 1a-1.

Sector B - The 'paired compasses' technique using implements having a triangular cross-section was present in Level 2c and was producing triangles extending into commas in Level 2d. Fig. 7.32 shows double, superimposed triangles resulting from an inaccurate return (fig.7.32,3,6), which originates a pseudo-carving effect (figs.7.31,8; 32,5). But there were also sherds where the implement was placed in the dots of the previous row with utmost care, resulted in rows of dots, and even triangles, arranged in a regular pattern, that was very much like comb impressions (fig. 7.31,3).

The maximum concentration was in Level 2b where 245 sherds were decorated with this implement (thirty-eight were so small that the pattern could not be clearly observed). The different manners of utilizing the implement was quite clear; a simple, oblique and shallow impression resulted in small triangles evenly spaced or scattered over the surface (fig.7.34,2,5), while a deeper oblique impression resulted in stabs (fig.7.34,4). The deep 'pair of compasses' impression is found on the greatest number of sherds (180, corresponding to nearly 75% of the total). When the punch is applied vertically to the surface there is no paste shifting and the resulting effect is a series of deep, circular pits that are evenly arranged in carefully spaced rows (10 sherds had coupled rows) (figs.7.35,5; 36,6). The resulting effect is very similar to that produced by comb impressions (figs.7.35,1;36,8). A peculiar feature may be observed in one sherd where the use of a tool having a triangular cross-section resulted in impressions of small triangles in two linked rows (fig.7.34,8). When the implement is used over an impressed motif, it results in triangularly-shaped stamps with paste shifting to produce a "stab and drag" effect (seven sherds) (fig.7.34,1). The rows of impressions may be very closely spaced, and this results in the 'honeycomb' type resembling a woven pattern (figs.7.35,2;36,5,7). In a few instances the impression is not accurate, as may be noted in three sherds in particular (figs.7.34,7;9.6,8). An arrangement of concentric rows of dots on vessel bases (fig.7.35,6), was a characteristic element.

A similar variety of decorative motifs were observed in Level 2a: shallow oblique impressions that result in small triangles (fig.7.38,1); eighteen sherds in which the returned technique produced double, divided impressions (fig.7.39,4); "star" shaped widening of the margin (fig.7.39,2); closely spaced rows giving the 'honeycomb' effect (fig.7.38,7); closely spaced and irregularly shaped impressions (fig.7.38,8). Some sherds show accurate use of the same technique, with no paste shifting (figs.7.38,6;39,3). Three sherds feature the dotting arranged over linked rows (fig.7.38,4).

The sherds coming from Level 1a presented the same characteristics. One had shallow oblique impressions resulting in horizontal dashes over a band of dots (fig.7.42,4). Two sherds had shallow oblique impressions resulting in small and slightly impressed triangles. On two sherds impressions were in two linked rows (fig.7.42,5,7). The sherd shown in fig.7.42,7, had the rows spaced out by a protrusion, due to the shift of the paste resulting from the 'compasses' technique. A deep impression is visible in sixteen fragments, where the returned technique is carefully employed, resulting in an evenly spaced arrangement of dots in well-aligned rows.

In Level 1, the recourse to this type of implement was only found on three sherds, one of which with a rim where it was used with a 'carving' effect (fig.7.44,1,5)(9).

The general frequency index, calculated in respect of Levels 2d-1a, was 50.4%; the maximum frequency being in Level 2b (61.5%) and tending to decrease in the upper levels, settling on 36.4% in Level 1a.

Two Toothed Implements with Geometric Cross-Section: Quoin

This implement has two square or sub-rectangular ends, giving impressions that are not clearly defined and which are quite distinct from the actual dots of the 'poinçon'. The remarks made with reference to the poinçon apply to this implement and different motifs result from different applications of it. While a vertically applied impression gives rise to even rows of small squares or rectangles, when the implement is applied obliquely the result is a series of deep triangles, much deeper than those resulting from the poinçon, giving a clear "carving" effect. The latter is particularly effective when the implement is used with the 'compasses' technique, turning, without much accuracy on the pivot-points of the preceding row. A magnifying lens shows a clear shifting of the paste, with the impression contours lacking any neatness.

Sector A - A number of sherds provide evidence of the use of the returned impression in Level 2c (fig.7.30,3,4). One had widely spaced impressions in rows divided by a slight protrusion due to the paste shift (fig.7.30,7). Level 2b was characterized by deep stamps of sub-rectangular shape that appeared at the surface. These were made using the 'divider' technique (fig.7.33,5,7,9). One sherd showed the implement used obliquely; this results in triangular motifs arranged over rows starting from a central spot (fig.7.33,11). A sherd from Level 2a had elongated impressions arranged over coupled rows separated by spaces (fig.7.37,4) and another a continuous, honeycomb, pattern obtained by going over the impressions in closely spaced rows (fig.7.37,8). A sherd coming from Level 2 features a very rough use of the returned 'compasses' technique (fig.7.40,9) and another had impressions over a double row, delineated by deep dotting. The different shapes of the impressions can be ascribed to different orientations of the implement (fig.7.40,4).

The use of this implement is not documented in the Upper Phase.

While the frequency in Level 2c was 6.6%, it rose significantly in Levels 2b and 2a (27.3% and 20.9%, respectively), to decrease once more to 6% in Level 2. The general index of this implement's frequency is 14.9%. It should be noted that the circular and rectangular two-toothed implements jointly account for 55.6% of the decoration.

---

9) In this case it was obtained with a single-pointed implement.

Table 7.4 - Uan Muhuggiag, Wadi Teshuinat. Absolute and percentage frequency of ceramics (according to implements used for decoration).

	SECTOR A													
	1	1a	Total	%	2	%	2a	%	2b	%	2c	%	Total	%
Two-toothed implements	1	-	1	0.5	29	43.3	22	45.9	32	72.8	22	73.2	105	56.6
- w. circular cross-section	1	-	1	0.5	25	37.4	12	25.0	20	45.5	20	66.6	77	40.7
- w. geometric cross-section	-	-	-	-	4	5.9	10	20.9	12	27.3	2	6.6	28	14.9
Spatula	2	1	3	1.6	6	8.9	15	31.3	4	9.1	3	10.0	28	14.9
Plain comb	-	-	-	-	-	-	-	-	-	-	-	-	-	-
Fileté comb	7	-	7	3.8	2	2.9	1	2.0	1	2.2	-	-	4	2.1
Cord impressions	2	-	2	1.1	-	-	2	4.2	1	2.2	-	-	3	1.6
Composite decorations	1	-	1	0.5	6	8.9	4	8.3	3	6.8	4	13.4	17	8.9
Undecorated	-	171	171	92.5	24	35.9	4	8.3	3	6.8	1	3.3	32	16.9
Total	13	172	185	100.0	67	99.9	48	100.0	44	100.0	30	99.9	189	100.0

S E C T O R   B

	1	%	1a	%	2	%	2a	%	2b	%	2c	%	2d	%	Total	%
	3	5.1	41	62.2	23	35.4	110	58.9	308	77.5	22	52.5	5	33.4	509	66.9
	3	5.1	24	36.4	13	20.0	88	47.1	245	61.6	15	35.8	4	26.7	389	50.4
	-	-	17	25.8	10	15.4	22	11.8	63	15.9	7	16.7	1	6.7	120	15.5
	4	7.0	14	21.2	23	35.4	40	21.4	22	5.5	2	4.7	3	20.0	104	13.5
	-	-	-	-	-	-	-	-	-	-	1	2.3	1	6.7	2	0.2
	3	5.1	5	7.6	6	9.3	5	2.6	5	1.2	5	12.0	4	26.7	30	3.9
	4	6.9	-	-	5	7.7	6	3.2	20	5.0	2	4.7	-	-	33	4.3
	-	-	6	9.0	2	3.1	15	8.0	40	10.1	7	16.6	2	13.2	72	9.3
	44	75.9	(82)		6	9.2	11	5.9	3	0.7	3	7.2	-	-	23	3.0
Total	58	100.0	65	100.0	65*	100.1	187*	100.0	398*	100.0	42	100.0	15	100.0	773	100.0

*Unclassifiable sherds 2=15; 2a=35; 2b=42

Sector B - In Level 2d, the only evidence of this implement is on a sherd which however shows quite clearly that the 'compasses' technique had been used (fig.7.31,5). In Level 2c, one may note the characteristic honeycomb (fig.7.31,12) and "star"-shaped (fig.7.31,10) patterns, quite similar to those obtained with the use of a poinçon. A sherd made in two pieces - that may be related to a small globular bowl with an everted rim (figs. 7.28,5; 32,1), clearly shows an inaccurate use of the divider technique, since the implement seldom returned roughly to the pivot point in the preceding row; the impressions were made by using a corner of the tool. Another sherd had two rows or impressions separated by a groove (fig. 7.32,2). More sherds (sixty-three) were recovered from Level 2b. They confirm that different impressions depended upon the angle at which the tool was applied, vertically (rectangular, square and sub-square) or obliquely (triangular). The decoration of four sherds used the 'compasses' technique to produce two rows (fig. 7.34,6). A sherd where the implement had been used obliquely had elongated impressions (fig. 7.34,3) and the returned technique was so accurately performed on another sherd as to achieve an impression closely resembling that of a comb (fig. 7.35,4). The same technique was used on another sherd, from the base of a vessel, to imitate a basket pattern (fig. 9.4,5). In Level 2a, the use of this implement is reasonably frequent, commonly with the 'compasses' technique. Accurate impressions in rows are clearly separated sometimes even by a slight ridge (fig. 7.39,8). In Level 2, evidence of this implement comes from ten sherds (fig. 7.41,4,6), one with a honeycomb pattern (fig. 7.41,1). Level 1a yielded seventeen specimens characterized for the most part by a closely spaced impressions (fig. 7.42,1,9).

No evidence of the implement utilization was found in Level 1.

The use of the implement gradually increased from Level 2d to Level 1a, where it reaches the peak frequency of 25.8% and where its use appears to be similar to the spatula (21.2%) and the poinçon (36.4%). The general index in Sector B is 15.5% and together with the 'poinçon' it accounts for 65.9% of the fictile decoration.

Spatula

This flat implement with a straight or curved edge is mainly used with a rocking movement. Depending on the curvature of its edge, it may produce zigzag curves, which, depending on the way it is applied, will be more or less closely spaced. It should be borne in mind that if the implement overlaps the row of grooves preceding, this results in a continuous and closely spaced woven pattern (fish-scale, or *guilloché*). The same type of decoration can result in a carving effect which is used rather frequently on the very short necks with everted rims.

When a spatula with a rocker movement is found in the decorating of rims, the implement is used with other implements (usually, two-toothed ones). The specimens where the spatula was utilized with other implements were included in the "compositely decorated sherd" class. This title was suggested because the implement the spatula is coupled with is not always recognizable.

Sector A - In Level 2c the spatula seems used on only three sherds (fig. 7.30,6). The four samples from Level 2b were indeterminate but twelve sherds from Level 2a showed more details. While one small sherd shows

Fig. 7.30 - Uan Muhuggiag, Wadi Teshuinat. Level 2c, Sector A. Impressed pottery. 1, 2, 4, 5, punch; 3, 7, quoin; 6, 8, spatula; 9 composite implements (2:1).

simple, vertical impressions (fig. 7.37,7), a rather well-spaced rocker movement is detectable on four (fig. 7.37,5). Seven sherds show the same deep and closely-spaced rocker impression as the guilloché type (fig. 7.37,2,9), that is quite close to carving (fig. 9.3,1). While spatulae are exclusively used with the rocker movement, there are many differences in the impressions caused both by implement edge curvature and the way of holding the implement. One may observe specimens characterized by shallower (fig. 7.40,8) and deeper impressions in guilloché patterns and carving (fig. 7.40,1,2,7).

In the Upper Phase there was little evidence of the implement. Level la yielded a single specimen (figs. 7.29,5; 43,5) while Level 1 yielded two sherds, one, a piece of rim with a rhomb-patterned decoration (fig. 7.43,1,6).

The general index for spatulae in the Lower Phase in Sector A, was 14.9%. It was 10.0% in Level 2c and 9.1% in Level 2b. A noteworthy increase is reported in Level 2a (31.3%) while it was back to 8.9% in the subsequent level.

Sector B - The sherds from the basal levels (2d-2c) were too small to allow any significant observation to be made (fig. 7.31,2,6,9), but a few comments may be made on the sherds coming from Level 2b, even though they are still quite small. A sherd from a small vessel with an undifferentiated rim, was decorated by a rocking spatula, the grooves being spaced out on the vessel body and quite close to the rim (fig. 7.36,1). A larger sherd, possibly part of a globular vessel with an undifferentiated rim, was also decorated on the body and in the upper part of the rim by a rocking spatula.

The greatest number, forty sherds, came from Level 2a. Ten sherds were characterized by a thin woven motif obtained by the shallow impression of a short-edged spatula (fig. 7.39,1,5). Other sherds featured widely spaced zigzags (figs. 7.39,7; 9.3,2); in one of these the decoration began well below the rim, leaving an undecorated area. A sherd from a slightly everted rim was characterized by a deep and closely spaced rocker impressions (figs. 7.29,1; 38,2). Only rocker points may be detected in another sherd (fig. 7.39,6). There were twenty-three examples of spatula decoration in Level 2. One sherd was part of a small vessel with a significantly everted rim, the under side of the rim and the walls were characterized by deep, rocker impressions (fig. 7.41,2). Four everted rim sherds, two from a single vessel with a short neck, show a beautiful deep rocker impression (fig. 7.41,5) which gives a 'carving' effect (fig. 7.41,8). Four other sherds have shallower impressions giving a woven-mat motif (fig. 7.41,3). Fourteen specimens of spatula decoration from Level la are fully similar. Eleven of them are from the upper section of thin-walled vessels (group 3) where the spatula was allowed to rock with closely spaced movements (fig. 7.42,2); the point on which the implement was rocked was clearly visible in one of them (fig. 7.42,8).

The spatula was only represented by four specimens in Level 1 (fig. 7.44,3,7,10,11).

The general index, calculated with respect to the Levels 2d-la, is 13.5% and quite similar to that from Sector A. Even though the use of the spatula is very low in the base levels, it is commoner in Level 2a reaching its highest level of utilization (35.4%) in Level 2. In Level la it drops

Fig. 7.31 - Uan Muhuggiag, Wadi Teshuinat. Level 2d (1-6), Level 2c (7-12). Sector B. Impressed pottery. 8,11, punch; 4,5,10,12, quoin; 2,6,9, spatula; 1,7, plain comb; 3, composite implements (2:1).

Fig. 7.32 - Uan Muhuggiag, Wadi Teshuinat. Level 2c, Sector B. Impressed pottery. 3,5,6, punch; 1,2,4, quoin (2:1).

Fig. 7.33 - Uan Muhuggiag, Wadi Teshuinat. Level 2b, Sector A. Impressed pottery. 1,3,6,8,10, punch; 5,7,9,11, quoin; 2, spatula; 4, fileté comb (2:1).

Fig. 7.34 - Uan Muhuggiag, Wadi Teshuinat. Level 2b, Sector B. Impressed pottery. 1,2,4,5,7,8, punch; 3,6, quoin (2:1).

Fig. 7.35 - Uan Muhuggiag, Wadi Teshuinat. Level 2b, Sector B. Impressed pottery. 1,2,5,6 punch; 4, quoin; 3, fileté comb (2:1).

back to roughly the same value as that of Level 2a.

Plain Comb

There are still doubts as to the actual presence of the combs with multiple teeth that produces dots in straight lines at the site. In fact, the motif resulting from the use of the comb may only be differentiated from the pattern produced by two-toothed implements, when one can show the recurrence of straight sequences with the same number of serrations and jointing points. This is extremely difficult with sherds as small as most of those that were recovered. Thus, even though the presence of this implement is not definitely ruled out, the use of a plain comb is only mentioned in connection with two sherds coming from Sector B which, owing to their general features, have an intrusive character. The rest of the sherds decorated with dotted straight lines have been ascribed to two-toothed implements (10).

Sector B - The existence of this implement must be accepted in Levels 2d and 2c, that is at approximately the base of the deposit.

Level 2d yielded a small rimmed sherd characterized by an extremely compact, thin paste that differed from the others in this level. The pattern, which unfortunately does not allow an easy reading, shows comb decoration on the wall, associated with a rocker spatula decoration on the rim (fig. 7.31,1). In view of its paste and the arrangement of the decoration, it may be likened to the Dotted Wavy Line style in the Ti-n-Torha horizon. A sherd from Level 2c, another small specimen, has similar paste and thickness characteristics; it lacks a rim and only the comb impressions in wide spaced rows may be recognized (fig. 7.31,7).

"Fileté" Comb

It is an implement the serrations of which were obtained by binding with a cord. Only few sherds provide evidence of its presence but document both the rigid and the flexible varieties. It was always used in rather irregular rows, producing grooves with inside notching, and resulting in a woven-mat motif in the case of flexible combs. Whenever the rigid variety was used, the result was a series of grooves with inside notching arranged in clearly differentiated rows with shifting of paste.

Sector A - The implement has a very low incidence in the Lower Phase with only four specimens in Levels 2c-2 (general index 2.1%). It is absent in Level 2c while in Levels 2b and 2a it is present with one specimen of the rigid- (fig. 7.33,4) and the flexible- type (fig. 7.37,6) in each. Two extremely small sherds decorated with the rigid variety were present in Level 2.

In the Upper Phase, while it is not represented in Level 1a, it used over 50% (7 sherds) of the decorated products in Level 1. This frequency

10) This reappraisal and the absence of combs in this ceramic industry should nonetheless be further checked and looked for in other areas, since it implies an important change in the decorative style attributed to these groups.

Fig. 7.36 - Uan Muhuggiag, Wadi Teshuinat. Level 2b, Sector B. Impressed pottery. 5-8, punch; 2, quoin; 1, spatula; 3,4, plant impressions (2:1).

must be related to the total change in the nature of the ceramic production in this upper horizon. There is a different decorative repertoire that, as we have already seen - is verifiable in different ways. Out of the seven, special reference should be made to one from a carenated bowl of good-quality paste, which was mat red on both the inner and outer surfaces. The decoration was arranged in oblique lines and covered the whole vessel body as well as the everted rim (fig. 7.29,7); another fragment was decorated in the same style with mat red surfaces on very dense paste about 10 mm thick (fig. 7.43,2). Finally, there was a small sherd with thin impressions in very close rows (fig. 7.43,4).

Sector B - The frequency is slightly higher than in Sector A; the general index, calculated with respect to Levels 2d-1a is 3.9%. The use of the implement is documented from Level 2d, which yielded a sherd that, in view of its paste, belongs to the Dotted Wavy Line context. Most of the specimens are extremely small wall sherds that provide no guide to the shape of the original vessel. There is evidence of the use of both the rigid (Levels 2b, 2, 1a: figs. 7.35,3; 41,7), and the flexible variety (Level 2a: fig. 7.38,3,5).

In Level 1, this implement is documented by three examples, of which two were from the rigid type (fig. 7.44,9) and from the flexible type. The frequency of the filetê comb is high, especially in comparison with different implements. There is no evidence of the significant difference between the Upper and the Lower Phase which was observed in Sector A.

Cord impressions

This is a flexible category that includes different types of impressions that are obtained through the use of a variety of vegetable artifacts. Firstly, there are the cord-impressed motifs that are the most characteristic element of the ceramics of the Upper Phase. This type of "woven-mat" decoration is very wide-spread among the Saharan ceramics of a later period. Secondly, there is the use of stems or small, fissured reeds, resulting in a series of crescents. Fibres, even if accidentally impressed on the surfaces, give rise to slight dots or streaks. We have previously referred to two-toothed implement impressions that reproduces basketry but true or matt basket impression (roulage sur natte) is unknown in the Uan Muhuggiag complex.

Sector A - There are only three examples in the Lower Phase, one in Level 2b is doubtful and two are in Level 2a (fig. 7.37,3).

No specimen was discovered in Level 1a, but in Level 1 the technique is found on two sherds, one of which, being large, allowed the impression to be accurately noted (fig. 7.43,7).

The general index is 1.6% in the Lower Phase while the two specimens in Level 1 represent 15.4% of the decorated series.

Sector B - In this area there was a greater number of cord- and plant-impressed decorations. Level 2c yielded two doubtful specimens but in Level 2b, two specimens show genuine cord impressions while a series of small crescents, obtained by stem impression, were recognized on nine sherds (fig. 7.36,3,4). In levels 2a-2, while total eleven sherds showed plant impressions, they are too small to allow an identification of the type of

Fig. 7.37 - Uan Muhuggiag, Wadi Teshuinat. Level 2a, Sector A. Impressed pottery. 4,8, quoin; 2,5,7,9, spatula; 6, fileté comb; 3 cord and 1 plant impressions (2:1).

vegetable element used.

In Level 1, two sherds, one from the vessel wall (fig. 7.44,2), showed distinct cord impressions; a third specimen was more doubtful (fig. 7.44,8). A further sherd clearly shows impressions obtained with a stem (fig. 7.44,6).

The general index is 4.3% in the Lower, and 6.9% in the Upper Phase.

Composite decorations

This category is represented for the most part by rim sherds with only a small parts of the vessel walls. As was previously pointed out, the class is necessary because while the decoration of the rim with a spatula is always quite recognizable, this is seldom true of the type of implement used in the underlying decoration. Without this group the percentage frequency of spatulae would be exaggerated. The spatula is certainly associated with two-toothed implements and filetē combs.

Sector A - On seven sherds from Levels 2c and 2b the spatula was combined with two-toothed implements (fig. 7.30,1,5). This combination is not easily analysed. On a sherd, the decoration most likely involved a spatula together with a flexible, filetē comb (fig. 7.30,9). Four sherds from a type of small globular short-necked pot were recovered from Level 2a; their decoration involved the use of a spatula in combination with a quoin (fig. 7.28,3). Level 2 yielded six sherds of short necked vessels with everted rims that show impressions of spatulae on the rim and divider technique quoin impressions on the neck (fig. 7.40,6).

A single example of composite decoration was recovered from Level 1.

The general index of this type of decoration is 8.9% in the Lower Phase with a peak frequency in the basal Level 2c (13.7%).

Sector B - This decorative style has much the same frequency as in Sector A (general index 9.3%), and is commonest in Level 2b (forty specimens: 10.1%). Twenty-seven rim sherds show combinations made with spatulae and two-toothed implements. Eleven decorative wall sherds were similarly decorated (fig. 7.28,6,7).

Composite decoration had the same relatively high frequency in Level 2a, (fifteen sherds, 8.0%) (figs. 7.28,4; 29,3; 38,2). Level 1a yielded six specimens; three decorated at the rims and the rest on the walls.

Undecorated

Sector A - Undecorated sherds were rare (7 sherds) in the lower part of the stratigraphy: one in Level 2c; three in Level 2b and four in Level 2a. A sudden increase in Level 2 yielded twenty-four sherds (35.9%).

In one later level (1a) they were very common (171) but all belonged to three vessels. Only a single, spatula decorated sherd was reported. Of the three vessels, the first (forty-six sherds) was, quite likely, a bowl with a convex rim, undifferentiated from the vessel body, with slightly bulging walls; the base was probably rounded. The paste belonged to

Fig. 7.38 - Uan Muhuggiag, Wadi Teshuinat. Level 2a, Sector B. Impressed pottery. 1,4,6-8, punch; 2, spatula; 3,5, fileté comb (2:1).

group 1 and was approximately 10 mm thick. Its surface was polished and brilliant, with black areas on red. The shape of the second vessel could not be reconstructed (five sherds) but the rim was decorated with incised straight lines encrusted with white. The paste belongs to group 2 and its thickness ranges around 5 mm, with a mat black surface. The third group of 120 sherds belonged to a spheroidal bowl. The sherds were very small and the rim is of everted type, delimited by a cordon . The paste belonged to group 2 and its thickness ranges from 5 mm in the bottom of the bowl to 3 mm in the walls. The red surface was burnished and showed many spatula strokes, but was black in both the core and the inner surface.

Undecorated sherds were not found in Level 1.

Sector B - There are no undecorated sherds in Level 2d, and few in the levels directly above (2c-2) but the extremely sharp increase in Level 1a calls for a similar reasoning as that made with reference to Sector A. All the sherds in Level 1a belong to two pots. One of these belonging to group 1 (average thickness 7 mm) and has a rough, reddish grey surface. The other one was thinner (group 2, 3 mm thick) with a homogeneous black surface.

3. THE SITE: TECHNOLOGICAL AND ECONOMIC CHARACTERIZATION

There are a number of reasons that determined the importance of Uan Muhuggiag. Firstly, there is its excellent stratigraphy with C14 dates ranging from the 7th to the 4th millennium B.P., secondly its wealth of different kinds of archaeological remains: stone; ceramic and organic materials; faunal samples; palynological and other botanical remains.

All these elements contributed to distinguish two decidedly differentiated archaeological horizons. From the stratigraphy we ascertained that occupation started with small and closely spaced lenses, very rich in hearths; two or three hearths may be observed on the floors of individual lenses (fig. 7.6). The shift from one to the other lens occurs almost imperceptibly and is, at time, marked only by compact surfaces that flakes off as soon located.

A very evident break was observed in Sector A, where the early phase was abruptly closed by a collapse of the back wall due to extreme temperature changes. The entire character of the archaeological and sedimentary remains from that time (subsequent to 3770 B.P.), appears quite different.

In Sector B this break did not appear so clearly from the sedimentation, but a soil horizon mixed with vegetable remains and cattle dung that was clearly observed in Sector A (Levels 1a-1, fig. 7.5) was present, although much thinner, only in Level 1 of Sector B. Therefore, the archaeological and palaeofaunal data of Level 1a was assigned to the Lower Phase. This means that the later occupation, starting from 3770 B.P., concentrated on the western section, only marginally affecting the eastern section, where Sector B was opened.

The break observed in the stratigraphy was confirmed by definite changes in the artifactual typological series.

Fig. 7.39 - Uan Muhuggiag, Wadi Teshuinat. Level 2a, Sector B. Impressed pottery. 2-4, punch; 8, quoin; 1, 5-7, spatula (2:1).

The Lower Phase

The flaked stone assemblage is numerically poor and the difference between Sector B and Sector A (1020 pieces and 203, respectively) is evidently related to the narrowing of the excavated area in the latter. Comparing the totals with those of the Ti-n-Torha East collection, Uan Muhuggiag may be considered as a non-residential site. Coarse sandstone was the major raw material, and was unquestionably responsible for the poor character of the industry.

In Sector B, retouched tools were only 18.52% of the flaked stone total and the tools/debitage ratio was approximately 1:4. In Sector A, the percentage frequency of tools is rather higher (39.9%) and the tools/debitage ratio approximately 1:2. As far as flakes s.s. are concerned, the type from single platform cores predominates while primary flakes rank second; even early-stage core-preparation flakes are frequent. Rejuvenation and core preparation techniques are seldom represented. Ninety-six core-trimmings are reported in Sector B as against seventy-six cores (less than two core-trimming elements to one core). The core-trimming elements from Sector A were twenty to thirteen cores, quite a similar ratio. There is a very high incidence of chips, this class forming 27.9% of the whole cores/debitage group in Sector A and 37.6% in Sector B. On the whole cores were infrequent (10.6% and 9.1% in Sector A and Sector B, respectively). As for core-types, the most frequent was the single platform core, but even patterned and unpatterned multiple cores were quite frequent.

The retouched tools show differences from those at Ti-n-Torha East and Two Caves but a correlation with the Torha North site (Levels II-I). There was a similar contraction of classes and a disappearance of both the geometric microliths and the microburin technique. As stated beforehand, blades were kept separate from the other debitage - and included among tools - to allow a comparison with the Ti-n-Torha facies. The blade index was very low (34.6% in Sector A and 46.0% in Sector B) and lacked the most significant Ti-n-Torha types, such as backed blades/bladelets and truncations. There was instead an increase in the pieces having less specific typological characterization, such as perforators, scaled and continuously retouched pieces and side-scrapers. A high frequency of notches/denticulates as well as arrow-heads was observed.

Due emphasis should be laid on the completely unusual presence of a few backed bladelets in the deepest levels of the sequence. The single straight backed and pointed bladelet with rounded base, which was recovered from Level 2c of Sector B (fig. 7.9,3), was fully comparable with similar examples coming from Two Caves. Further specimens, even though less finished, were the partially backed bladelet (fig. 7.9,6) and the proximal fragment of a backed bladelet (fig. 7.7,2), also from Sector B. Finally, it is worthwhile mentioning the fragment of backed bladelet that was the only item of this type recovered from Sector A (Level 2b: fig. 7.11,3).

A type of triangular, naturally sharp pointed at the distal end, 'blade-like' flake appears instead to be more common. Owing to its characteristic shape, it recalls some Levallois points and, at times, shows a retouched base. It appeared in Sector B from Level 2b onwards while in Sector A points of this kind, or quite similar, are the only ones to have come to light since no real arrow-heads have been recovered. Arrowheads of Hugot's types only appeared in Sector B in Levels 2d, 2c and 2b and represented 7.9%

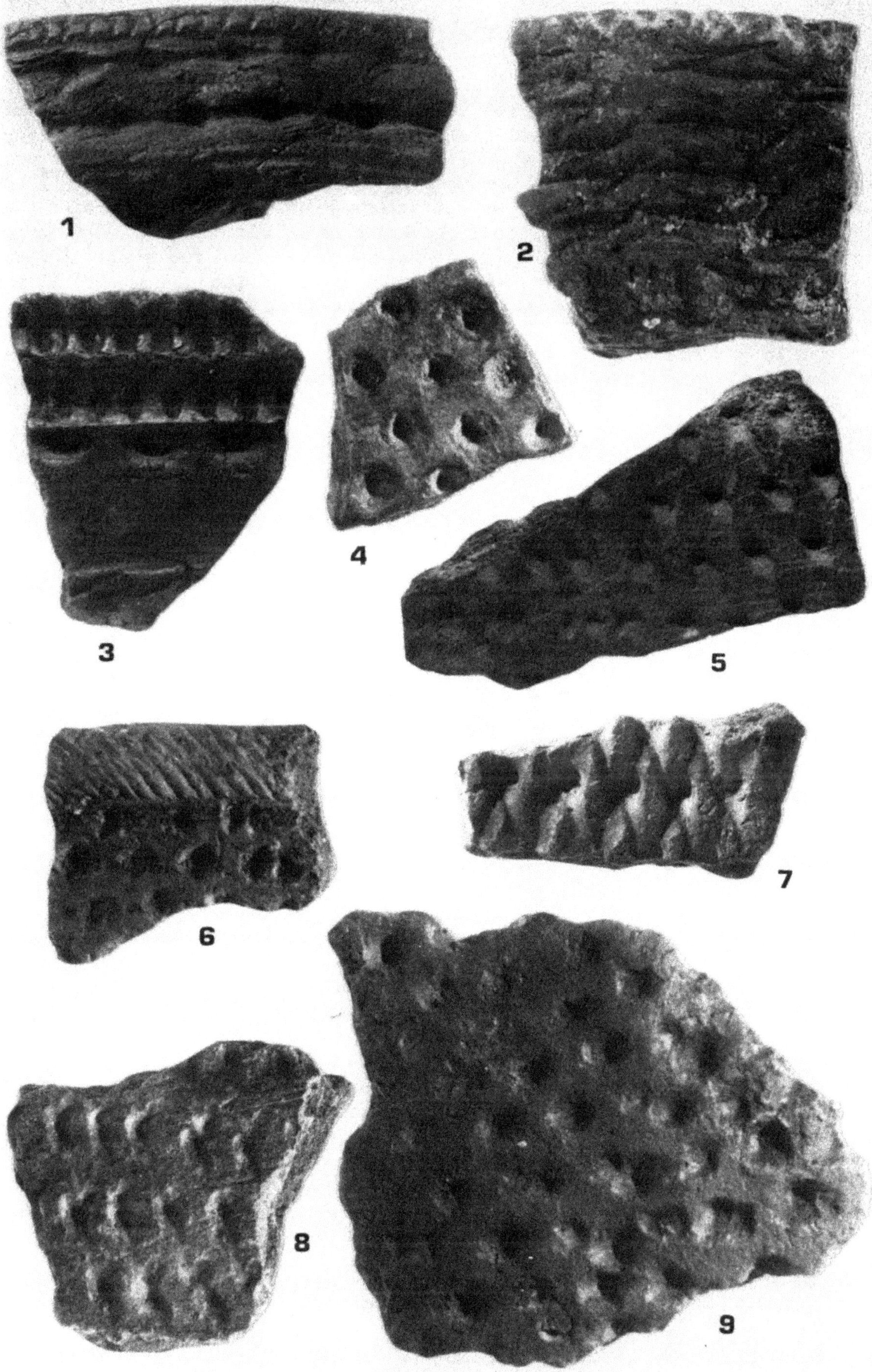

Fig. 7.40 – Uan Muhuggiag, Wadi Teshuinat. Level 2, Sector A. Impressed pottery. 5, punch; 9, quoin; 1,2,7,8, spatula; 6, composite implements (2:1).

of tools; out of the fifteen items only eight are of western-central Sahara well-defined types. According to the typology of J. Hugot (1957), the triangular flat-based, straight sided type (A1) is the best represented, followed by the hollow-based (A18) and leaf-shaped (C4). All the other types recovered, convex-based and leaf-shaped, pointed at both ends (types B3 and C1 respectively), are represented by single specimens. Leaf-shaped and triangular arrow-heads may indicate specific relationships with the western-central Sahara where they are known to be wide-spread. The Torha East sequence could allow a local development of these implements (Barich 1974: 141) apart from a few late elements, such as the leaf-shaped point with serrations that might really be an imported element.

In conclusion, the flaked products at Uan Muhuggiag may be said to have elements of continuity in the local tradition of the Tadrart, for the pointed and backed tools and the bifacial arrow-heads are found in the earlier Ti-n-Torha facies. At the same time, associated with them, one can observe the development of the group of large and crude tools, unstandardized and rough, which belong to the Later Neolithic complexes of the Sahara.

No real polished stone industry has been found, which appears poor even at Two Caves. The scarceness of the group of grinding-implements confirms the limited interest in gathering activities already observed at Torha North. This conflicts with the extensive evidence at Torha East where both lower grinders and handstones were in great number (Barich 1974: Tab. III). At Uan Muhuggiag the 12 specimens all belong to the lower horizon and none were observed in the Upper Phase.

The bone artifacts, although not numerous, include specimens that bring Torha North to mind (fig. 10.4). They are extremely well-made and were clearly intended for the manufacture of pottery (smoothing, decorations). Ostrich eggshell beads and unworked fragments, although not frequent, represent a further element of similarity with Ti-n-Torha.

The ceramic assemblage allows the most significant comments. In both Sectors, Levels 2d to 2 (11) are characterized by blackish (5YR 3/1 - 5YR 2.5/1) or dark reddish brown (5YR 3/3) ceramics decorated with impressions made by two-toothed implements, spatulae, and cord-wrapped (filetés) combs. The decorative patterns often involve the combination of two implements. A number of sherds are not decorated but have smoothed, if not burnished surfaces. The ceramic paste allowed to differentiate three groups that may not be considered as actual "classes" to be distinguished, since the mix of components appeared to be rather uniform (see the study by A.M. Palmieri, Chapter 8).

Ceramics using decorative techniques were analysed by the criteria used for Ti-n-Torha; the implements employed were taken as the starting point for analysis (figs. 7.45-47). This preliminary level of classification fits within the study of general implement utilization and proved the 'divider' technique suggested by I. Caneva. This technique allows covering with decorations of broad areas within a limited time and maintains a homogeneity in the decoration that is unattainable with a simple stamp.

---

11) In level 2d of Sector A, where the excavation was stopped before the bedrock finds were meaningless (one ceramic sherd and three chips).

Fig. 7.41 - Uan Muhuggiag, Wadi Teshuinat. Level 2, Sector B. Impressed pottery. 9, punch; 1,4,6, quoin; 2,3,5,8, spatula; 7,10, fileté comb (2:1. No. 5 is 1:1).

Fig. 7.42 - Uan Muhuggiag, Wadi Teshuinat. Level 1a, Sector B. Impressed pottery. 5,7, punch; 1,4,9, quoin; 2,6,8, spatula; 3, fileté comb (2:1).

Fig. 7.43 - Uan Muhuggiag, Wadi Teshuinat. Level 1a (3-5), Level 1 (1, 2, 4,6,7). Sector A. Impressed pottery. 3, punch; 1,6, spatula; 2,4, *fileté* comb; 7, cord (2:1).

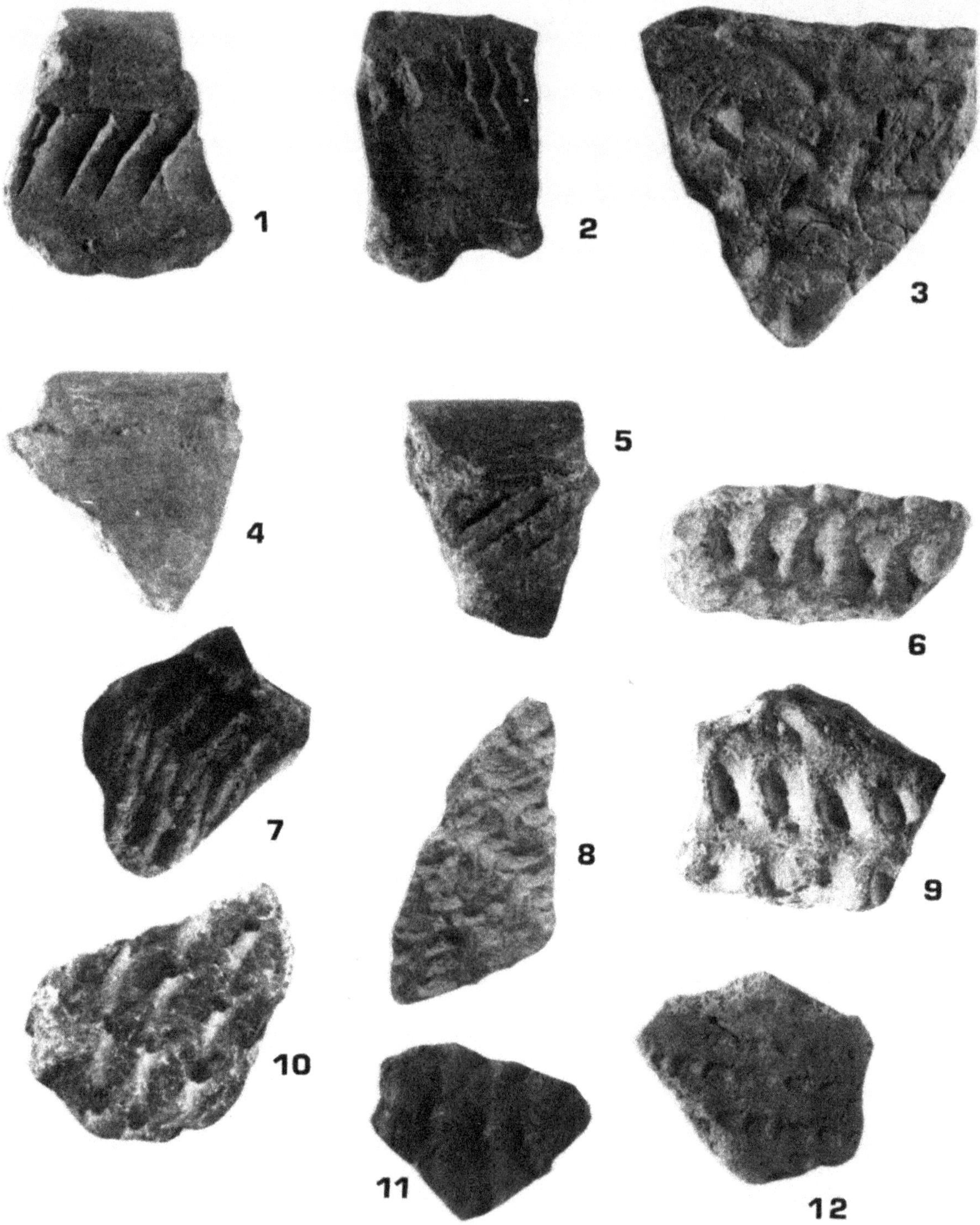

Fig. 7.44 - Uan Muhuggiag, Wadi Teshuinat. Level 1, Sector B. Impressed pottery. 1,5, punch; 3,6,7,10,11, spatula; 9, fileté comb; 2,8 cord (2:1).

A reappraisal of the presence of multi-toothed combs impressions showed that there was practically no difference in the patterns produced by a comb and a skillfully used two-toothed implement. The presence of comb decoration could only be unquestionably ascertained by observing the points of juncture of the comb impressions, and since many sherds available to us were small and therefore defied this verification, it was deemed advisable to restrict comb utilization until it could be re-considered in additional samples.

It proved possible however to show the use of two-toothed implements on 55.6 and 65.9% of the decorated sherds in Sectors A and B, respectively (fig. 7.45). The comb technique was definitely ascertained on two rimmed sherds with very compact thin paste, differing from other ceramic sherds in the complex. On both counts they can be related to specimens of the Dotted Wavy Line complex of the Ti-n-Torha type. It should be pointed out that both fragments were associated with arrow-heads and (in Level 2c, Sector B) the straight backed and pointed bladelet with rounded base, probably representing an imported element in the otherwise homogenous context of Uan Muhuggiag.

The conflict with the method used in the classification of central-Saharan ceramics is only an apparent one (12). It is well known that in the earlier facies of the Dotted Wavy Line complex, there is significant evidence of the use of the comb, both for the single impressions and in rocker technique ones. The hypothesis that even simpler implements (punches, quoins) might have been used in similar techniques and can be recognized gives greater plausibility to the ways in which the general decorative scheme was obtained. Furthermore, it enables a better definition of the finds by removing subjective elements (we are dealing with such definitions as: 'rows of dots'; 'deeply impressed triangles'; 'chevrons'; etc.).

The validity of this hypothesis was checked in different pastoral contexts, such as Wadi Athal (figs. 10.6,7) and Ti-n-Torha North, Levels II-I (fig. 10.3) in the Acacus; Adrar Bous (Smith 1980: figs. 18.2; 18.8) and Adrar Chiriet (Roset 1978) in Niger; and, in the Tibesti, Bardagué and Bardai (Gabriel 1981: 202, Photo IIc, IId). The lower horizon of Uan Muhuggiag agrees in a number of aspects (lithic, bone, ceramic) with other pastoral contexts in the Tadrart Acacus. Further afield there are similarities at the sites in the central and western Sahara dating from about the 6th millennium B.P. (Niger, Adrar Bous: 5780±500; the Hoggar, Méniet: 5400±150). The latter association is made on the basis of ceramics and the polished bone and ostrich eggshells that go along with them. The lithic evidence of the region has poor characterization and only general comparisons are possible. As previously pointed out, bifacial arrow-heads are the only reliable element for a comparison among different complexes.

In my opinion attempts to draw parallels with distant regions should not neglect the chronological factor, so that the arrow-heads present in the Acacus would appear to be among the most ancient types. The types observed at Uan Muhuggiag are consistent with the local tradition that, found in the

---

12) It is the same method that I followed in the ceramics classification of the Ti-n-Torha East and Ti-n-Torha North complexes, using the guidelines provided by H. Camps-Fabrer (1966).

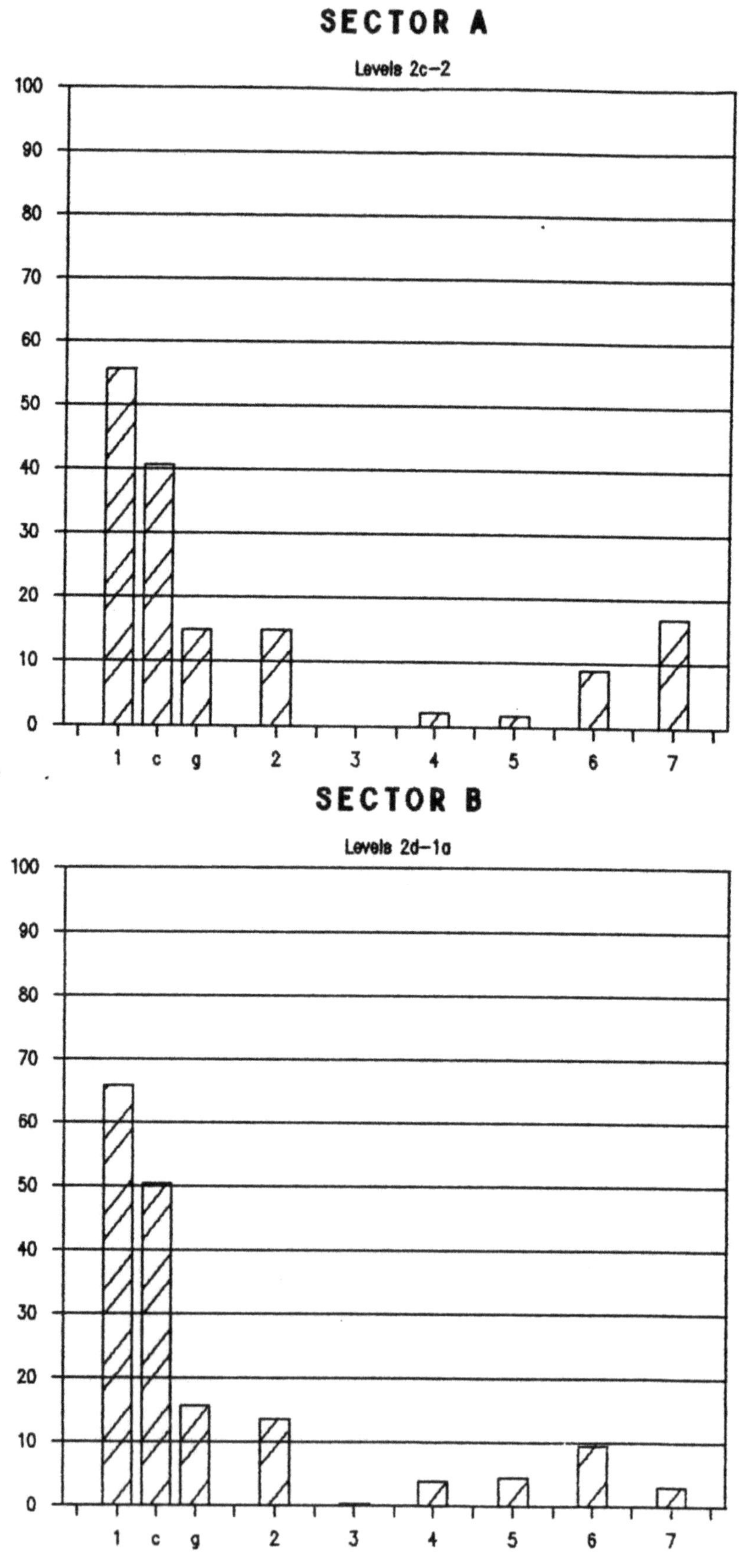

Fig. 7.45 - Uan Muhuggiag, Wadi Teshuinat. Sectors A and B. Absolute frequencies of implements for ceramic decoration. Key: 1 = two-toothed implements with c (circular) or g (geometric) cross-section; 2 = spatula; 3 = plain comb; 4 = fileté comb; 5 = cord/plants; 6 = composite; 7 = undecorated.

Upper Level of Torha East (CI), is continued in Torha North (V-I). These types allow us to recognize a derivation from the microlithic technique that results in a reduction in the size as well as in the high standard of the bifacial manufacture. An external origin might be identified only in the Kiffian complex that, through its dates from around 7500 B.P., might offer a possible ancestory. The Kiffian sites, that A.B. Smith (1976: 192-93) defines as antecedent to the local Tenerian and Temet complexes (Roset 1983: 132, fig. 13) show an association of types similar to that observed at Uan Muhuggiag. Furthermore, just as at Uan Muhuggiag, it seems that the microlithic arrow-heads have their antecedent in the Epipaleolithic techno-complex of the Wadi Greboun (Clark 1976) (13).

The Upper Phase

In the second phase of occupation, Sector B Level 1a is still markedly in use. Its total artefact frequencies fall within the mean values observed in the Lower Phase. The persistence of preceding characteristics suggests a continuation in this level from the Lower Phase. In Sector A (Level 1a) evidence is extremely poor owing to a landslide. The two sectors were both in use by Level 1 when, on the whole, the occupation of the site was greatly decreased.

In the lithic industry, sandstone is still being used but,at the same time, there was a reduction in tool-types; thus, the artifacts were, for the most part, undifferentiated types that, apart from end-scrapers, may be considered to be simple flakes with irregular retouch. No grinding-implements were found.

The polished bone industry includes some well-manufactured punches similar to those of the earlier phase. Some large fragments with rough polishing, including one with notching on one side were found. Ostrich eggshell beads were still being made.

The ceramics were completely different from that the underlying levels, in respect of both paste and decoration. Paste was still composed with smoothed or polished surfaces; its colour was always red. Surfaces were, for the most part, plain but when decorated, there was a marked preference for patterns obtained with cord-wrapped (filetés) combs making woven-mat motifs. Spatulae were still being used to a limited extent but punches were almost absent. Even this part of the sequence fails to provide little indication of the shapes of the vessels; one bowl with a carenated profile (fig. 7.29,7) could be recognized.

If the Lower Phase at Uan Muhuggiag may be associated with the pastoral facies already known in the Tadrart, the Upper Phase is quite new. Nothing allows us to think that this change occurred as a gradual shift since there are no significant elements of connection between the lower and the upper phases, nor is the evidence presently available to us sufficient to suggest any similarity with other compasses. It represents the latest occupation visible in the stratigraphy (since 3770±200 B.P.) and defies comparison with other excavated deposits. The carenated bowl and the woven-mat decoration

---

13) See in particular the long lived Ounan Point, peculiar to Greboun but found both at Temet (Roset 1983) and Adrar Bous III (Tixier 1962).

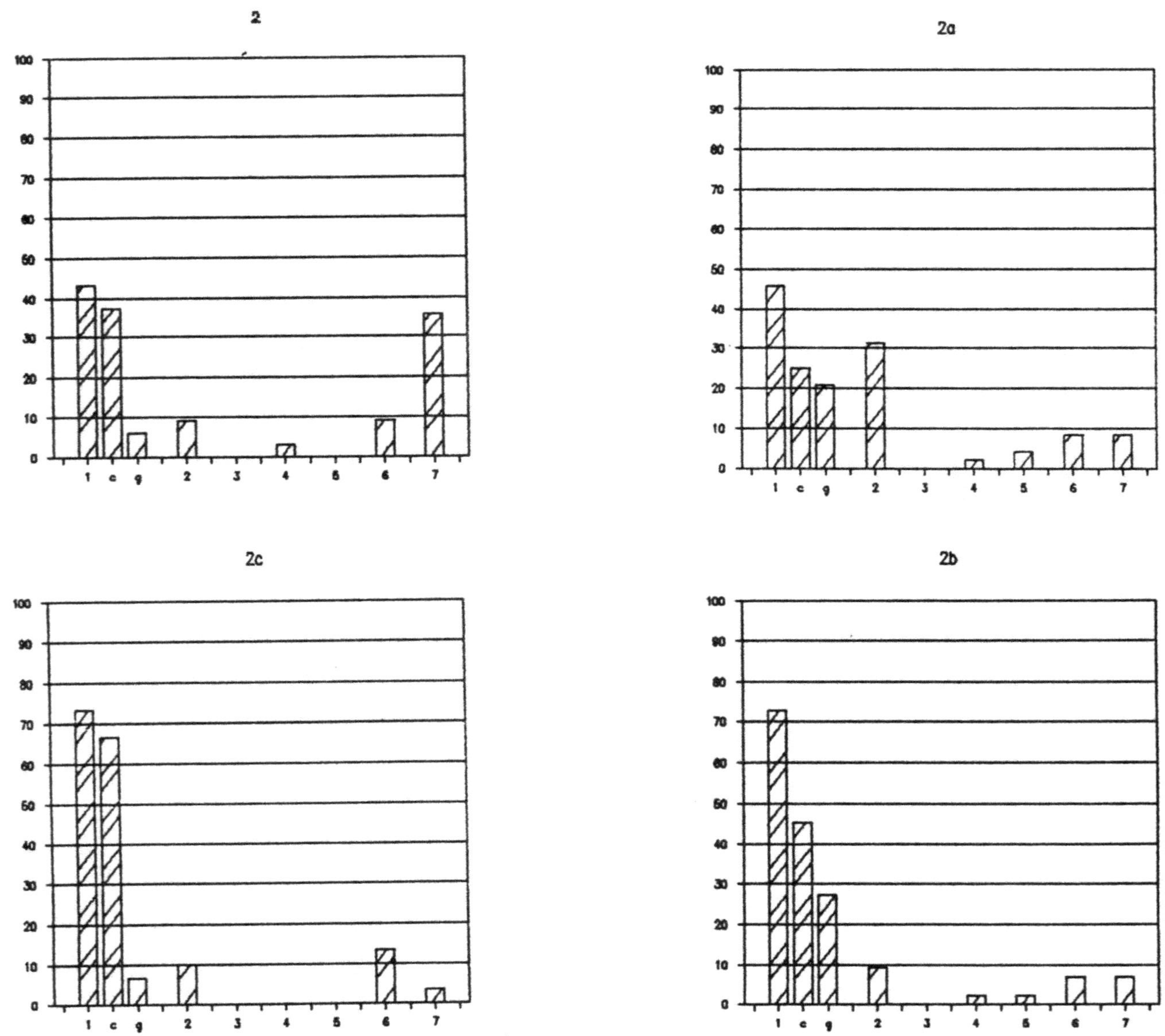

Fig. 7.46 - Uan Muhuggiag, Wadi Teshuinat. Sector A. Percentage frequency of implements for ceramic decoration level by level.

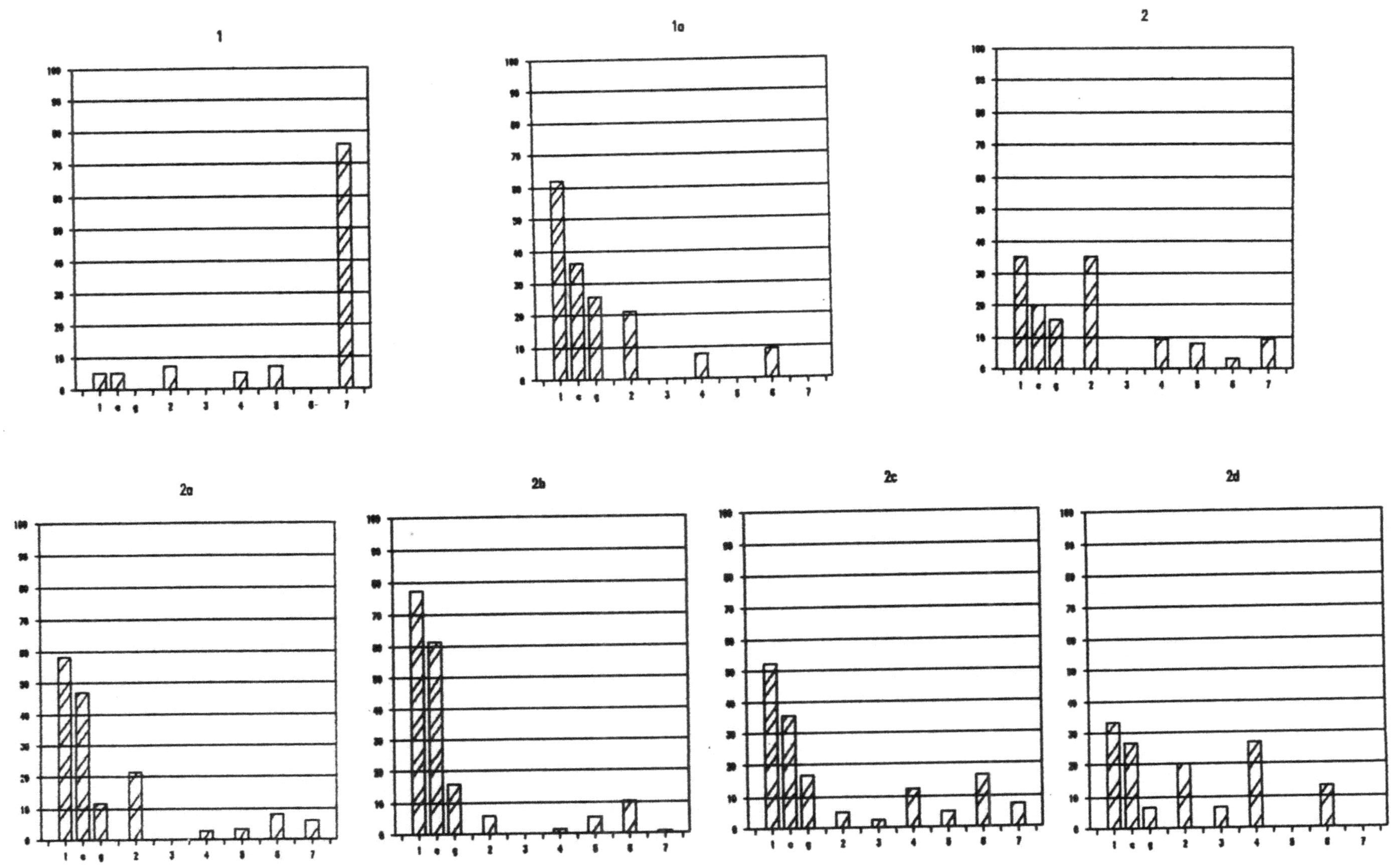

Fig. 7.47 - Uan Muhuggiag, Wadi Teshuinat. Sector B. Percentage frequency of implements for ceramic decoration level by level.

suggest a very faint parallel to the Proto-Berberian ceramics of the coastal area (cf. Camps 1974a: 342 et seq., Pl. XXV, 3).

Vegetation and Climate

The beginning of the deposit may be traced back to an extremely moist period (prior even to the 7th millennium B.P.) with a plant-cover typified by Acacia-Panicum, shrubs and Typha (already reported in Pasa, cf. Pasa, Pasa-Durante 1962) suggesting closeness to water. Most of the observations made by Pasa related to Trench IA, opened in the eastern side of the deposit in a position adjoining our Sector B. In his lower series (hearths X to VI), he identified an optimum climatic period with high humidity, shown by the levels of Typha and confirmed in the new excavations. Very rich pasture were postulated, while the faunal determinations showed the contemporary presence of cattle and wild species and, therefore, of the existence of a mixed pastoral/hunting economy.

No dates are yet available for this period, corresponding to the initial habitation (Levels 2d, 2c). As shown by the summary Tables, the artifacts of this period are, on the whole, much scarcer than in the subsequent series. These base levels included a few elements of the Ti-n-Torha facies, such as straight backed and pointed bladelets and the two sherds of Dotted Wavy Line type. Even the microlithic arrow-heads, quite different from the blade-like pointed flakes that appear from Level 2b, should be included among these elements. Gautier in his archaeozoological reconstruction, stressed a higher incidence of cattle than in the upper phase sequence. This should be related to a greater availability of both water and pasture.

Above hearth VI (in the Pasa's stratigraphy) the pollen spectra showed an impoverishment of plants including the disappearance of Typha. The sediments, consisting of sand and muds, led the author to suggest an arid phase. Caprids became increasingly more important among the faunal remains. This arid interval should correspond to Levels 2b-2 of the stratigraphy uncovered in the 1982 excavations. These levels represent the time when the site was more intensely inhabited, recently dated to 6035±100 and 5350±200 B.P.

Starting from Level 2, the previous plant complex was replaced by one dominated by Artemisia Herba Alba and A. Campestris, Aristida and scarce Ciperaceae. Typha reappears, although in a limited way. Moving upwards in the sequence, Aristida becomes increasingly more abundant. The sediments changed showing an increased component of coarse windblown sand. This situation - to be interpreted as a progressive deterioration of the climate, bringing about a curtailment of pasture and an increase in small livestock - tends to degenerate in the upper part of the sequence. Pasa observed, in his trench IL sequence (western area) the same accumulation of eolian dune sands with abundant silts mixed with fragments of sandstone from the vault of the shelter, that we observed in the excavation of Sector A, Level 1a. The uppermost level (Level 1) showed a moderate reappearance of vegetation related to a slight moisture increase; Typha definitely disappears and the plant cover is dominated by Artemisia, Aristida and Chenopodiaceae. The prevailing fauna in the latter context is represented by caprids, with goats predominant over sheep and representing 90% of the whole faunal complex (see Gautier, Chapter 12, this Volume).

Considering the faunal composition as a whole, A. Gautier reports 92.4%

to the domestic animals and only 4.2% to the wild ones (Barbary sheep that were so important at Ti-n-Torha East). This clearly points to the fact that an important change has occurred and that pastoralism has turned into the major activity of the group. It is important to note that Gautier reports that most of the domestic livestock is made up by small flock animals (14). Indeed he emphasizes that the cattle/caprid ratio is 1:3. A thorough examination of the sequence (Table 12.3) shows that while in the two base Levels (2d-2c) the cattle/caprid ratio is 1:1, in Level 2b a sudden change occurs. The ratio returns to 1:1 in 2a but caprids again become predominant in Level 2.

On the basis of this evidence, supported by the archaeological data (less intense habitation, archaic typological elements, presence of small millstones witnessing the persistence of grinding activities) one may reasonably affirm the existence of an initial habitation phase, with its own peculiar characteristics, with the 7438±220 dating obtained by Pasa confirmed. The Final Phase may be pastoral occupation in which goats become the major breeding animal. It should be recalled that on the back wall of the shelter is a scene depicted in the Ti-n-Lalan style where F. Mori identified the latest type of herders among those that frequented the Acacus.

Owing to the variety of evidence of its occupation, Uan Muhuggiag must have been a site of considerable importance. This was evidently determined by its location in one of the central wadis, its visible but well-protected position close to water. All these elements contributed to make this site a point of return for the pastoral groups that moved within the Acacus massif, thus determining the formation of an outstanding archaeological sequence and the well developed pictorial complex.

---

14) The prevalence at Uan Muhuggiag of small livestock, contrary to what was observed at Ti-n-Torha North, may be related to the thicker plant cover reconstructed in the former site, as against the more open vegetation of Torha North.

CHAPTER 8

CHEMICAL ANALYSIS OF THE ACACUS POTTERY: A PRELIMINARY ESSAY

By Alberto M. PALMIERI*

Archaeological research in the last decade is characterized by a remarkable increase of the number of studies devoted to ancient ceramic technology. Archaeological interpretation, therefore, can now rely on a series of additional data concerning source areas of raw materials, technological features of different cultural contexts and detailed definitions of different phases or schools within broadly known contexts.

This new perspective becomes particularly interesting when we deal with the most ancient pottery, which is often undifferentiated and conservative typologically over long periods and extensive areas. In this respect the Sahara seems to be one of the regions where such data would supply most useful information, provided that they are systematically gathered from every excavation. Unfortunately, this is not the case. Only a few fragments from Amekni (Hays, Hassan 1974) and the ceramics from the Tibesti (Gabriel 1973), were analysed. In addition, this restricted panorama involves the petrographic aspects, rather than the quantitative elemental composition of the wares.

The aim of the present study, therefore, was to test the validity of chemical analysis applied to the specific context and to provide the first, albeit preliminary, reference for future work.

As far as clay supply is concerned, preliminary indications can be gathered by observing the geomorphological structure of the Tadrart Acacus. The whole massif does not present great lithological variations: a sandstone series of the Upper Gotlandian and the Lower Devonian, with lateral outcrops of Tenezzuf clay. Sandstone strata are often interbedded with clay layers which are the best supply for present day ceramic production (Sindelar et Alii 1980). The alluvial plain of the Wadi Auis and other large valleys, with their bulk of loose sediments, may also have been a source of raw material.

The sample analysed was 24 fragments of pottery typologically selected, coming from different prehistoric sites of the Tadrart Acacus (1). They can be considered as belonging to three cultural contexts, radiocarbon dated respectively to ca 3700, 6000 and 8000 years ago. The most ancient context (Samples 6, 7 and 8) is characterized by dotted wavy line and related

---

* Istituto per le Tecnologie Applicate ai Beni Culturali, C.N.R., Rome, Italy.

1) Coming from the Excavations and Surveys made by the Libyan-Italian Joint Mission for Saharan Research.
Information on the archaeological context can be found in Barich, Mori 1970; Barich 1974 and this Volume.

pottery, belonging to Epipalaeolithic hunter-gatherer groups. Potsherds come from the site of Uan Muhuggiag (surface), from Fozzigiaren and from the area of Wadi Ti-n-Torha. The second context (Samples 3, 4 and 5) comes from the well known "pointillée" pottery, widespread all over the Sahara, among the Neolithic pastoral communities. The third (Samples 1 and 2) represents the latest development of the pastoral cultures, with red undecorated pottery. All potsherds belonging to the pastoral phase come from the site of Uan Muhuggiag (see fig. 1.1).

ANALYTICAL METHODS AND EQUIPMENT

The present study involves the quantitative analysis of major trace elements through XRF. 21 elements have been analysed.

The XRF - equipment used was an automatic spectrometer Siemens SRS 200 with ten position sample changer, controlled by an Ericsson Personal Computer for setting of instrumental parameters. The raw data (pulse numbers for each spectral line) have been related to an external standard and corrected for matrix effects by means of the same Ericsson P.C. (according to the well known method by Franzini and collaborators (1975)). The CPU time needed was 15 s and the dosage of 21 elements in ten samples took about 10 hours. The X-ray tube used was a Siemens AG CR 61 (3000 W) with chromium target. The voltage and current varied in function of the single element and its concentrations, from 30 kV and 80 mA for Na, Mg, Al (more difficult to detect) to 55 kV and 40 mA for the other elements. Lower voltage and lower current was used for Ca, Ti, Fe, K, easily to detect when present in high concentration.

Three ceramic samples analysed by the Laboratorium voor Aardkunde of the University of Gent (Belgium) (2) have been used as reference standards.

Major elements were measured by flow counter; trace elements by scintillation counter.

The counting time varied between 10 s (Ca) and 200 s (Mg, Mn). All spectral lines were measured in vacuum.

All incrustrations, adhering sediments, etc. were mechanically removed from the samples; afterwards they were dried at 105 °C, powdered and pressed without adding a binding agent by a Herzog HTP 40 pellettising tool.

Data due to L.O.I. (loss on ignition), needed to compensate for the matrix effect, were evaluated on a 0,5 g powdered sample in a muffle furnace at 960 °C for one hour (3).

The complete results of the analysis are given in Table 8.1. Among the 21 elements analysed, 12 have been selected as the most significant variables, i.e. the major elements SiO_2, Al_2O_3, Fe_2O_3, MgO, CaO, Na_2O, K_2O and the trace elements Sr, Nb, Zr, Rb and Ba. Major elements may give

2) Results kindly provided by dr. Paul de Paepe.

3) I am indebted to Dr V. Francaviglia for his substantial contribution in sample analysis.

Table 8.1 - Tadrart Acacus. Quantitative elemental composition of ceramics.

	SiO_2	Al_2O_3	Fe_2O_3	MnO	MgO	CaO	Na_2O	K_2O	TiO_2	P_2O_5	S	Cl	Nb	Zr	Y	Sr	Rb	Ba	Th	Pb	La
S1C1	54.83	19.59	6.96	0.11	2.40	0.78	0.65	3.56	0.97	0.25	117	1550	9.2	169	22.3	97	125	892	10.4	22.3	55.7
S1C2	55.13	18.73	6.82	0.11	2.50	0.83	0.78	3.47	0.95	0.21	272	3652	10.5	139	17.1	110	114	839	3.9	22.1	59.5
S1C3	74.56	11.28	4.48	0.05	2.34	1.82	1.76	3.46	0.75	0.21	980	2905	11.5	335	15.4	240	129	1730	11.5	39.2	38.2
S1C4	86.26	8.78	3.28	0.03	0.91	0.80	0.35	1.52	0.57	0.09	150	645	8.9	301	14.5	101	53	436	7.7	17.4	51.1
S1aC1	75.76	12.75	3.16	0.03	1.04	0.94	0.58	2.86	0.69	0.29	201	3042	6.7	233	14.2	100	52	547	13.8	24.6	2684
S2C1	56.24	19.83	7.24	0.13	2.56	0.85	0.75	3.56	0.99	0.23	302	2866	6.5	151	17.7	119	127	887	13.9	21.3	70.1
S3C1	60.44	15.33	6.82	0.08	2.22	1.54	1.05	2.91	0.91	0.52	527	3451	12.9	325	22.2	140	66	574	13.0	19.9	704
S3C2	67.35	12.91	5.48	0.06	2.41	1.09	1.53	3.50	0.75	0.89	751	3525	9.4	240	16.2	122	66	558	6.9	27.7	49.3
S3C3	70.20	13.70	6.06	0.08	1.92	1.40	0.88	2.29	0.96	0.20	423	4063	11.8	295	22.9	130	65	534	7.8	27.6	68.9
S4C1	60.28	15.49	6.95	0.07	2.40	0.71	1.22	2.87	0.73	0.22	690	2749	10.4	284	21.9	100	74	492	12.7	24.6	57.4
S4C2	63.36	15.75	7.36	0.09	2.40	0.92	0.79	1.94	0.90	0.21	367	1181	16.2	392	26.7	139	68	482	9.7	32.4	57.9
S4C3	60.35	15.11	8.55	0.09	2.69	1.58	0.96	2.07	0.70	0.17	491	3046	8.3	250	17.4	123	57	381	9.4	17.4	45.4
S5C1	64.01	14.94	7.20	0.09	1.98	1.24	0.86	2.17	0.92	0.23	184	2111	14.8	343	24.6	129	70	586	12.1	28.8	66.0
S5C2	63.03	15.58	7.26	0.06	2.26	0.68	0.54	1.69	0.79	0.23	182	263	9.9	246	17.3	104	73	469	14.2	24.6	60.0
S5C3	59.21	15.74	7.90	0.10	2.06	0.78	0.64	1.76	0.94	0.37	209	578	9.3	244	14.3	102	58	469	12.4	25.0	71.9
S5C4	55.76	17.73	6.92	0.06	1.94	0.79	0.52	1.28	0.82	0.24	86	210	0.0	126	0.0	66	44	434	5.1	28.7	52.1
S6C1	51.42	22.38	6.38	0.04	1.65	0.89	2.89	3.18	0.38	0.12	161	496	9.1	340	25.9	171	132	2813	19.4	22.1	220
S7C1	40.15	17.56	13.65	0.10	5.35	4.85	1.47	2.37	1.58	0.48	304	1029	16.8	277	31.6	244	89	999	12.2	23.5	25.0
S7C2	45.87	23.52	6.32	0.04	2.07	0.92	1.62	3.22	0.41	0.10	125	517	8.9	193	24.0	132	159	1143	18.0	43.3	240
S7C3	41.71	16.39	14.75	0.13	5.42	3.92	1.59	2.20	1.92	0.50	137	895	19.4	313	31.1	220	81	985	11.3	19.5	32.3
S7C4	45.29	21.83	6.96	0.04	3.06	1.25	1.24	3.66	0.52	0.22	81	1534	9.5	217	24.1	166	177	1246	18.0	37.3	149
S7C5	51.74	20.78	6.38	0.05	1.85	1.51	1.19	3.86	0.52	0.13	174	1185	11.5	297	24.0	122	175	1741	14.4	45.4	99.3
S8C1	47.71	22.87	7.53	0.06	3.65	2.27	1.60	2.84	0.61	0.18	144	762	8.9	290	22.4	463	125	4357	11.9	33.1	89.4
S8C2	48.17	24.23	7.86	0.05	1.84	2.01	1.49	2.97	0.63	0.15	79	195	12.6	296	21.5	433	125	4571	11.0	29.3	278.1

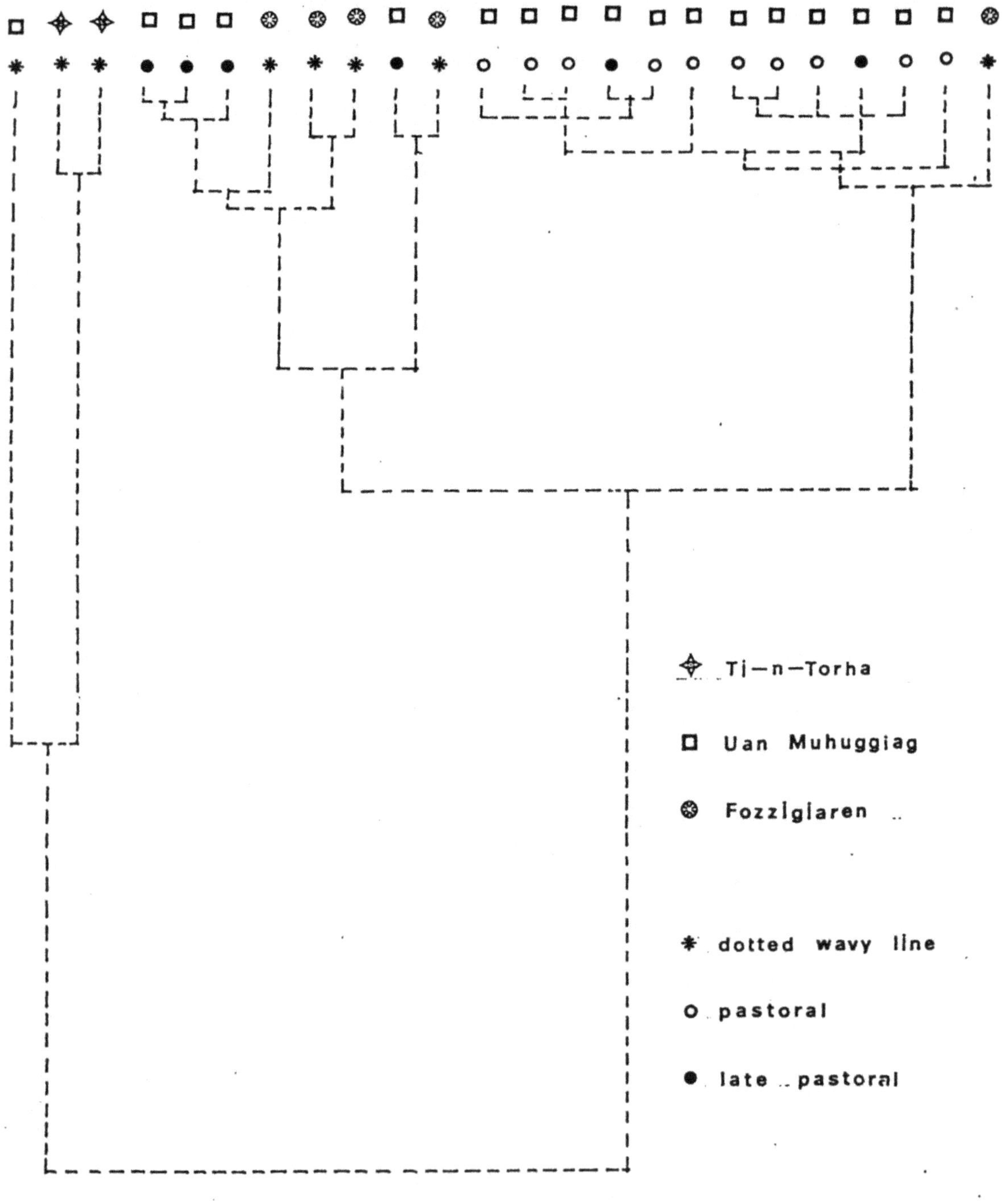

Fig. 8.1 - Tadrart Acacus. Dendrogram of cluster analysis by the rate of trace elements.

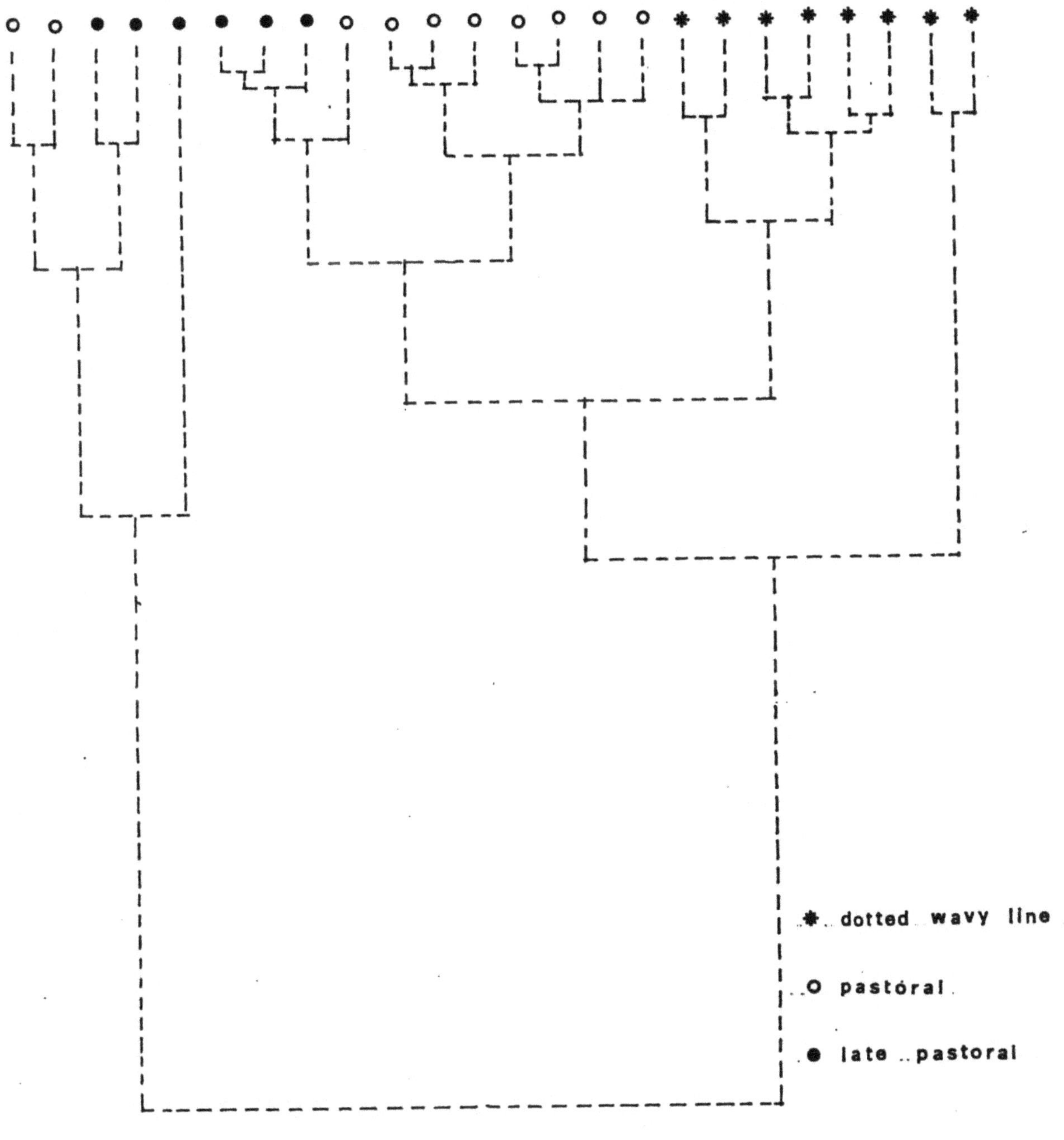

Fig. 8.2 - Tadrart Acacus. Dendrogram of cluster analysis by the content of major elements.

information on tempering material, therefore, on source areas, as well as technology, whereas trace elements give indications as to the geochemical environment, reflecting very specific local features.

Due to the variability inherent in prehistoric pottery, where the domestic production can not avoid some fabric differences a certain fluctuation in the results was expected. Taking this into account, our attempt to obtain an objective classification of the specimens with the cluster analysis gave fairly satisfactory results. The method examines the similarity coefficients in successive cycles where elements with gradually lower correlations are grouped. The subsequent fusions of groups, according to the decreasing similarity coefficient, are illustrated in a computer based graphic dychotomic dendrogram. The program is the DENDR, provided by the Institute of Physics of the University of Rome "La Sapienza" (Bietti et Alii 1978).

The 12-element dendrogram does not give very significant groupings, owing to the combined effects of the two kinds of elements. The trace elements dendrogram (fig. 8.1) shows four clusters at a sufficient similarity level: 1) an isolated sample representing the oldest ceramic context of Uan Muhuggiag, 2) the small Ti-n-Torha group, 3) the Fozzigiaren dotted wavy line context, and 4) the Uan Muhuggiag pastoral phase, these last two mixed with the latest pastoral aspects of Uan Muhuggiag. It seems, therefore, that a significant relationship exists between the selected elements and both the geographical and chronological distribution of the specimens. The latest phase of the Uan Muhuggiag sequence seems to be the only unclassifiable group.

The major elements dendrogram provides the best grouping (fig. 8.2). Sherds are divided into two very distinct clusters, one of which contains two small groups belonging to the pastoral phase. The other cluster includes three main groups, characterized by a good similarity level, which correspond to the three contexts under examination. The pastoral phase is closely correlated with its latest aspects. Again, the latter, unlike the others, seems to escape any classification system. Since SiO_2 and Alk seem to be the most significant elements for classification, a binary diagram was built to emphasize their relationship (fig. 8.3). The two main groupings suggested by the major elements diagram, thus are better seen. The latest pastoral phase is not well characterized by these variables or by the others examined, as already seen. Silica concentration seems in general to increase from the most ancient periods (45-50%) to the recent ones (60-65%), with a tendency, in the most recent aspects, towards a greater variability.

CONCLUSIONS

A certain difference between the wares of the three contexts was already visible to the naked eye. The aim of the analysis, therefore, was to quantify the difference, to identify its nature and to explain it in terms of manufacturing technology or raw material sources.

Clay appears to have specific characteristics in each site, suggesting a local manufacture and a sufficient standardization of pottery making, at least for the two ancient periods. On the contrary, a greater variability in both provenance and fabric characterizes the latest pastoral phase. The phenomenon could be tentatively ascribed to a greater mobility of the groups

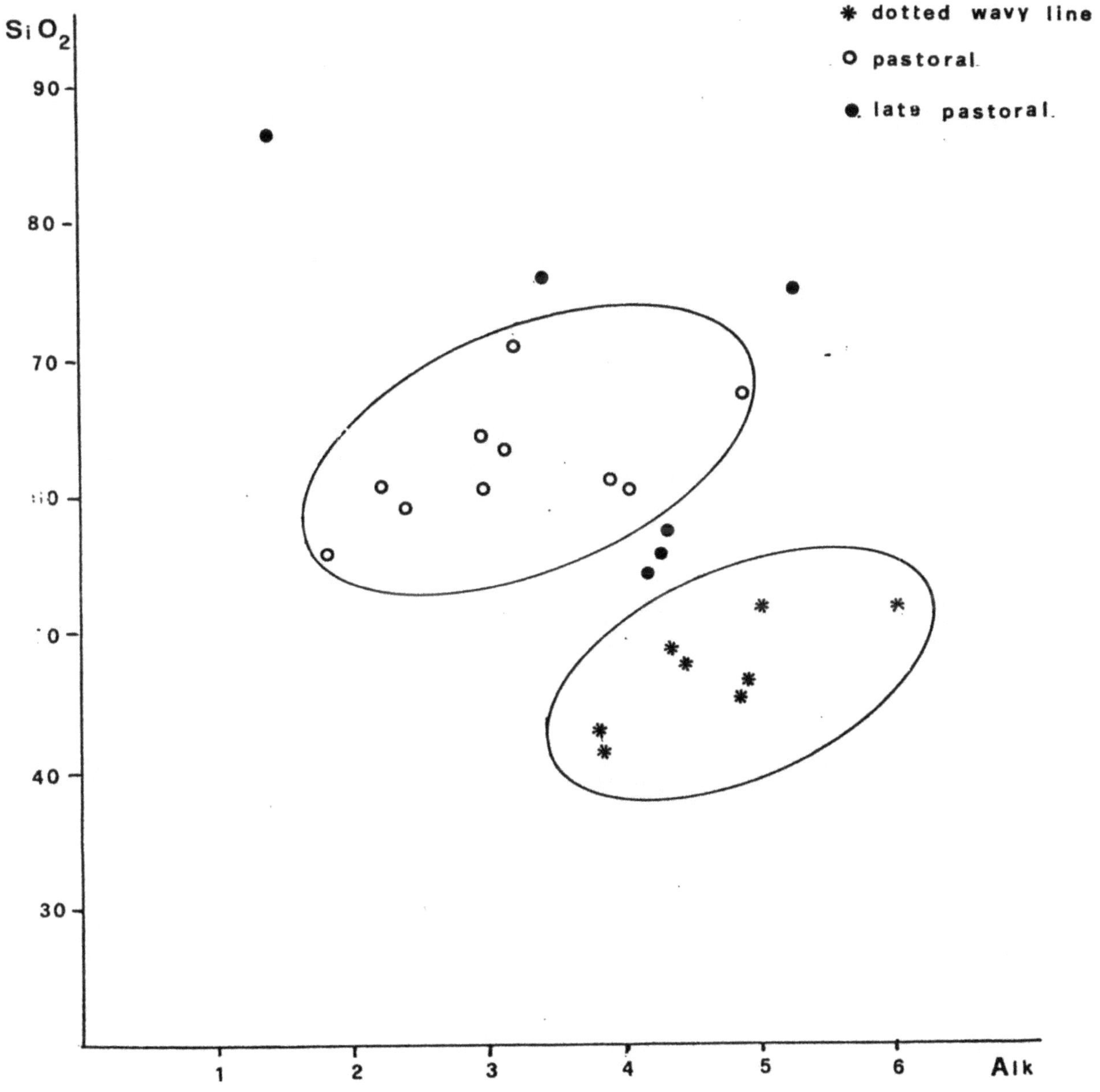

Fig. 8.3 - Tadrart Acacus. Ceramics clustering based on the SiO_2/ Alk ratio.

involved with respect to the others. Great mobility, due to the increasing severity of the ecological conditions, may mean great dispersal of small groups with different traditions and access to scattered clay sources.

Two homogeneous clusters are shown by the ratio SiO_2/Alk, i.e. the main factors of variation. The dotted wavy line related context and the earliest pastoral aspect, in fact, are well characterized. Since SiO_2 is linked to the quartzitic sand component, the low level of this element in the most ancient context would suggest the use of natural clay without any additional temper. Such a level, around 50%, appears indeed to be characteristic of the clay layers interbedded with sandstone in the central Libyan regions (Sindelar *et Alii* 1980). The increased SiO_2 rate in the later contexts, therefore, would suggest a technological change, with sand added to the clay. In support of this assumption we observe the stability of the SiO_2 content within the two cultural contexts, even if the clay appears to have a different chemical composition according to its provenence. In addition, a very similar trend was recently observed in the cultural development of the middle Nile Valley, in the Sudan, between the ceramic production of the locally corresponding dotted wavy line and pastoral groups (Francaviglia, Palmieri 1983).

The results, in conclusion, emphasize a cultural-chronological subdivision rather than a geographical one, which is to say that an intentional rather than a natural change characterizes the subdivision.

REFERENCES

BARICH B.E. 1974 - La serie stratigrafica dell'uadi Ti-n-Torha (Acacus, Libia). Per una interpretazione delle facies a ceramica saharo-sudanesi, Origini, VIII: 7-184.

BARICH B.E., MORI F. 1970 - Missione Paletnologica Italiana nel Sahara Libico. Risultati della Campagna 1969, Origini, IV: 79-142.

BIETTI A., RAMBALDI A.E., e ZANELLO I. 1978 - DENDR - Un programma Fortran di "cluster analysis" per applicazioni archeologiche, paletnologiche e paleoecologiche, Quaternaria, XX: 49-86.

FRANCAVIGLIA V., PALMIERI A.M. 1983 - Petrochemical Analysis of the "Early Khartoum" Pottery: a Preliminary Report, Origini, XII: 191-205.

FRANZINI M., LEONI L. e SAITTA M. 1975 - Revisione di una metodologia analitica per fluorescenza X, basata sulla correzione completa degli effetti di matrice, Soc. Ital. di Miner. e Petrogr. Rendiconti, XXXI, 2: 365-376.

GABRIEL B. 1973 - Vorgeschichtliche Keramik aus dem Tibesti (Sahara), Berichte der Deutschen Keramischen Gesellschaft, 50/8: 261-67.

HAYS T.R., HASSAN F.A. 1974 - Mineralogical Analysis of "Sudanese Neolithic" ceramics, Libyca, XXII, 157-165.

SINDELAR J., JADRNICEK, P. 1980 - Studies of Some Clays in Libya, in The Geology of Libya, Vol. III, eds. Salem M.J., Busrewil M.T., Academic Press: London, pp. 993-999.

CHAPTER 9

POTTERY DECORATION IN PREHISTORIC SAHARA AND UPPER NILE: A NEW PERSPECTIVE

By Isabella CANEVA*

The problem of classifying prehistoric archaeological materials in the Nile Valley and deciding which elements are relevant for specific classification has been dealt with previously by the author on several occasions (Caneva 1970, 1983, 1984).

The need of more significant typologies has led the author to emphasize technological procedures rather than morphological aspects in pottery classification. Technological procedures, indeed, represent the synthesis of the operator's investment in terms of labour, time, care and taste.

In the central Nile Valley, the classification of prehistoric pottery on the basis of decoration techniques rather than decorative motifs or tools provided groupings with a remarkable chronological significance. A trend towards the predominance of <u>rocker</u> (including the "dotted wavy line" motifs) as opposed to other techniques was present from the early Mesolithic (7500 B.P.) to the later Mesolithic (6500) and Neolithic (5500), cultures. A further development of this trend was also recognized, which leads to the exclusive use of a two-toothed instrument in later Neolithic aspects (including the later "dotted wavy line" motifs).

A comparable trend can be seen in the chronological development of Saharan pottery. A certain parallelism can be followed especially in the shift from the <u>rocker comb</u> impression to the compass-like application of a two-toothed instrument. This shift is particularly emphasized in both areas by the evolution of the "dotted wavy line" decoration, obtained first with the <u>rocker</u> technique and later with the <u>alternately pivoting</u> stamp.

As far as Uan Muhuggiag is concerned, its pottery appears to be decorated with techniques belonging to the later phase of the aforementioned trend. The most common is the "return" technique, which consists in the compass-like application of a two-toothed tool, going over the last line of dots, thus ensuring equal spacing between all the dots of the pattern. This results to be the most common technique of decoration all over the Sahara during the 7-6 millennia B.P., including most of the so-called "punctations".

Despite similarities and parallelism, substantial dissimilarities are also recognizable between the Saharan and the Nilotic areas. Many elements point, therefore, to a normal framework of exchange and contacts between adjacent regions, sometimes influencing the local production, but never in terms of predominance of one region over the other.

The elaboration of true and proper typological structures, which are

* Dipartimento di Scienze Storiche, Archeologiche e Antropologiche dell'Antichità, University of Rome "La Sapienza", Via Palestro 63, Rome, Italy.

subsequently compared with a view to reconstructing local cultural sequences, determining relationships between different areas and establishing general regional chronologies, makes the choice of basic elements for consideration one of the most difficult problems of any study.

When dealing with ceramics the precise aims of classification may be defined as follows:

a) To differentiate the various ceramic elements, reconstruct the complexity of the ceramic production and assess chronological variations.

b) To recognize significant elements which permit easy comparisons among contexts.

c) To have a clear view of the amount of labour involved in the production, the time required, the special care needed, etc.

d) To show the importance of certain activities over others. This is normally related to the shapes of the pots and, therefore, is not considered relevant in contexts where shapes can not be reconstructed.

As may be seen from this very short general list of goals, the basic premise in selecting typological elements (or associations of elements) is that the choice must come after observing the material and formulating the specific research aims, and not decided upon in advance.

In any case, the typological elements must have the following properties:

a) strongly differentiating attributes in the specific context, so as to produce a clear-cut characterization of groups being compared;

b) a significant correlation with the intentions of the potter;

c) an easy means of being recognized, so that anyone can repeat the same analysis without significant variations and at any time.

It is noted that these requisites are difficult to fulfill simultaneously in a single element; they are more often found when grouping several elements either together or at different levels of study. In this case, the structural order of the headings (division into main groups and sub-groups) can be decisive for either grouping the data and making them significant, or dispersing them, even if in theory it should be possible to read the list in different ways.

In prehistoric Saharan ceramics the available elements are limited, on the one hand, by the poor state of conservation of the material which makes it impossible to reconstruct complete shapes or complex decorative patterns on the surface of pots and, on the other hand, by the great homogeneity in both the make up of the ceramic paste and the surface treatment.

The only analysable element without laboratory aid is, therefore, the decoration. There are, however, different ways in which the decoration can be evaluated, namely:

a) the instrument used;

b) the decorative patterns achieved;

c) the technique adopted to produce the decoration.

It is difficult to separate these aspects, one from the other. The visual impression obviously takes all three into account. However, the rationalization of the visual impression must give them a sequence. These sequences must follow a logical pattern which is not necessarily either the succession in time of the various acts performed or the content of the "visual effect" (in this sense, as already mentioned, the elements are inseparable), but reasoning based on the quantity of information which each element possesses.

These three aspects of the decoration were adopted for the first Saharan ceramic typology (Camps-Fabrer 1966; Camps 1968), and then, conventionally, by all archaeologists who subsequently dealt with the subject, virtually without further questioning. The sequence in which these aspects were considered is usually that set out above, whereas in classifying Sudanese ceramics it is the motifs that are considered first (Arkell 1949; 1953; Haaland 1981; Hays 1974; Mohammed Ali 1982). However, the author feels that the approach calls for two remarks: the first is general and theoretical and concerns the order followed, the second is practical and particular and concerns the analysis of decorative techniques.

1. THEORETICAL COMMENT

The classification of decorative motifs is not based on syntactic arrangement, which can only be reconstructed in extremely rare cases, but on the optical impression given by the individual decorative elements. This, by the way, is not easy to describe and is extremely subjective both for the author and for the reader of the description. The grouping into "classes" is particularly subjective (the motifs often overlap classes), so that there is a risk of either having groups which are too extensive and generic (dotted patterns, wolf's tooth, etc.) or going to the other extreme and separating almost every fragment. In all probability, the craftsman was subjective in using motifs generally accepted by the community and varying his execution of it from one time to the next.

The same may be said for the implements used, for which there were no standardized materials, shape or size permitting them to be associated specifically with decoration motifs. Tools were probably made or collected individually and were, therefore, merely personal objects. Therefore, this classification is no longer based on real objects present, but on objects whose existence and physical aspect must be assumed. This procedure, then, although necessary for formulating assumptions about descriptive details, does not attain the value of a useful classification, since it becomes as subjective as the decorative motifs.

By considering the technique used before taking the other elements into account, the following advantages are achieved:

1) the technique is a <u>fact</u>; and does not lend itself to individual subjective interpretation; it can be recognized in the same way by anybody and is easily described;

2) it provides a great deal of information, giving an immediate realistic reconstruction of a moment in the past, with indications as to the types of gestures made, the position of the hands, the way the pot was held, the care taken in executing the operation, and the duration and cost of the latter;

3) the technique is very distinctive even with respect to the decorative motif which is made by it and is not detached from it. On the contrary, often the motif can only be done in one particular way (the smocking pattern for instance). The technique is probably the first thing that an old craftsman taught a young apprentice, together with all the motifs it can produce and the different ways to enrich and vary them. Each craftsman then developed his personal version of the patterns;

4) the technique reconstructs a "general trend" (a specific way of treating the material), which conveys one organized meaning to the variability of the motifs present. It, therefore, goes beyond a detailed attribute list of myriad of different executions, by accomplishing an operation which is a synthesis of what is simply seen (1).

As a result of the preceding remarks, which are applicable to many categories of archaeological material, there is a dichotomy as to the different kind of information conveyed by a) the technical and b) the morphological aspects of the materials. It has been decided, therefore, to adopt for Saharan and Sudanese prehistoric ceramics a criterion for classification based on the techniques used and then make internal sub-divisions which take into account the other aspects mentioned, in the following order:

- Technique of decoration
- Type of instrument used
- Resulting decorative morphology

This method, applied to Sudanese prehistoric ceramics (Caneva 1983, 1984) yielded highly satisfactory results both for the reconstruction of a sequence and for the internal differentiation of cultural aspects. Chronologically speaking, this made it possible to identify a technological trend which relates all cultural phases from the early Mesolithic to the late Neolithic, a trend which appears to follow similar phases also in the Sahara. In terms of an internal differentiation, for the late Neolithic, it made it possible to differentiate two sites whose ceramics, both decorated with motifs which would have been conventionally classified as dotted lines, had in one case been executed exclusively with comb impressions and in the other with rocker impressions using a double-pointed instrument.

---

1) The importance of the technique used for ceramic decoration compared to the specific shape of each instrument is stressed by Camps (1968:127), even though the type of intrument then influences his typology far more than the technique.

As previously pointed out, even in a classification of motifs considering technical aspects in the internal subdivision, it would have been possible to observe the differentiation between the two sites and, similarly, although with greater difficulty due to the dispersion of data, it would have been possible to reconstruct the trend supporting the entire development of ceramic decoration in the Nile Valley. However, since the techniques were the more distinguishing elements, it was logical to give preference to a criterion of classification which emphasized them more than the other elements.

2. PRACTICAL OBSERVATIONS

In previous studies of the decorative techniques used, analysis was mainly carried out without considering the logic behind the information visually recorded. One of the most surprising cases is that of the well-known dotted wavy line decoration: it was imagined that a half-moon shaped toothed stick was used, directly applied to the pot, alternating the position of the curve upwards and downwards (Camps-Fabrer 1966: 446). This mechanical movement was repeated on the entire surface of the pot, for thousands of impressions arranged in hundreds of rows all perfectly parallel. Assuming this to be so, one has to marvel at the dexterity of the craftsman who operated in such a way as to make it impossible to find any trace of joining the juxtaposed impressions and at the same time obtaining perfectly parallel lines.

In most cases, however, the curves are not repeated in exactly the same shape all along the line, nor is the dotting of the different curves forming the strip uniform, but it is less dense in the external, convex parts of the hand, and both these features are unexplainable in a repetition of impressions using the same instrument. The lines are then normally grouped in bands of 5 or 6, as are the straight dotted lines that usually form the entire decorative pattern above and below the undulated bands. Finally, an account must be taken of the absurdities of the suggested procedure in terms of manual dexterity and time required for the levels of precision achieved. What was proposed was a procedure which was not suggested by any real element; the individual elements themselves, i.e., dots and curved lines, were abstracted from the context in which they repeatedly recur. If, on the contrary, one succeeds in showing the relationship between these elements (parallelism even in assymmetry, grouping of lines, position, number and spacing of each dot with respect to the dots of the lines above and below, frequency of impressions in modules or panels, etc.) and endeavors to give a logical explanation of the entire operation in terms of time and ease of execution, then one will obtain a certain number of elements which will be rational and provide a significance for what is seen.

This type of observations, together with a careful analysis of the material from the foregoing point of view, leads to proposing a procedure which has so far been confirmed (or, at least, not contradicted) in all cases examined in both Sudanese and Saharan contexts. This consists of making fans of dotted impressions with a rocker movement of a toothed comb (fig. 9.1,_5_). With this method the dots on the lines closest to the point of the fan are the closest together whereas on the outer lines they are more spaced, as in all rocker patterns. This produces perfectly parallel patterns, accounts for the difference between one curve and another on the same line and ensures a perfect repetition on successive lines. It also accounts for the grouping of lines in bands, wavy or otherwise and finally for the more widely spacing of the dots in the arc farthest from the centre. Laboratory tests using this system gave good results in a surprisingly short time (fig. 9.1,2) (2).

A similar case, which arose from the observation of the Acacus material and was resolved following the same perspective of showing the "relationship" between the various elements, concerns the direct application of combs (plain, "filetés", etc.) in parallel horizontal rows. Although the existence of this technique is totally logical, widely observed in Sudanese contexts and probably to be assumed, in certain cases, also in Saharan contexts, its recognition must be supported by a very specific series of relationships between the elements. Basically speaking, the following characteristics must be encountered in the decoration (fig. 9.2,1):

a) presence of a "module" with specific characteristics (for instance irregular spacing and shapes), repeated alone along the line and in the various rows;

b) presence of interruptions, even if well joined, between one impression and another along the line;

c) minimum spacing of the holes along the line;

d) irregular (unavoidable) spacing and rows not perfectly parallel.

If these characteristics are not encountered, there is good reason to doubt that the decoration considered was executed with a comb normally impressed in the clay. If this is the case, then one should look for the following characteristics:

a) unbroken lines;

b) impressions with varied spacings in the row;

c) rows with impression of different shapes and sizes;

d) equal distances between perfectly parallel rows;

---

(2) Good evidence of the validity of our proposal were also found in published materials: for the less densely dotted convex waves cf.: Caneva 1983: fig. 15,3; Arkell 1949: pl. 79; Addison 1949: pl. XCIV,C, 1, 4; Bailloud 1969: fig. 2,3,15; Courtin 1969: fig. 21, 5, fig. 25; Gabriel 1981: photo Ic,d; Roset 1983: fig. 5,2,4; Hugot 1963: pl. XVIII a,h, pl. XIX 4,2; Camps 1968: pl. XVI, 1-3, pl. XVIII, 5; Barich 1974: fig. 30, 1, fig. 33,3,7; Arkell 1949: pl. 102.1. For the series of fan impressions cf.: Bailloud 1969: fig. 2,A (last row); Courtin 1969: fig. 25; Gabriel 1981: photo Ic; Roset 1983: fig. 5,2,4; Barich 1974; fig. 33,6,8. For repeated rows of variously shaped waves cf.: Caneva 1983: fig. 15,2; Roset 1983: fig. 5,4; Barich 1974: fig. 30,9. For clear evidence of rocker comb impressions cf.: Caneva 1983: fig. 15, 5-7, fig. 13,4; Arkell 1949: pl. 72,5; Bailloud 1969: fig. 2,A.

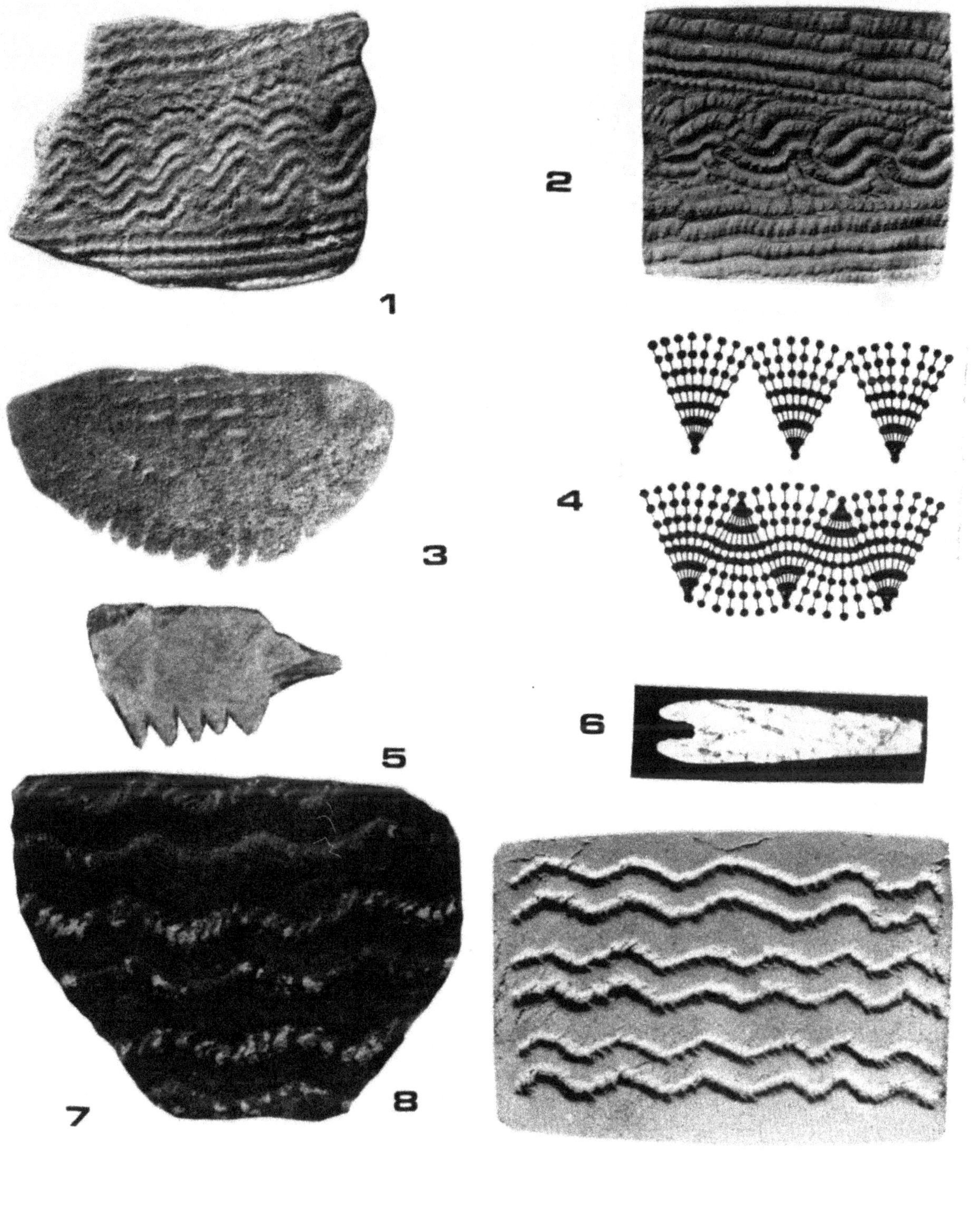

Fig. 9.1 - Dotted wavy line decoration. 1–5, potsherd, scheme, replication and combs connected with the "Mesolithic" type; 6–8, potsherd, comb and replication connected with the "Neolithic type (sherds and combs are from Sudan) (1:1).

e) impression placed regularly in an intermediate position with respect to the row above and below;

f) spaced rounded tooth impressions.

If some of the above characteristics are present (a, b, c, d), the pattern was probably made vertically, with a comb used pivot fashion (3) (fig. 9.2,4-5). If all are present, they were probably made with double pointed tools using compass-type movements, going over the holes already made for each new pair of rows (fig. 9.2,2). However, if d) is not present then the same technique has been used but without going over the holes a second time (fig. 9.2,3).

Another case which is frequently misinterpreted is that of a decoration formed by a series of opposed triangles or "v"s (fig. 9.3,3). Very often, in both Saharan and Sudanese contexts, it is considered as a series of normal impressions made with a triangular-section stick (or quoin), in the opposite direction in every row (Haaland 1981; Roset 1978). It must be noted, however, that this type of impression, which in some cases recurs, gives a very different effect, due to the inclined plane of impression and to the less acute angle (fig. 9.6,1).

Arkell already pointed out the symmetrical nature of the pattern and recognized in it the effect of a double-pointed instrument, making impressions in pairs of lines (1953: 71).

In the Sahara almost all what is called "poinçonnage" decoration appears to be produced by a double-pointed instrument. The technique is very simple, as we have seen already. It consists in imparting a rocker movement to a double-pointed stick or (in the case of triangles) to a short convex edged spatula with a central notch, along a horizontal line. This system accounts for a great number or reciprocal relations between points or groups of points, observable on a surface generally defined as dotted: pairs

---

(3) A good example is the fragment from Amekni (Camps 1968: pl. XIX, 1) with two and a half bands of rocker decoration on 12 rows (5 with simple oblique dots, 2 with double dots - on the middle of the pattern, where the dots are gone over twice, thus doubling the number of impressions - 4 with simple thin horizontal dots) corresponding to the 12 teeth of a comb, maybe _fileté_, applied vertically with a pivoting impression. The same pattern is visible in the other fragment illustrated in the same plate (No. 4) showing a rocker pattern on 5 rows (two bands of which are represented): 2 rows of rounded dots at the beginning and end of the pattern, 2 inner rows of bigger dots most of which have been gone over and so close that they are almost completely amalgamated. We cannot believe that 5 different combs were used to make this pattern. Moreover, this appears as a typical rocker pattern, with twice as many dots in the middle rows as there are in the outer rows, where the comb pivots and, therefore, makes only one dot for each two zig zags.

Fig. 9.2 – "Punctations". 1, comb impression (from Geili, Sudan); 2, "return" technique (from Uan Muhuggiag); 3, alternate pivoting stamp (from Geili, Sudan); 4–5, rocker stamp (from Saggai 1, Sudan) (1:1).

of parallel lines, equidistant impressions, matching positions of the latter (which are always intermediate with respect to those of the other row), regular repetition, on alternate rows, of different impressions corresponding to the different shapes of the points. By going over the holes with the point of a compass with a fixed opening, evidence of the extremely regular effect (never attainable by simple punching) was obtained.

3. POTTERY DECORATION AT UAN MUHUGGIAG: DESCRIPTION OF THE TECHNIQUES

Except for the plain rocker, the pottery of Uan Muhuggiag showed at first sight two contrasting features: great general homogeneity and different details on every fragment. It, therefore, appeared almost impossible to build up a typology other than that based on recognition of the use of various pointed sticks impressed in the paste at various angles, both features being of such a personal nature that they were unable to provide any significant grouping.

The attempt to recognize clusters or particular schemes from the uniform pattern of punctation, was, in most cases, very frustrating: all the impressions were, without any apparent reason, very regularly spaced. However, the extreme regularity itself appeared suspicious, suggesting that very special techniques and tools were used, to warrant both the equidistance and the "smocking" distribution of the impressions. The rocker technique could never produce this effect, even when the teeth of the comb are regularly spaced and shaped, because the vertical zigzag forms horizontal lines with differently spaced dots. As far as the normal impression of combs or single points is concerned, this technique can only accidentally produce the pattern, but it is not a guarantee. The solution was offered by rare cases of inaccuracy or errors, which suggested that double pronged instruments were used as compasses pivoting alternately on the points, and in every new pair of lines going over the dots in the last line (figs. 9.5,6). This way both the perfect spacing and the endless exact reproduction of the scheme were warranted.

Therefore, from a technical point of view, the two main groups which were first evident in the ceramics from Uan Muhuggiag (rocker and punctations) could be further divided as follows:

Plain rocker

We call rocker what in the french literature is called "impression pivotante". In this case, the tool used is not a toothed comb, but the plain, thin edge of a spatula or a bivalve shell, applied to the surface of unfired pots and moved zig-zag fashion, lifting first one end of the edge and then the other, alternatively. The result is a continuous line, like an incision, forming various kinds of zig-zag, according to their spacing, as well as to the form and inclination of the tool. The holder of the tool inclined it towards the direction of the movement, so that the impression is normally sharp and deep on one side (where the movement is arrested) and shaded on the other side, from where the tool was dragged.

This technique was first described by Arkell (1953: 72) for the Middle Nile Valley, where it is known in the Shaheinab culture (but completely unknown in the earlier contexts), and by Arambourg and Balout for the Sahara (1952).

Fig. 9.3 - Uan Muhuggiag, Wadi Teshuinat. Impressed pottery. 1,5, "fish scale"; 2, plain rocker; 3, alternate pivoting stamp ("vees"); 4, "return technique" (2:1. No 1 is 1:1).

At Uan Muhuggiag tools with particularly thin sharp edges seem to have been used, applied in two very different ways:

a) horizontal bands of curved, lightly impressed zig-zags. They are often employed for rim or lip decoration and they represent the most common application (fig. 9.3,2);

b) vertical bands of zig-zags in which the tool is so deeply impressed in the paste that it raises backwards a great mass of clay. This decoration occurs very often on certain kinds of everted rims (fig. 9.3,1).

"Fish Scale"

It is so defined by Arkell (1953: 73) as a particular form of plain rocker decoration, in which no space is left between the bands, the edge of the tool always pivoting on the bottom points of the already made band.

The effect is a lozenge network, which recalls a sort of net slinging which is still in use in Africa to carry vessels (Camps-Fabrer 1966: 466).

Rare in the Sudan, it is well known in the Sahara, where it was called "décor guilloché" (Arambourg, Balout 1952: 292).

The applications are the same as in the foregoing pattern, but they are usually found as body decoration rather than rim or lip decoration.

Alternate pivoting stamp

This technique is a kind of extreme simplification of the older rocker technique: a double pronged object is applied to the surface of the pot and moved horizontally, pivoting alternately on one point and then on the other.

The result is a pattern of pairs of dotted lines. In each pair the dots of one line are equidistant and intermediate from the dots of the other line, which only happens by chance between lines of different pairs.

The dots have various forms, according to the shape of the tool. Often the points of the same tool differ one from the other, providing a lively effect of horizontal alternation of different lines (fig. 9.4, 2) The motifs vary also depending on the way of pressing and moving the tool, sometimes with lateral dragging in the direction of the movement, sometimes with greater pressure on one point. The latter produces the impression of the lesser tapered part of the point, giving the dot a triangular form, as at the top of a zig-zag design. Very rarely in this context, but frequently in the middle Nile Valley and also in parts of the Sahara, for instance in the Niger (Roset 1983: fig. 7), both points give the impression of opposed triangles or vees (fig. 9.3,3).

An effect of horizontal pattern is typical of this technique, and the closer the dots are on the line, with respect to the space between the lines, the greater is the effect (fig. 9.2,3). In some cases the pattern looks as if it was made with a normal comb. However, the difference between these two techniques is normally recognizable, since a comb is supposed to have small, quadrangular, closely spaced teeth, if it is made by cutting the

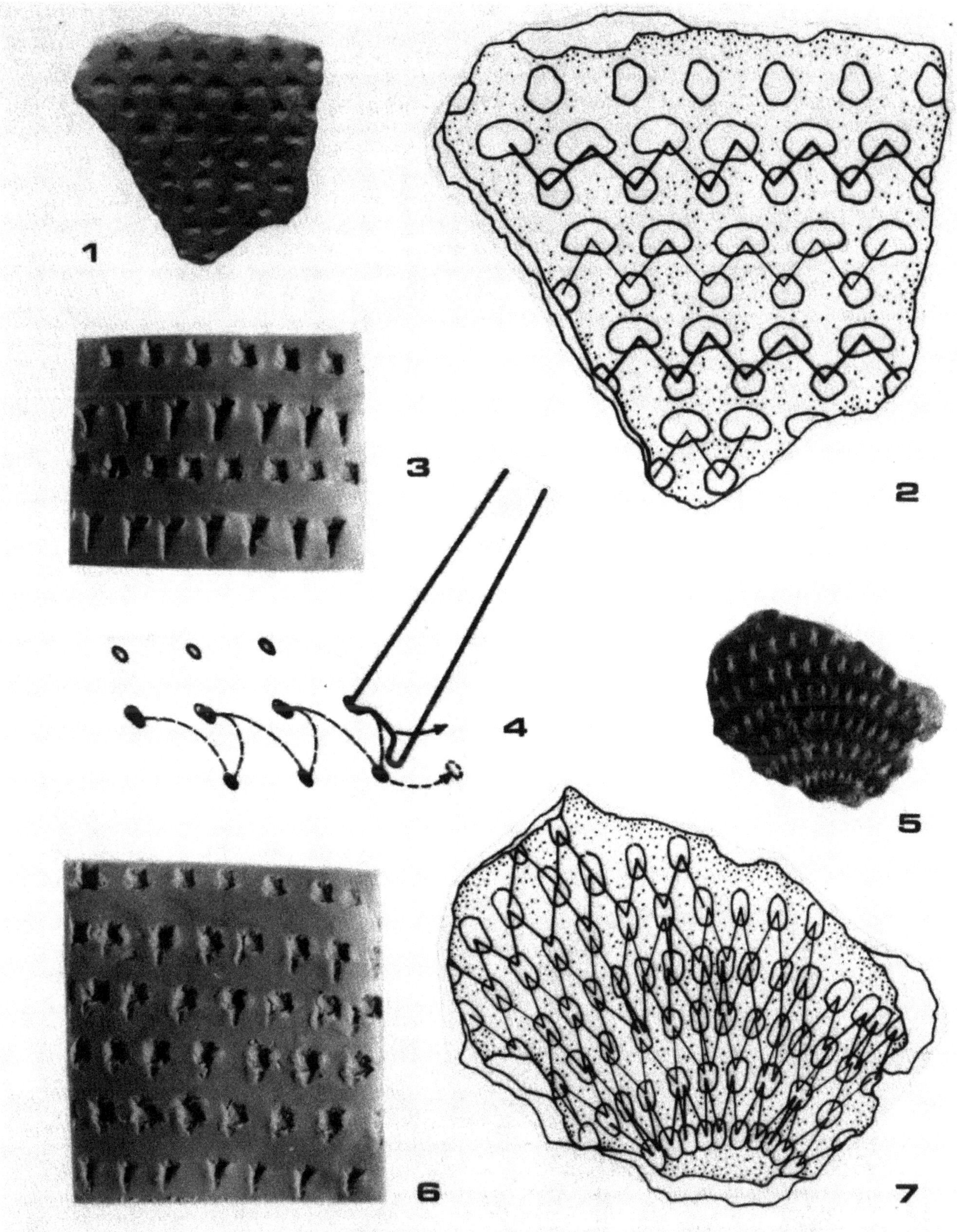

Fig. 9.4 - Uan Muhuggiag, Wadi Teshuinat. Impressed pottery. 1-3, alternate pivoting stamp (sherd, design, replication); 4-7, "return" technique (scheme, sherd, replication, design) (1:1. Schemes are 2:1).

frontal or lateral edge of a spatula or stick. This is the kind of comb which has been found so far either of bone, stone or potsherd. The distinction between the two techniques, which, as already said, also relies on other elements, has appeared in some cases to be of very important significance.

The so-called "basket" motif appears also to be obtained with this technique (or with the following), not with the impression of vegetal fibres - the latter technique ("roulage sur natte"), in fact, gives very different effects: soft and shading impressions, with regular schemes but very irregular pressure on the paste (cf. the pot published by Hugot 1963: pl. XX; cf. also Hurley 1979). In the concentric pairs of circles at the bottom of the vase, the progressive reduction of the inner circle is obtained making the compass pivot twice on the same dot whenever the two points of the tool fall along the same vertical line (fig. 9.4,7).

This technique, recognized in some cases by Camps-Fabrer (1966:467) and suspected by Savary (1967:863), was never taken into consideration for Saharan contexts. It was, on the contrary, well known in the Sudan, first described by Arkell in 1949, but not recognized thereafter.

"Return" Technique

This is, in a general sense, the same technique, but without spacing the pairs of lines, i.e., for each pair of lines the compass goes back to pivot on the dots of the last line impressed ("return" technique), ensuring a perfect equidistance between the dots on the whole dotted surface, providing a well balanced smocking effect (fig. 9.4,4-7). In many cases the pattern is so perfectly performed that there are no clear technological indications. However, as already said, the regularity itself was for us a good indication. In addition, in most cases little imperfections were noticed which led to the discovery of this technique: the dots are irregularly shaped, deformed by repeatedly going over them, often with different shaped points (figs. 9.5,2;6,2); where the compasses did not go over exactly the same point, additional dots are present, intermediate and extraneous to the pattern (fig. 9.5,1) more frequently, inside the hole, the clay is seen to be raised by the second impression, which partially filled in the previous one. If the rim, i.e. the beginning of the pattern is present, a first line of small regular dots can be noticed which is different from all the others, because it is the result of a single impression (fig. 9.6,1) (Barich, Mori 1970: fig. 12,2,6,11).

This technique, unknown in the middle Nile Valley, is the most widespread for decoration in the Saharan Neolithic. Similar, in principle, to the rocker "guilloché", it was nonetheless never taken into consideration to account for the impressed punctations, which were always considered to have been made with normal impression of combs or single points.

The last two techiques are apparently behind all the decorations known as "punctations". They are characterized by deep impressions, sometimes evident also on the back of the potsherds.

Despite the simple character of the foregoing technological picture, the comprehensive decoration group is still rich and complex. An internal subdivision, classifying the different patterns and tools, therefore, would be useful, but it is difficult to achieve.

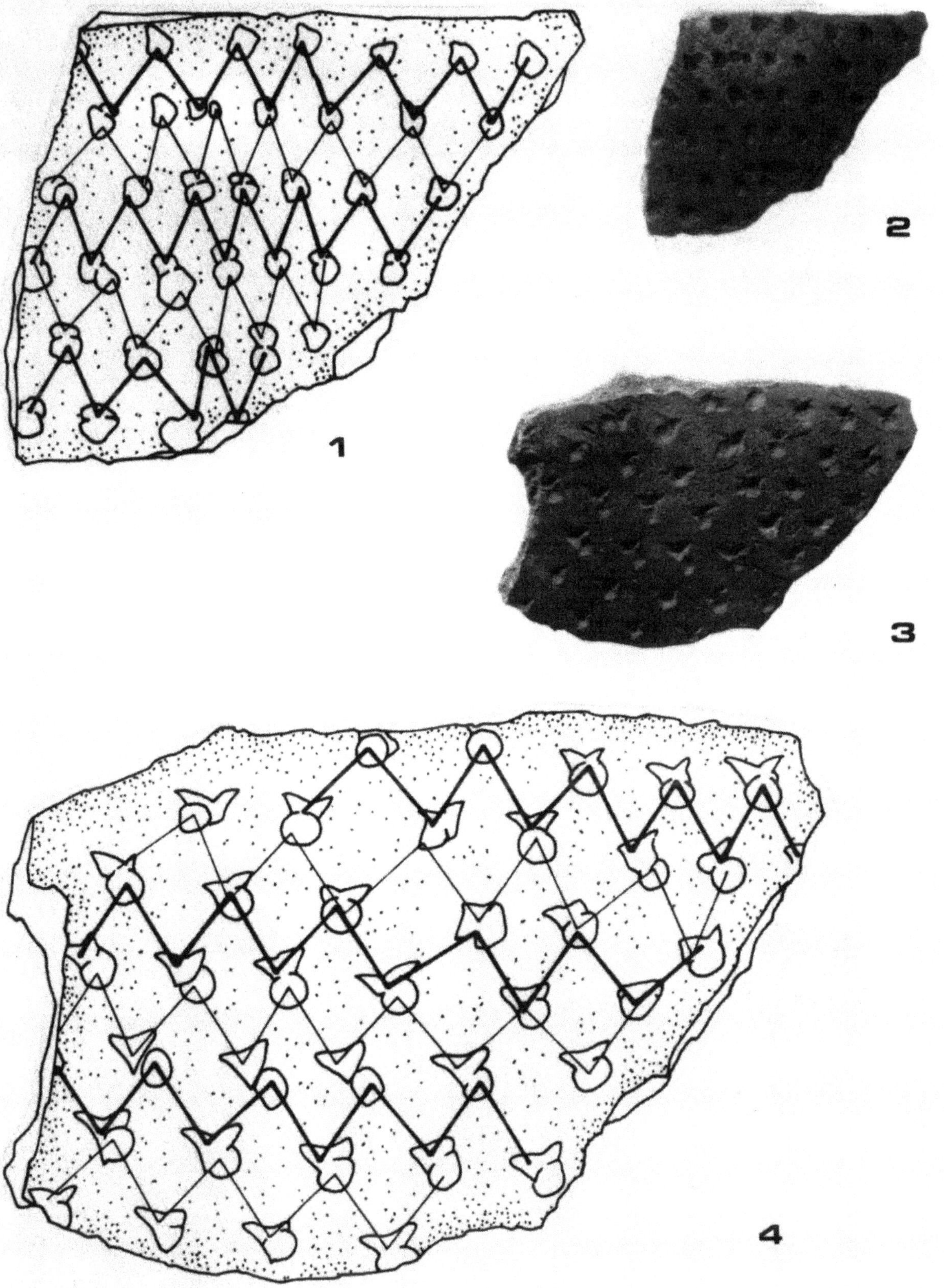

Fig. 9.5 - Uan Muhuggiag, Wadi Teshuinat. Impressed pottery. "Return" technique. Potsherds and schemes showing deformed and additional holes due to repeated stamp (1:1. Schemes are 2:1).

It must be said, in general, that within the four groups the motifs result from the use of different tools, whose nature and aspect will never be completely described and listed (suffice it to think of the endless possibilities provided by vegetal or animal objects). In addition, the inclination, the intensity and regularity of pressure applied to the tool and the spacing of the impressions could produce such different effects in the same kinds of decoration that the relationship between them is completely hidden, as already noticed for instance by Camps-Fabrer (1966a:147). These features are, therefore, the result of such accidental movements that they appear rather as shades, often mixed on the same vase, fading one into the other and, in any case, characterize almost every fragment. Hence, they will not be considered in this chapter, the aim of which is to formulate a new perspective of ceramic classification. For a detailed analysis of patterns the reader is referred to the complete description of the Uan Muhuggiag materials by B.E. Barich in chapter 7, this Volume.

4. THE TECHNIQUES OF DECORATION AT UAN MUHUGGIAG: FREQUENCY DISTRIBUTION BY LEVELS

The table (Table 9.1) showing the frequency of the techniques of decoration contains the four groups, a separate heading of "varia" with rare cases of rocker combs, and a category of unclassifiable sherds, which includes fragments decorated with alternate punctations in which the "return" technique was not clearly understandable (4). Therefore, when clustering the whole group of punctations, in contrast with the group of plain rocker, unclassifiable sherds are included in the first group.

It is to be noted that, as far as the plain rocker is concerned, the presence of this decoration on the rim has been considered as an additional occurence.

A detailed analysis of the occurrences is difficult and not fully reliable because of the scarcity of the materials, except in one case (less than 50 units per level). Any conclusion has, thus, a merely indicative value.

Sector B provided more material and a greater stratigraphic extension. Both Sectors, however, show the same trend, especially in their lower levels. A gradual increase of the rate of the rocker technique can be seen throughout the stratigraphic development (levels 2d, 2c, 2b, 2a in B;

---

(4) In most cases, from the extreme regularity of the pattern, we must infer that the "return" technique was applied. The simple alternately pivoting tool, in fact, gives less regular patterns.

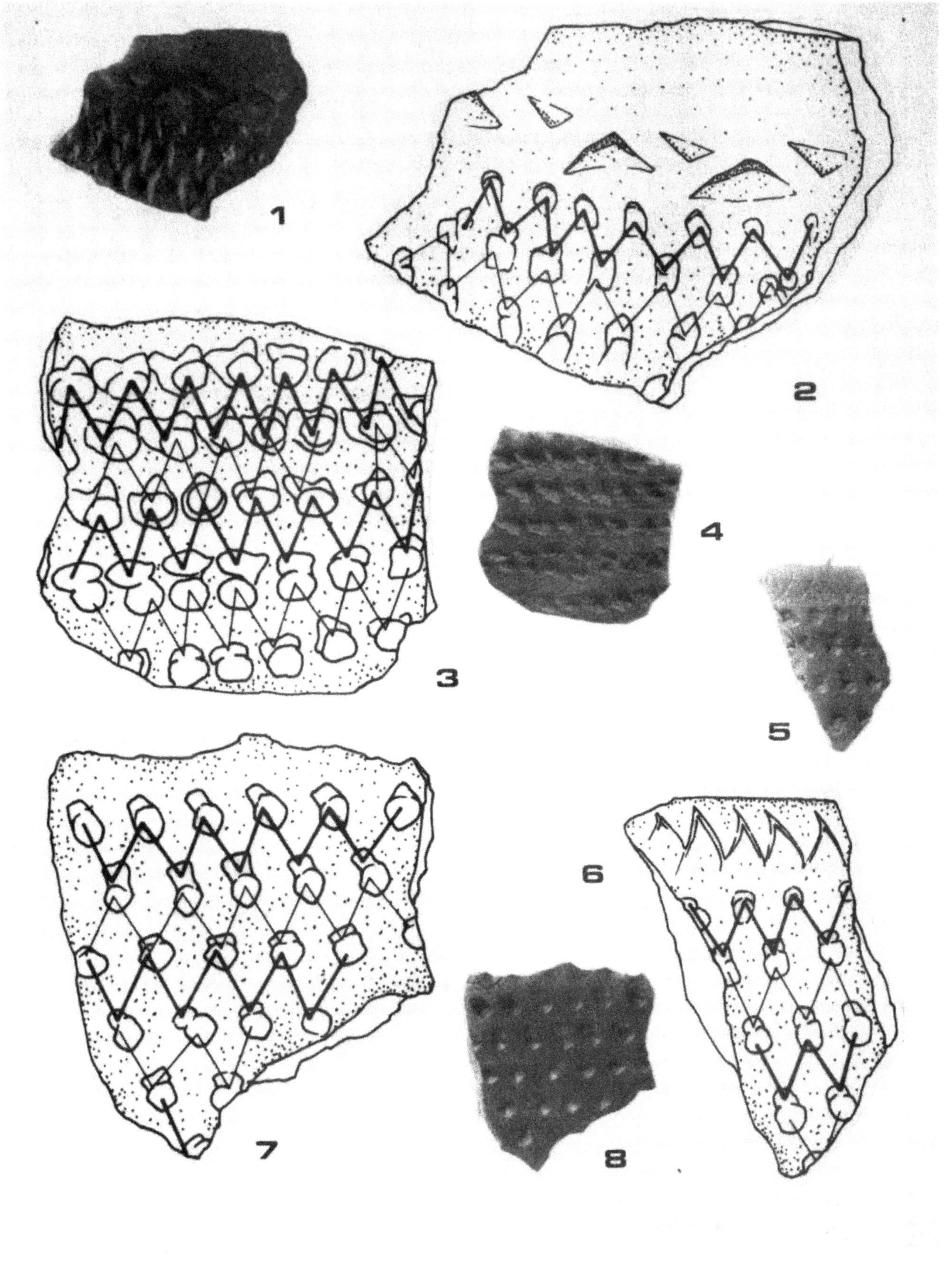

Fig. 9.6 - Uan Muhuggiag, Wadi Teshuinat. Impressed pottery. "Return" technique. Potsherds and schemes showing deformed and additional holes due to repeated stamp (1:1. Schemes are 2:1).

levels 2c, 2b, 2a in A) maintaining, in both Sectors, almost the same ratio between the two main groups.

In the highest levels, however, there is an inconsistency between level 2 of Sector A and the corresponding 2 in Sector B. The latter is probably to be seen as part of the underlying level (i.e. 2a), since the inversion of the trend, with an exceptionally high rate of alternate punctations, characterizes level 2 in sector A and level 1a in sector B. This is probably just a matter of labelling, not the result of a different situation. Actually, the two Sectors compare very well, which confirms the reliability of the analysis.

Generally speaking, it can be said that the "return" technique is the most common in the site throughout the chronological development. It also prevails on the simple alternate stamp. The increase of the plain rocker technique accordingly appears to follow a phase of "return" technique ("fish scale"). In other words, the diffusion of this technique could be linked to its possibility of providing kinds of smocking patterns like the "return" technique. The same situation seems to be attested at Wadi Athal (Barich, Mori 1970); at Ti-n-Torha a similar situation is evidenced in the lack of rocker technique in the lower two levels, while this decoration accounts for 13-18% in the latest contexts (Barich 1974: tab. X).

COMPARISONS AND CONCLUSIONS

As far as we can gather from lists and descriptions relying on a different typological system, the foregoing techniques of decoration are typical of the Neolithic Sahara, where almost in every region they chronologically follow contexts characterized by the dotted wavy line ware (<u>sensu</u> Arkell; cfr. note 5) and by the rocker comb decoration.

As far as the Saharan occurrences are concerned, however, we refer to the comparisons made by B.E. Barich on the basis of the complete contexts, limiting the analysis here to the most eastern, non Saharan, evidence.

As far as the Nile Valley is concerned, the alternate impression of two-toothed instruments occurs already in Mesolithic contexts, associated with wavy line, dotted wavy line and rocker comb decoration (Caneva 1983). Its rate appears to increase during the early Neolithic (Shaheinab culture) becoming then the exclusive impressed decoration during the late Neolithic (cf. Kadada: Geus, Reinold 1979).

The shift from the rocker comb to the pseudo-rocker double point is well evidenced in the pottery from central Sudan. A good example is provided by the evolution of the dotted wavy line decoration, which was first obtained with a comb used rockerwise (Early Khartoum and related assemblages) and later with double toothed tools alternatively pivoting (Shaheinab and related assemblages). The first one is always applied to unburnished pottery, associated with Mesolithic tool kit and is characterized by round waves of multiple line rows (= rocker comb); the second is always applied to burnished pottery, associated with Neolithic assemblages and characterized by short and sharp waves of pairs of dotted lines (= alternate pivoting

Table 9.1: Uan Muhuggiag, Wade Teshuinat. Frequency distribution of pottery decoration techniques.

Sector A

levels	2			2a			2b			2c					
techniques	N	%	%	N	%	%	N	%	%	N	%	%	N	%	%
Varia	1	3.4								1	5.5				
Alternate pivoting stamp	1	3.4		5	17.8		13	28.9		4	22.2				
Return technique	19	65.6	79	3	10.7	50	23	51.1	87	6	33.4	72			
Unclassifiable	3	10.4		6	21.4		3	6.7		3	16.7				
Plain rocker	5	17.2		5	17.8		6	13.3		4	22.2				
Fish scale				9	32.1										
Total	29			28			45			18					

Sector B

levels	1a			2			2a			2b			2c			2d		
techniques	N	%	%	N	%	%	N	%	%	N	%	%	N	%	%	N	%	%
Varia																		
Alternate pivoting stamp	4	15.1		2	7		4	10.8		19	18.3		2	10				
Return techinuqe	13	48.1	70	7	24	34	16	43.2	68	47	45.2	82	15	75		7	50	50
Unclassifiable	2	7		1	3		5	13.5		19	18.3					3	21.4	
Plain rocker	8	30		19	66		7	18.9		14	13.4		3	15		4	28.6	
Fish scale							5	13.5		5	4.8							
Total	27			29			37			104			20			14		

stamp) (fig. 9.1) (5).

The same applied (as far as the photographs show) in the Ennedi, where a kind of dotted wavy line with small, sharp waves, associated with rough triangular punctations, is found to overlap the rocker decoration and the classical dotted wavy line (levels B and A at Soro Kezenanga; Bailloud 1969: 38, fig. 3, A) (6).

In the middle Nile Valley the plain rocker follows the same trend: almost lacking in the Early Khartoum contexts (only one fragment is published in Early Khartoum, Arkell 1949: fig. 79, 3; not one was found at Saggai), it becomes a more and more common technique of decoration during the Neolithic phases. Often used as rim decoration, it is confined to the rim of otherwise undecorated vases in the latest cultures (Arkell 1953: fig. 31, 40-42).

However, other elements differ considerably in the two areas; not only paste composition and vessel shapes (as far as we can infer from such small fragments, rims are more varied in the Sahara) but also the decorative motifs, which are quite diversified and usually well spaced in the Sudan. Here, in addition, the designs are less deeply impressed into the paste, never visible on the inner surface. But the most impressive difference is the role that these techniques hold in the two areas: exclusive in the Sahara, medium in the Sudan, where they are always associated with other kinds of surface treatment, such as incision, brushing, red slipping, combing-and-washing (ripple ware), which give the ceramic production a well varied character in comparison with the stereotyped aspect of the corresponding Saharan production.

The "return" technique is almost unknown in the Nile Valley (except for some cases of "fish scale" motifs). Therefore, the particular effect of oblique alignments and lozenge spaces, which is so characteristic of the Sahara, is there completely unusual.

From this comparison there is, therefore, a general impression of parallel technological development through similar stages, which were already partially recognized by Arkell for the Khartoum province (1949) and by Bailloud for the Ennedi (1969). The sequence of these stages now appears to extend all over the Sahara: from wavy line to dotted wavy line, to rocker comb impressions, to alternate dots together with plain rocker.

These stages not only are more or less contemporary in both areas, but also are associated with broadly similar cultural contexts, i.e.

---

(5) The misconception concerning the dotted wavy line ware was already known by Arkell, who tried then to emphasize (apparently too late) the difference between the two kinds of this decoration recognized in the Sudanese prehistory (1972).

(6) These are the levels to be compared with Sudanese Neolithic cultures instead of the lower ones, where both archaeological materials and C14 datations perfectly correlate with the Early Khartoum cultures.

"Mesolithic" economy and tool kit, with documented hunting/fishing and gathering activities as far as the more ancient series (wavy line, dotted w.l. and rocker comb) is concerned, "Neolithic", pastoral economy, with reduced lithic tool-kit (both in number and variety, cf. Barich 1974) associated with the pottery having alternate dots.

The dates of the first group range between 6000 and 4500 B.C., both in the Sahara and in the Nile Basin (cf. Shabona, Sorourab II, Saggai), while the more recent group is dated 4000 in the Acacus, 3000-3500 in the Ennedi, 2500-2900 B.C. in the Nile Valley (Kadada). The Saharan dates appear to be somewhat earlier than the Sudanese, but this is due to the lack in the Sahara of something really comparable to the Shaheinab culture which is dated 4000-3500 B.C. In any case, the series is very incomplete for such an extensive area, and no reliable hypothesis can be developed on the origin of one or the other cultural complex (7). Moreover, a valid comparison should take into consideration the whole complex of elements forming a cultural context. The decoration of pottery is merely part of one of these elements. It has been considered an important one, often the only one available for study of the Saharan prehistory (Camps 1974: 235), but its content of information is only indicative of the relationship between the two areas considered.

This concept, always at the basis of Barich's work in the Acacus, was not widespread among Saharan prehistorians, as Maitre has many times stressed. He pointed out the extremely varied internal faceting of the Saharan Neolithic (he says "Neolithics") and the poor possibility of comparison between the two areas (1972). However, his perspective was biased: behind his obstinate defence of the Saharan autonomy, one can guess that there may be a doubt as to the autonomous cultural development of one or the other areas, and that the question is which one influenced the other. Moreover, he is involved in an incorrect analysis of the Sudanese contexts, mistaking Shaheinab for Early Khartoum. Hence his evidence, which is based on the chronological priority of Delebo (5250 B.C.), becomes a refutation once Delebo is compared with its right homologue, that is Saggai (5500 B.C.).

Actually, the close parallelism and the contemporaneity of the stages, as well as the local diversities, do not suggest any particular priority. On the contrary, they emphasize the possibility of long, uninterrupted reciprocal relationships between the two areas.

These relationships, as seen through the pottery, seem to have always existed, lasting until very recent times (cf. the identity between the pottery of the Nubian C group and that of the Iron Age in the Borkou; Courtin 1969). We assume, therefore, that such relationships must be considered in a normal network of contacts among fairly mobile groups inhabiting the same environment and practising the same activities, rather than as episodic events requiring a specific explanation.

---

(7) Camps is rather inclined towards a Saharan origin of the African "neolithisation" (1974: 224), but he bases this assumption on the comparison between the farming communities of the Fayum, which, of course, have a very recent date (4500 B.C.), and the Mesolithic sites of the Acacus, about 2000 years older.

The different local features, especially the different associations, are easily explained as autonomous developments and utilizations of broadly common schemes. The data to quantify its significance in the internal development of a single context are still lacking. In addition, the actual contexts are very few, since the findings are either sporadic or old in most cases. This situation, in fact, could account for the lack of actual syntheses of both the Sahara and the Sudan.

REFERENCES

ADDISON F. 1949 - Jebel Moya, Oxford University Press: London.

ARAMBOURG C., BALOUT L. 1952 - L'ancien lac de Tihodaine et ses gisements préhistoriques, Actes du II ème Congr. Panafr. de Préhist., Alger: 281-292.

ARKELL A.J. 1949 - Early Khartoum, Oxford University Press: Oxford.

ARKELL A.J. 1972 - Dotted Wavy-line Pottery in African Prehistory, Antiquity, 46/183: 221-222.

BAILLOUD G. 1969 - L'évolution des styles céramiques en Ennedi (République du Tchad), Actes du Premier Colloque Intern. d'Archéol. Africaine, Fort Lamy: 31-45.

BARICH B.E. 1974 - La serie stratigrafica dell'Uadi Ti-n-Torha (Acacus, Libia). Per una interpretazione delle facies a ceramica saharo sudanesi, Origini, VIII: 7-184.

BARICH B.E., MORI F. 1970 - Missione Paletnologica Italiana nel Sahara Libico. Risultati della Campagna 1969, Origini, IV: 79-142.

CAMPS G. 1968 - Amekni: néolithique ancien du Hoggar, Mémoire du C.R.A.P.E., 10 (A.M.G.: Paris), 230 p.

CAMPS G. 1974 - Les civilisations préhistoriques de l'Afrique du Nord et du Sahara, Doin: Paris.

CAMPS-FABRER H. 1966 - Matiére et Art mobilier dans la Préhistoire nord africaine et saharienne, Mémoire du C.R.A.P.E., 5 (A.M.G.: Paris), 574 p.

CAMPS-FABRER H. 1966a - Sur quelques techniques décoratives de la céramique impressionnée saharienne, Congrés Préhistorique de France, Compte Rendu de la XVIIIème Session, Ajaccio: 143-154.

CANEVA I. 1970 - I crescenti litici del Fayum, Origini, IV: 161-204.

CANEVA I. ed. 1983 - Pottery Using Gatherers and Hunters at Saggai (Sudan): Pre-conditions for food-production, Origini, XII: 7-278.

CANEVA I. 1984 - Typological Notes: the Sudanese Case, Paper presented at the II Intern. Symp. on the Late Prehistory of the Nile Basin and the Sahara, Poznan.

COURTIN J. 1969 - Le néolitique du Borkou, Nord-Tchad, Actes du Premier Colloque Intern. d'Archéol. Africaine, 1966: 147-159.

GABRIEL B. 1981 - Die Ostliche Zentral Sahara in Holozän - Klima, Landschaft und Kulturen (mit besonderer Berücksichtigung der Neolitischen Keramik), in Préhistoire Africaine, Mélanges offertes au Doyen Lionel Balout, reunis par Roubet C., Hugot H.J. et Souville G., A.D.P.F.: Paris, pp. 195-211.

GEUS F., REINOLD J. 1979 - Fouille de sauvetage à el Kadada (Soudan): la campagne d'avril 1976, C.R.I.P.E.L., 5: 7-159.

HAALAND R. 1981 - Migratory Herdsmen and Cultivating Women, Mim: Bergen.

HAYS T.R. 1974 - An Element of Nilotic Diffusion, South Afric. Archaeol. Bull., 29: 27-32.

HUGOT H.-J. 1963 - Recherches Préhistoriques dans l'Ahaggar nord-occidental, Mémoire du C.R.A.P.E., 1 (A.M.G.: Paris), 206 p.

HURLEY W.M. 1979 - Prehistoric cordage. Identification of impressions on pottery, Manuals on Archaeology, 3, Aldine: Washington.

MAITRE J.P. 1972 - Notes sur deux conceptions traditionnelles du Néolithique saharien, Libyca, XX: 125-136.

MOHAMMED ALI A. 1982 - The Neolithic Period in the Sudan c.a. 6000-25000 B.C., Monographs in African Archaeology No. 6: Cambridge.

ROSET J.P. 1978 - Poteries néolithiques du Ténéré. I, la région de l'Adrar Chiriet, Cahiers O.R.S.T.O.M., sér. Sc. Hum., XV, 4: 379-406.

ROSET J.P. 1983 - Nouvelles données sur le probléme de la néolithisation du Sahara méridional: Air et Ténéré au Niger, Cahiers O.R.S.T.O.M., sér. géol., XIII, 2: 119-142.

SAVARY J.P. 1967 - Eléments remarquables du Néolithique saharien, Bull. de la Soc. Préh. Franç., LXIV: 859-864.

CHAPTER 10

UAN MUHUGGIAG AND THE PASTORALISM IN THE CENTRAL SAHARA

By Barbara E. BARICH

The interpretation of the Uan Muhuggiag site can be linked with the results from sites in the Ti-n-Torha area. These two sample territories, put together, help explain cultural and the economic processes in the Saharan region between 10th and 6th millennia B.P.

The Uan Muhuggiag communities parallel those in Levels V-IV of the North Shelter in the Wadi Ti-n-Torha (figs. 10.1,2), which around the 8th-7th millennia B.P., showed a cultural impoverishment of which the general contraction of the lithic assemblage (which included the abandonment of the blade technique and its replacement with flakes) was an example. After the period in which, at the North Shelter, satisfactory climatic conditions still existed (see the pollen spectra, Chapter 13), climatic change was extreme enough to cause the abandonment of the site after the roof had fallen in (Level III).

The entire lower horizon of Uan Muhuggiag belongs to the same chronological period as the Ti-n-Torha North Shelter starting from Level IV, and its beginning (Levels 2d, 2c) might even be contemporary with the end of Level IV. This latter site was becoming inhospitable (1) due, it has been suggested, to an oscillation towards aridity. The two sites then developed in parallel throughout the 6th millennium B.P. (fig. 10.2 b; Table 10.1).

Here Uan Muhuggiag and Torha North, which we already referred to in Chapter 7, will be compared in a more analytical way. In the Ti-n-Torha pottery, from the terminal level of the East Site (CI) onwards, a diminution of the use of comb decoration is noticed. In the Levels V-IV of the North Shelter (the latest aspect of the facies itself, Barich 1974: 150), one may observe decorative motifs very similar to those from Uan Muhuggiag, Levels 2d-2b. It would seem reasonable that, if the comb was being regularly used with a pivoting movement, it is quite likely that its use was prolonged by using simplified or more easily available tools (2). This is a normal process, which would have speeded up (and I would dare say, undramatize), the production of vessels, that, from that time onwards, were only decorated

1) The phenomenon may be dated from 6500-6200 ca B.P.. This contrasts with the observations of E. Schulz as regards the continuity of the pollen spectra.

2) A comb requires a certain degree of manufacturing care whether it is made of clay, bone or it is obtained from a pebble of soft stone. Double-pointed implements are easier and faster to manufacture.

a

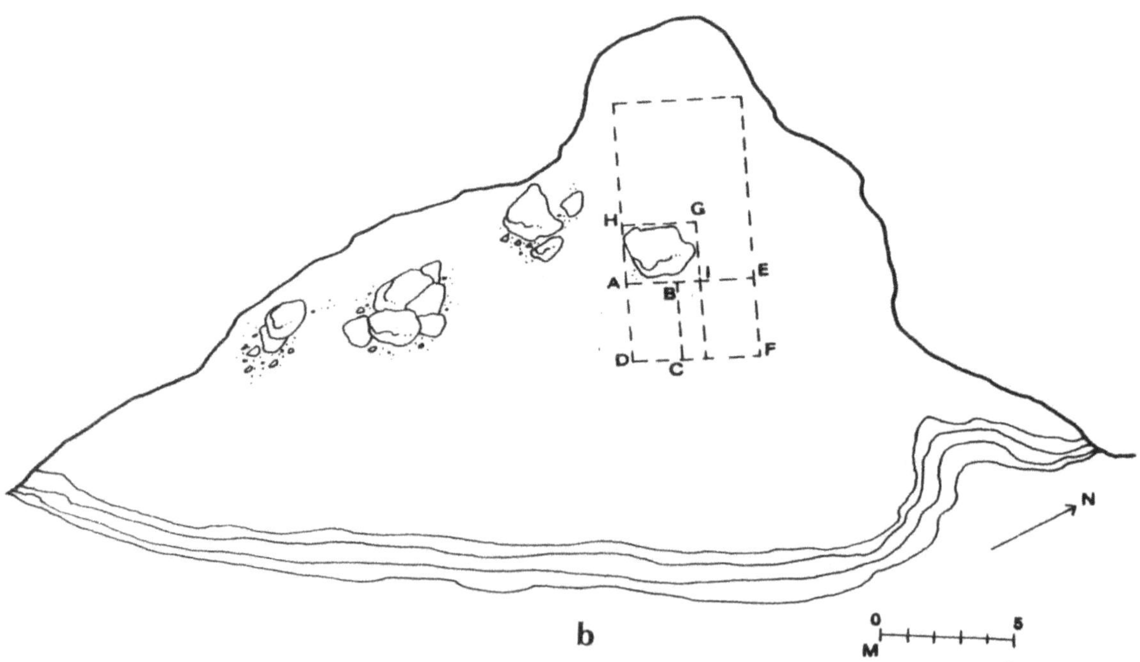

b

Fig. 10.1 - Wadi Ti-n-Torha. a: outer view of Ti-n-Torha North; b: layout of the deposit with the delimitated excavation area (after Barich 1974: fig. 16).

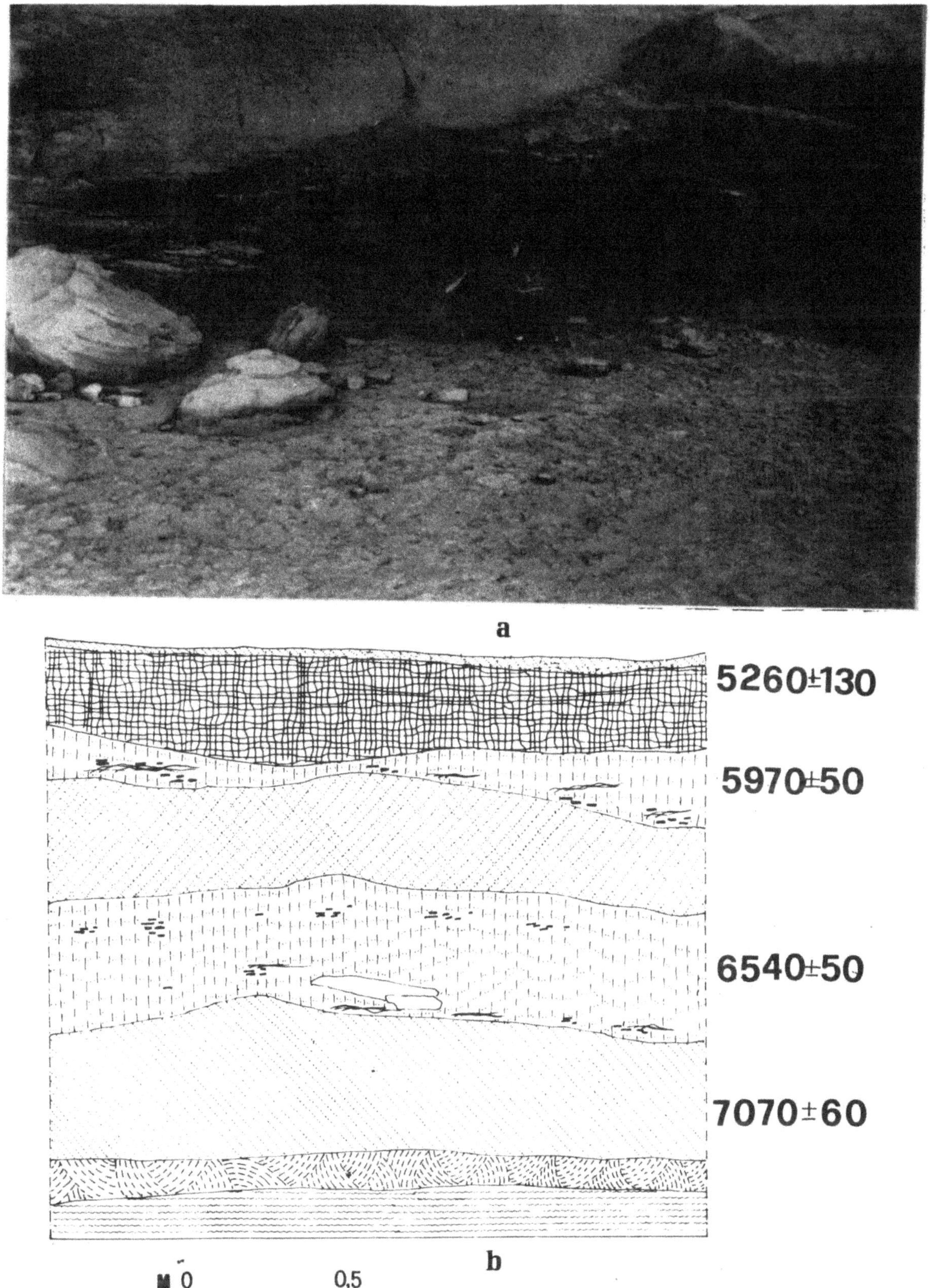

Fig. 10.2 - Wadi Ti-n-Torha. a: the North Shelter before the excavations; b: stratigraphy with radiocarbon datings of the levels.

by standard patterns. Maitre (1971) was not wrong when he pointed out that more recent periods witnessed a growing use of the punch and quoin instead of the comb, but he failed to stress that this was due to choice.

Table 10.1 - Chronological correlations between Torha North and Uan Muhuggiag.

TORHA NORTH	UAN MUHUGGIAG A	UAN MUHUGGIAG B
	Level 1a: 3770±200	
		LOWER HORIZON ↓
Level I 5260±130 5360±120 5600± 50	⟵⟶	Level 2b: 5350±200
Level II 5970± 50 ↑- - - - - -	Level 2: 6035±100	
Level III		
Level IV 6540± 50		
Level V 7070± 60 ↑- - - - - - - -	(7438±120 P1)	

What is the significance of this change? During the time of its earlier use ceramics still looked back to vessels supported by small nets which were copied in the prevailing "woven mat" patterns. The technological invention was emphasized by an extremely elaborate and imaginative decorative syntax. At that time, ceramics was used by semi-residential groups, living on lake shores, who preferred well-made products. The succeeding pattern (and my definition of "ceramica puntinata in serie" still applies since it stressed the peculiarity of these products), was employed by groups of herders, people accustomed to rapid movements who were not prompted by tradition to ascribe importance to the ceramic products. These groups kept on using ceramics, but simplified it, showing an increasing tendency towards punch and quoin techniques and repetitive patterns (fig. 10.3). Their stone tools were also quite limited, being predominantly nucleiform end-scrapers and side-scrapers on flakes, but having small arrow-heads and using pressure flaking already observed from CII onwards (fig. 10.4). They lacked the backed tools that characterized the wavy line complexes.

The main objection to the suggested continuity between the Ti-n-Torha and Uan Muhuggiag facies arises from the fauna: at Uan Muhuggiag, domestic livestock is 92.4% of the total and both cattle and caprids are present from the base levels.

Due emphasis was laid (Barich 1984a) on the preparatory rôle to domestication played by groups using wavy and dotted wavy line decorated

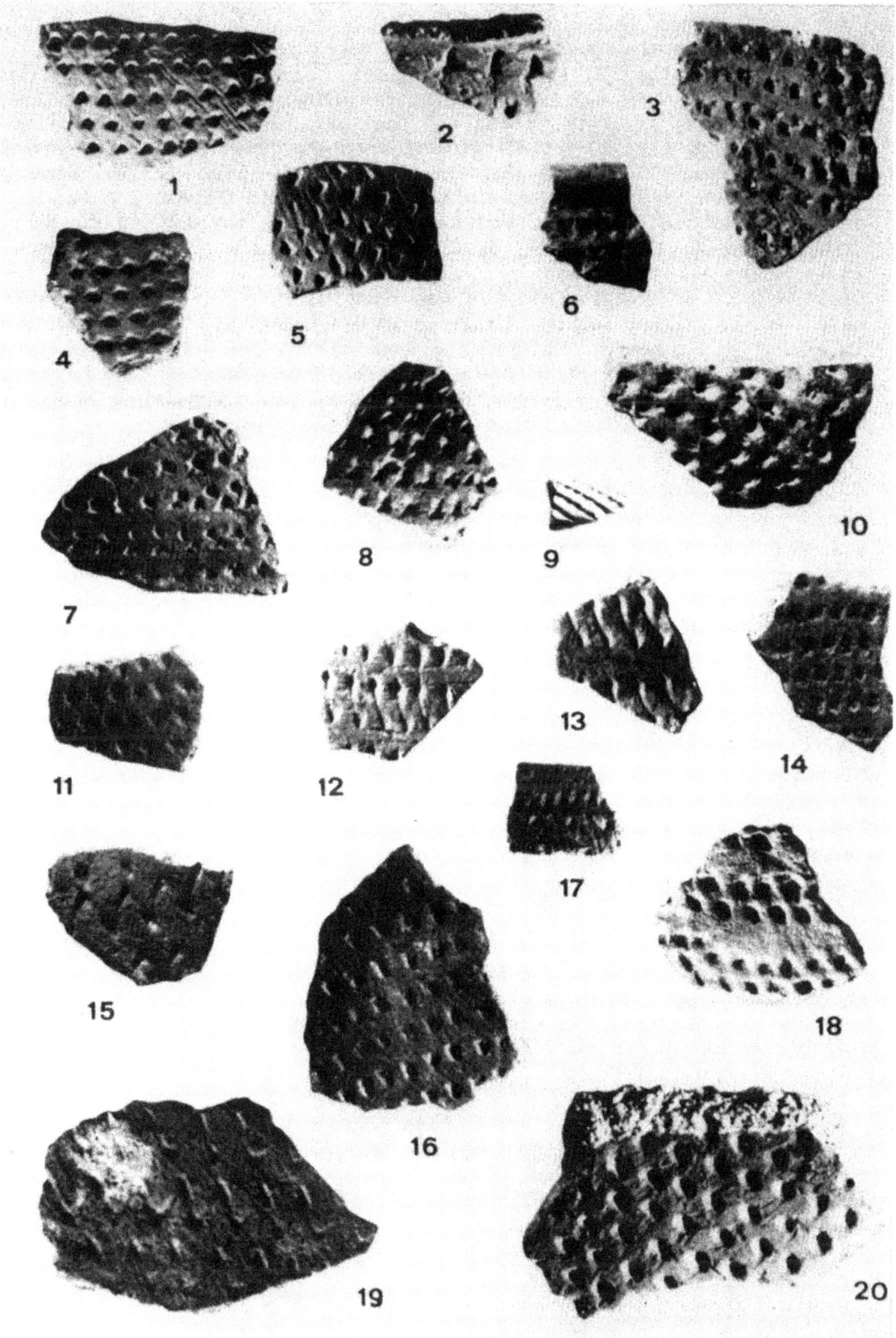

Fig. 10.3 - Wadi Ti-n-Torha. Pottery with impressed decorations from Ti-n-Torha North, Level II (1:1). (After Barich 1974: fig. 51).

ceramics. The semi-permanent settlements developed through their special, reduced nomadism, brought about by an intensive exploitation of the habitat. It included: a) hunting-fishing activities witnessed by both microlithic, backing-retouched types of tools, and bone harpoons and fish-hooks (Camps 1974a: 240, fig. 72); b) exploitation of wild graminaceous plants at Amekni and Méniet (Camps 1968: 186-188; Hugot 1963: 154-158); barley (six-row barley) and wheat (emmer wheat) in the area of Nabta (Stemler, Falk 1980: 393-99; Wendorf 1984:198); c) utilization of "large bovids" in the Nabta area and, at an earlier time, at Bir Kiseiba (Gautier 1980, 1984; Banks 1984). The acknowledgement by Gautier (Gautier, Van Neer 1977-82), now disavowed, of an archaic bovid in the fauna of Ti-n-Torha East, appeared sufficient to account for the situation at Uan Muhuggiag. Even though the intermediate stages in the passage from one to the other economy are difficult to reconstruct in detail, there is no question that, almost everywhere in the central Sahara, economic development appears to have the same diachronic evolution as that observed in the Tadrart Acacus.

Other Saharan Sequences

The Tenerian, as defined by A. Smith (1976), represents one of the most marked phenomena of central-Saharan pastoralism, preceded by an early ceramic Epipaleolithic facies (Kiffian), the appearance of which is considered prior to 7310±120 _ca_ B.P. It was found _in situ_ in the sandy basin of the Agorass-n-Tast (Adrar Bous, Niger) and its datings range between 5460±500 and 4440±120 B.P.

J.P. Roset reported from the region located directly to the East of the Aïr (Niger) the recovery of a group of Middle Neolithic ceramics, for which he suggested the same date as the Tenerian (Roset 1978). The region had a Dotted Wavy Line horizon that is well-documented by the stratigraphies of Tagalagal and Temet, dating from 9330/9370±130 and 9550±100 B.P. respectively (Roset 1983). Special significance was ascribed at Temet to stone prototypes of pottery, and to a lithic industry, with Ounan points, which corresponds to the Epipaleolithic horizon of the Wadi Greboun (Clark 1976).

The pastoral phase in the Tibesti has its climax between 5700 and 5300 _ca_ B.P., during an optimum climatic period (Gabriel 1981: 210) (3). Within the same area, a fauna dominated by wild types and a "large bovid" with a pure Dotted Wavy Line complex, was reported from Gabron, Enneri Dirennao (4).

In the Ennedi (Chad), L. Bailloud defined a sequence of two major occupation phases, at Délébo and Soro Kezenanga (Bailloud 1969). The transformation that came about in their ceramic decoration, is very similar to that observed between the Ti-n-Torha facies and the pastoral phase in the Acacus. The wavy line horizon of Délébo (_Zone_ IV) and Soro Kezenanga (Levels F-C) has classical Early Khartoum types, and at both sites, this is

---

3) It should be pointed out that Gabriel (1981) identifies an arid period between 5400 and 4100 _ca_ B.P. that corresponds to the evidence at Uan Muhuggiag.

4) Cf. Gabriel in Gautier (1984: 70). The remains under consideration are being restudied at the University of Tubingen.

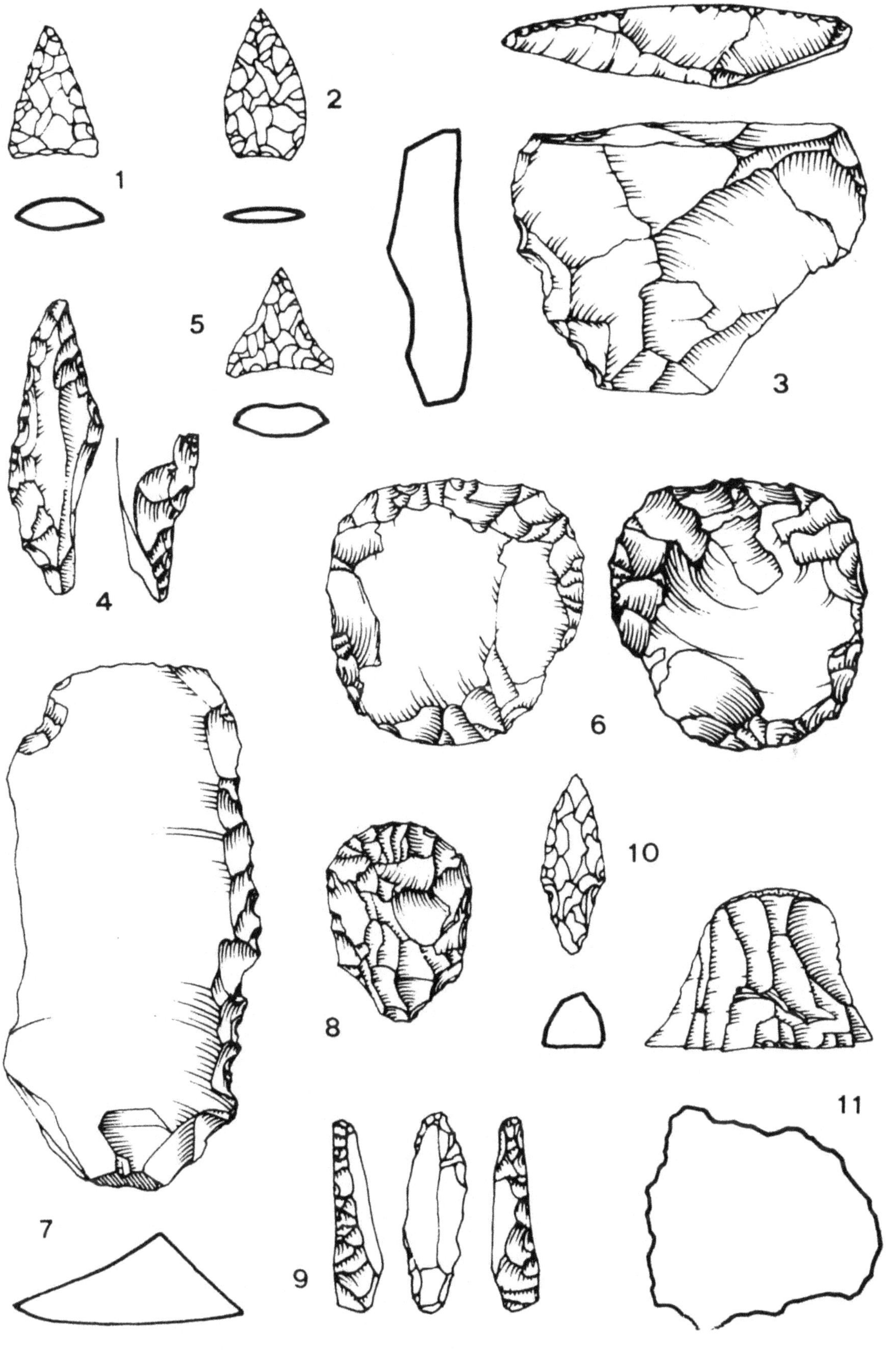

Fig. 10.4 - Wadi Ti-n-Torha. Lithic industry from Ti-n-Torha North. 1-4: Level I; 5-11: Level II (1:1). (After Barich 1974: fig. 47).

followed by a Dotted Wavy Line pottery of Shaheinab type: Délébo Zones III (7180±300 B.P.) and II (6900±300 B.P.); Soro Kezenanga Level C. In this latter site the dotted wavy line pottery is associated with rocker-stamped motifs and is subsequently followed (Levels B-A) by short wave, pointillée patterns, together with standard triangular impressions obtained by means of a punch. In these later levels, the fauna includes, in addition to antelope, gazelle and 'jackal', a "large bovid" (Bailloud 1969: 36-38).

This synthesis of the situation in the central Sahara shows that fully cattle-based economies are separated from the purely extractive, hunting-gathering economies by an interval of two millennia. From this originated the hypothesis, subsequently corrected, that the pastoralism might have been an imported phenomenon. I referred in Chapter 5 to the possible existence of a level, undifferentiated from a socio-economic point of view, in which proto-domestic experiments took place. This hypothesis was recently supported by the recovery of cattle at 9000 ca B.P. in the Bir Kiseiba area, and of barley and wheat (from 8100 to 7100 ca B.P.) at Nabta (5). According to Gautier, the presence of cattle at a stage of incipient domestication (6) in the Nabta-Bir Kiseiba area was the result of Saharan hunters reaching the Nile, where wild cattle appeared to have a special status since Qadan times (Gautier 1984:71). This model provides an explanation even for the situation in Saharan mountains. The hunters of the Early Holocene showed different economic developments through local adaptations. While a few groups settled within definite territories (such as at Ti-n-Torha East, Nabta E-75-6 upper level, Amekni lower level) other groups were more dispersed, in the same periods of favourable climate, and came in contact with groups that had already domestic herds.

No single direction of these movements suits the evidence. The Ti-n-Torha facies allowed a few lithic correlations to be established; the straight backed and pointed bladelets with modified base at Two Caves are related to the El Adam type Early Neolithic, while the elongated scalène triangles and small short crescents at Torha East (RSup-CII), may be related to the El Ghorab type Early Neolithic. However there is no correspondence between the ceramics of the two groups at this chronological level.

The comb-impressed ceramics decorated with the woven-mat pattern a Ti-n-Torha - and elsewhere - are in the Dotted Wavy Line tradition and appear in the Nabta-Bir Kiseiba area only in the Middle-Neolithic (Nabta E-75-8, lower layer; Bir Kiseiba E-79-2). A. E. Close (Chapter 3, this Volume) observing the correspondence in the Ti-n-Torha dating range, suggests that the RSup-CII occupation might have continued from an El Ghorab type Early Neolithic up to a Middle Neolithic phase.

These correlations should be interpreted as evidence of inter-relations that gave rise to exchanges and acquisitions on both sides. For instance, Nabta E-75-8, with its two layers related to the Middle and Upper Neolithic, may parallel the Torha East/Lower Torha North/Uan Muhuggiag

---

5) Nabta-Bir Kiseiba appears as a crucial area for domestic experiments. Similar evidence might come perhaps from the nearby basin of Gilf Kebir where R. Kuper has recently started work (cf. Kuper 1981).

6) Bos Primigenius f. taurus (cf. Gautier 1984:59).

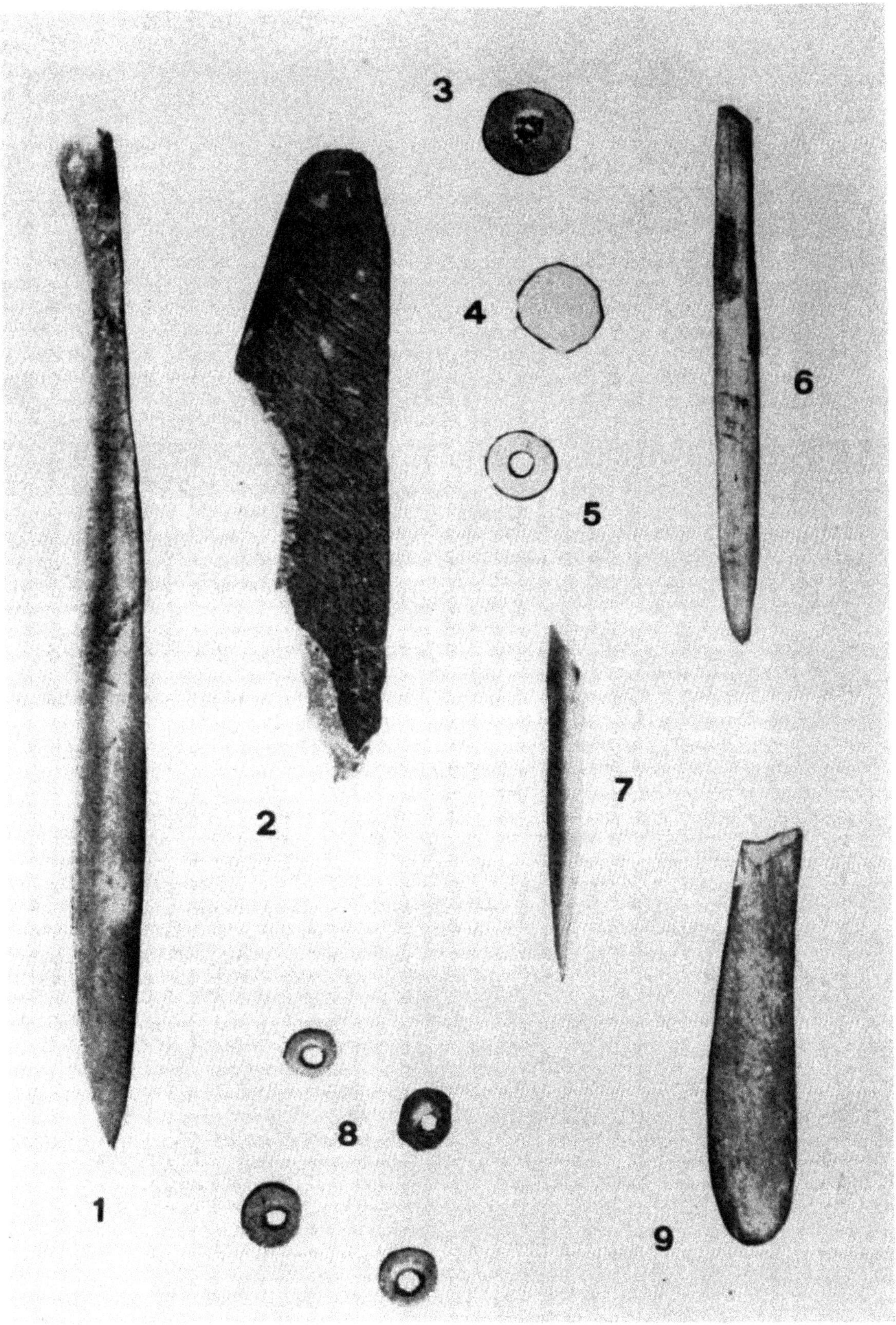

Fig. 10.5 - Wadi Ti-n-Torha. Bone industry from Ti-n-Torha North. 1,3-5,7: Level I; 2,8,9: Level II (1:1). (After Barich 1974: fig. 48).

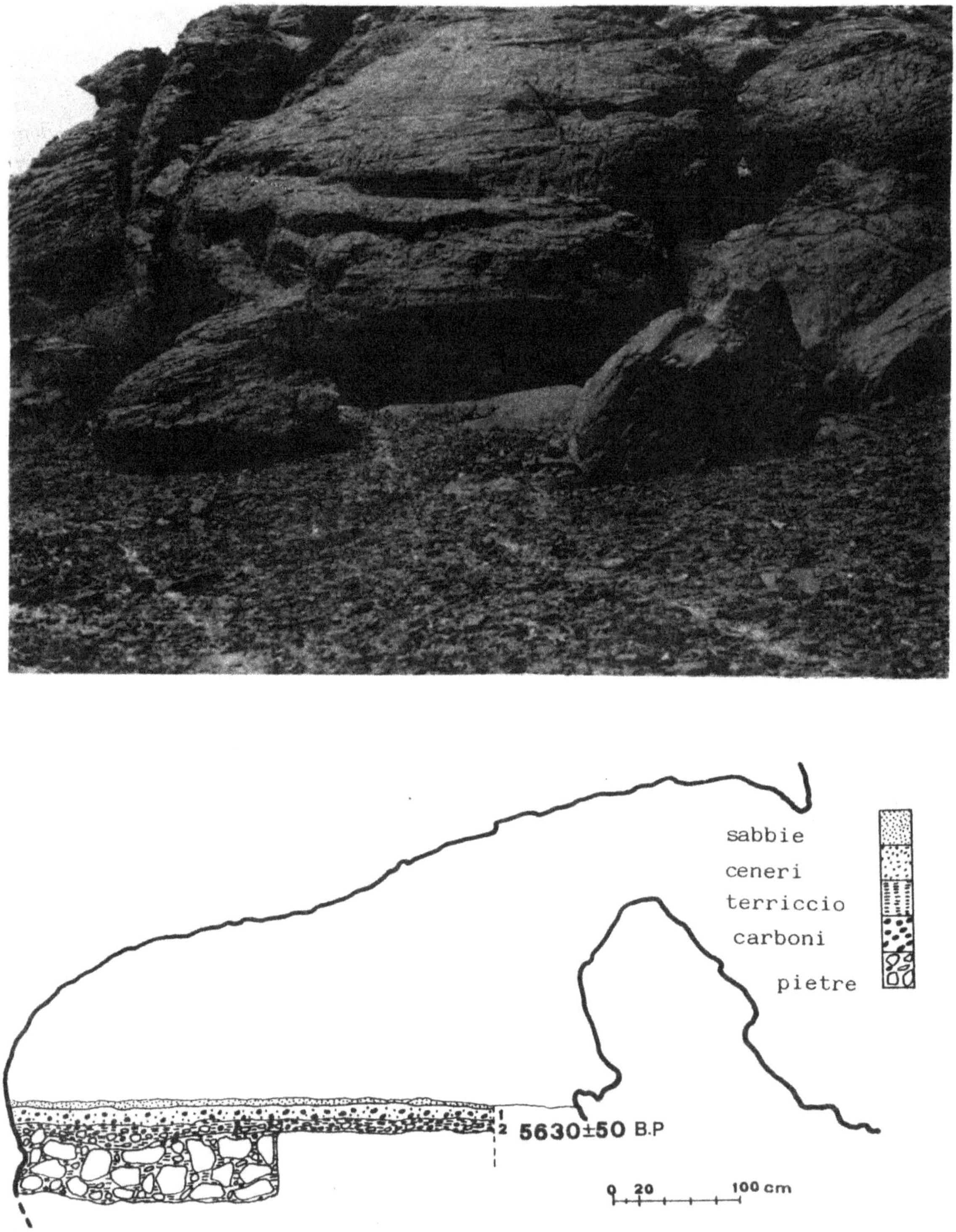

Fig. 10.6 - Wadi Athal. Outer view of the shelter and stratigraphy of the pastoral deposit (after Barich, Mori 1970: figs. 1,2).

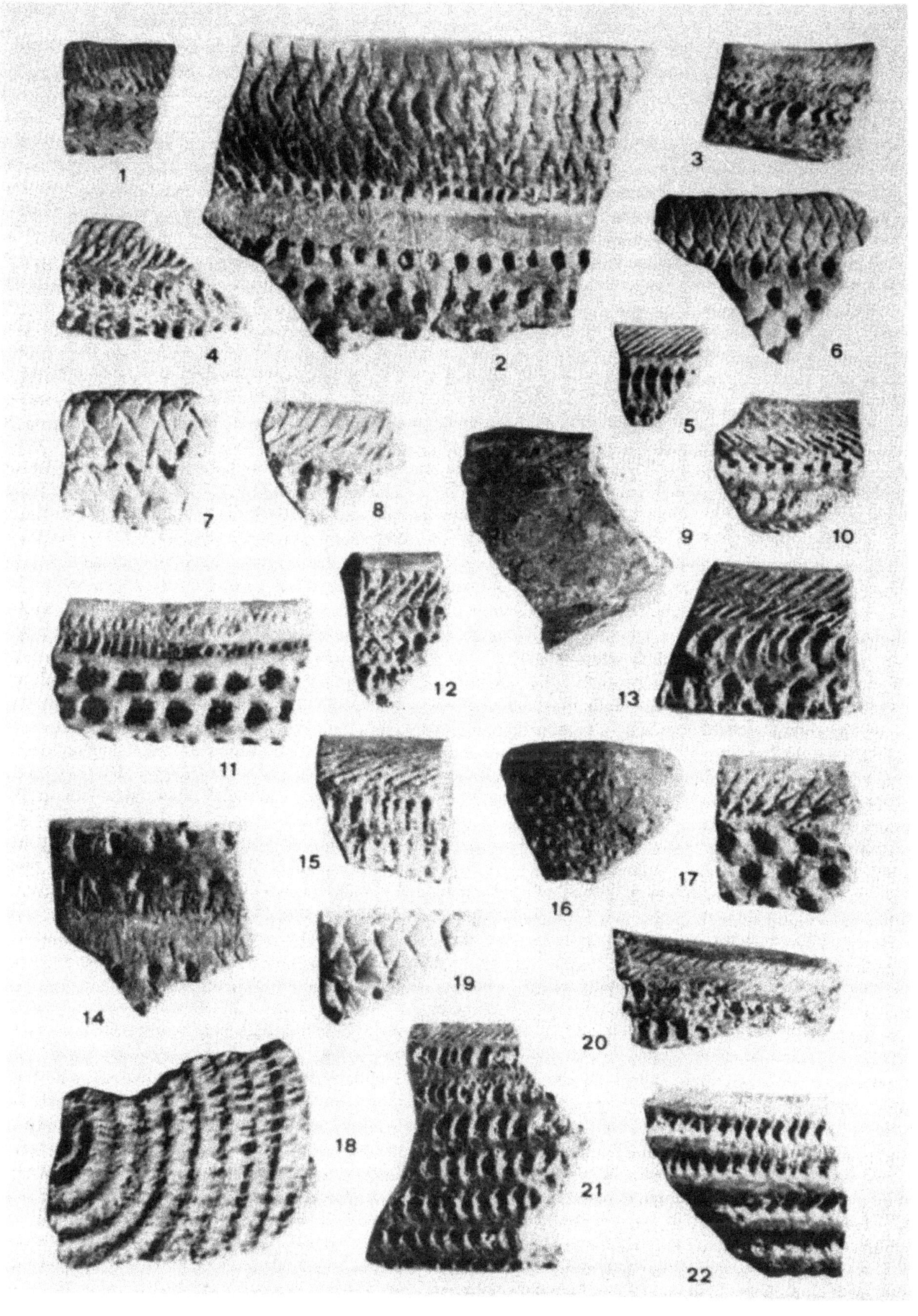

Fig. 10.7 - Wadi Athal. Impressed pottery from the Upper Level (1:1). (After Barich, Mori 1970: fig. 12).

sequences in the Acacus. Furthermore, only a pattern of the numerous interchanges through time can explain the introduction of domestic species and technological changes (7).

Environmental challenges should be viewed as recurring in these areas where, due to frequent climatic oscillations, human groups were forced to change their subsistence methods over and over again and the practically uninterrupted passage from a purely extractive to an animal husbandry economy had ancient beginnings. From a substratum of multiple activities, lacking specific orientation, a few groups specialized in animal domestication, living side by side for some time, with the groups of hunter-fishers. From an archaeological point of view, this is proved by the comparability of the finds previously mentioned in relation to the ergologic developments of Torha East, especially the synchrony of some of the Wavy Line ceramic complexes compared to others belonging to the standard punctated ceramics tradition (8).

Owing to its wealth of archaeological finds and the existence of an earlier complex which allows comparison with the first ceramic horizon of the region, the Uan Muhuggiag site is important. The late occupation phase there stretches the chronological span and shows a complex which is very different from its predecessors.

One may suggest that the complexity of the entire process, entailing different stages, does not rule out the possibility of non-local contributions in the later phases.

Gabriel (1984) has recently taken stock of the problem of the distribution of "stone places" in the Saharan plains from Serir Tibesti to Tanezrouft, ascribing them to groups accustomed to wandering from one water source to the other (9). These must have played a critical role in cultural exchanges. Some of the later complexes found in the Tassili (Camps 1974a: 245; 1974b), Tefedest (Maître 1971: 66 _et seq._) and even the Acacus, might indeed be ascribed to the coming of these groups to the massifs during their last phase (5300-3800 B.P.).

---

7) These observations do not rule out the possibility that the progenitors of the Saharan herds might have been the wild species known in the Maghreb at least since the Middle Paleolithic (Carter, Clark 1976: 492). In any event, the hypotheses made by A. Smith (1980: 491) about the introduction of cattle into North Africa have to be set aside on present evidence. But one must keep in mind the date of _Bos_ at the Cappelletti Cave (6530±250 B.P., cf. Roubet 1979) and, especially, the earliest appearance of caprids at the Haua Fteah remains (McBurney 1967).

8) By way of example, one may refer to the eastern sites of Khartoum, Shaheinab and Geili, dating from about the 6th-5th millennia B.P. and, therefore, contemporary with the "bovidian" developments further West.

9) Unlike the groups that lived in the massifs, who are supposed to have practiced a short-distance nomadism related to transhumance (Gabriel 1984: 395).

PART III: Studies on the Habitat and Chronology

CHAPTER 11

NATURAL RESOURCES AND PALAEOENVIRONMENT IN THE TADRART ACACUS: THE NON-CLIMATIC FACTORS DETERMINING HUMAN OCCUPATION

By Bruno MARCOLONGO*

1. INTRODUCTION

This study will deal with the applied geomorphological and geological aspects of the Tadrart Acacus region to acquire a better understanding of the influence - from a diachronic viewpoint - of environmental parameters on the type and evolution of various prehistoric settlements present in that area.

A decisive contribution to this preliminary analysis was given by satellite images (LANDSAT-3: figs. 11.1,2), both during direct survey planning and in the cartographic synthesis (Marcolongo 1982). These images allowed useful field work to be performed economically, and supplied the elements needed for a comprehensive physical and structural characterization of the entire Acacus chain, from the Libyan-Algerian border to the Serdeles oasis (Bartolomei 1966; Conant, Gourdazi 1967). They also formed the basis of topographic reference in locating phenomena and sites of interest.

2. OBSERVATIONS AND CONSEQUENT MAPS

The Tadrart Acacus mountain chain in the Libyan Fezzan is a good morphological example of "cuesta", high ground with a noticeable transverse asymmetrical profile where the eastern side is characterized by a gentle slope following the strata dip, while the western side has a much steeper slope set along a fault and fracture system. This asymmetry, conspicuously controlled by the overall, monoclinal-styled tectonic structure dipping toward the East, is of no merely formal interest since it involves a different development of the dissection pattern controlling erosion and drainage, thus affecting distribution, consistency and variation of groundwater reserves and, in the final analysis, natural resources in general. Obviously, it determines the susceptibility of the environment to the penetration and occupation by man, variously favoured by the presence of conditions that include, especially in arid environments and phases, high soil moisture and water-bearing aquifers with prolonged retention times.

These factors underline the major role played by non-climatic parameters (morphology, lithology, permeability, etc.) in the landscape evolution, with specific repercussions on the development of human communities (Rognon 1980). Indeed, even in times characterized by little or no precipitations, water resources stored at shallow depths create conditions favouring the permanence of a plant cover, however limited it might be, and the setting up

* Istituto di Geologia Applicata, C.N.R., Padova, Italy.

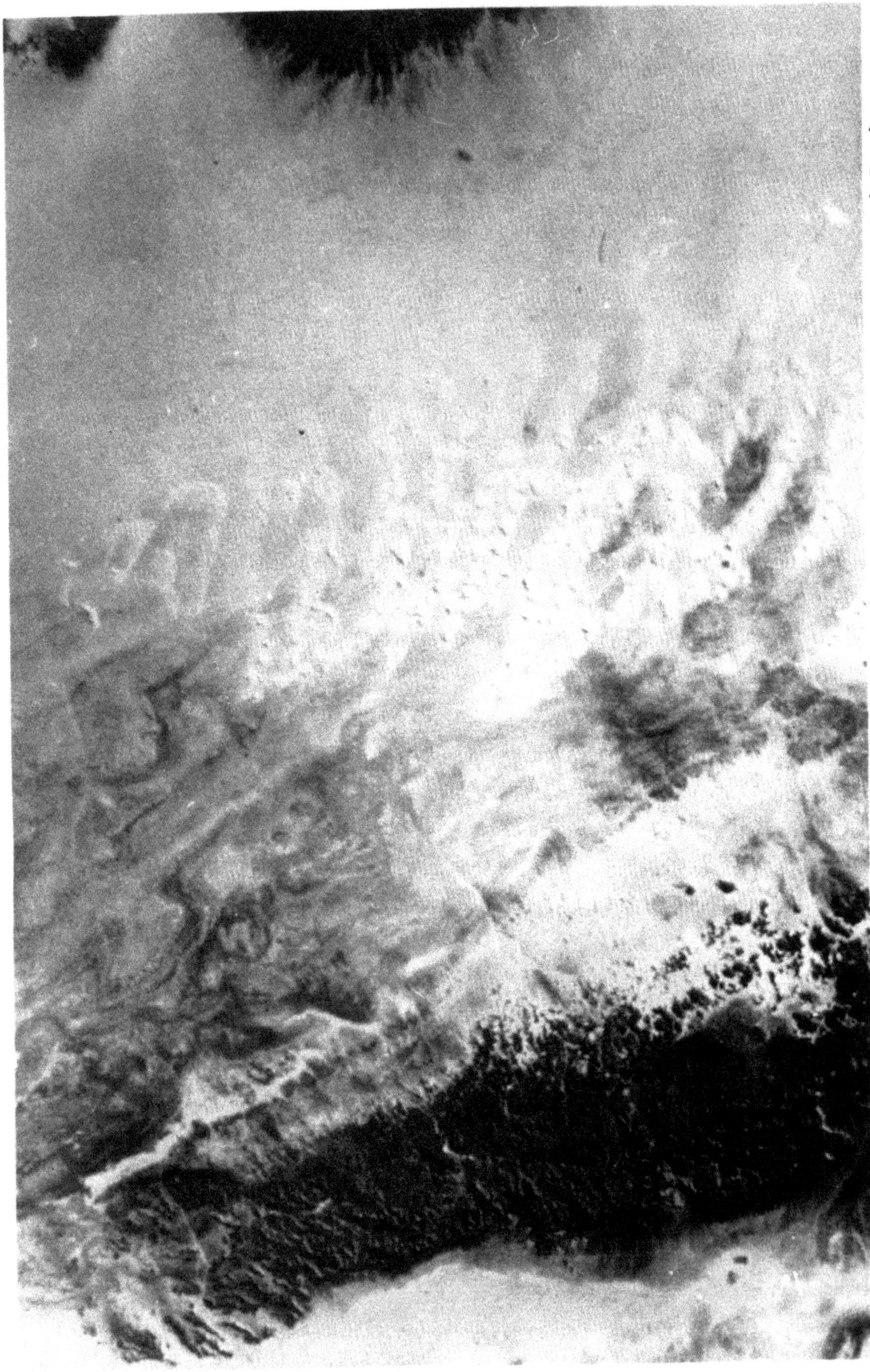

Fig. 11.1 - Tadrart Acacus. Landsat image taken in December 17th 1978, Band 7 (0.7-1μ).

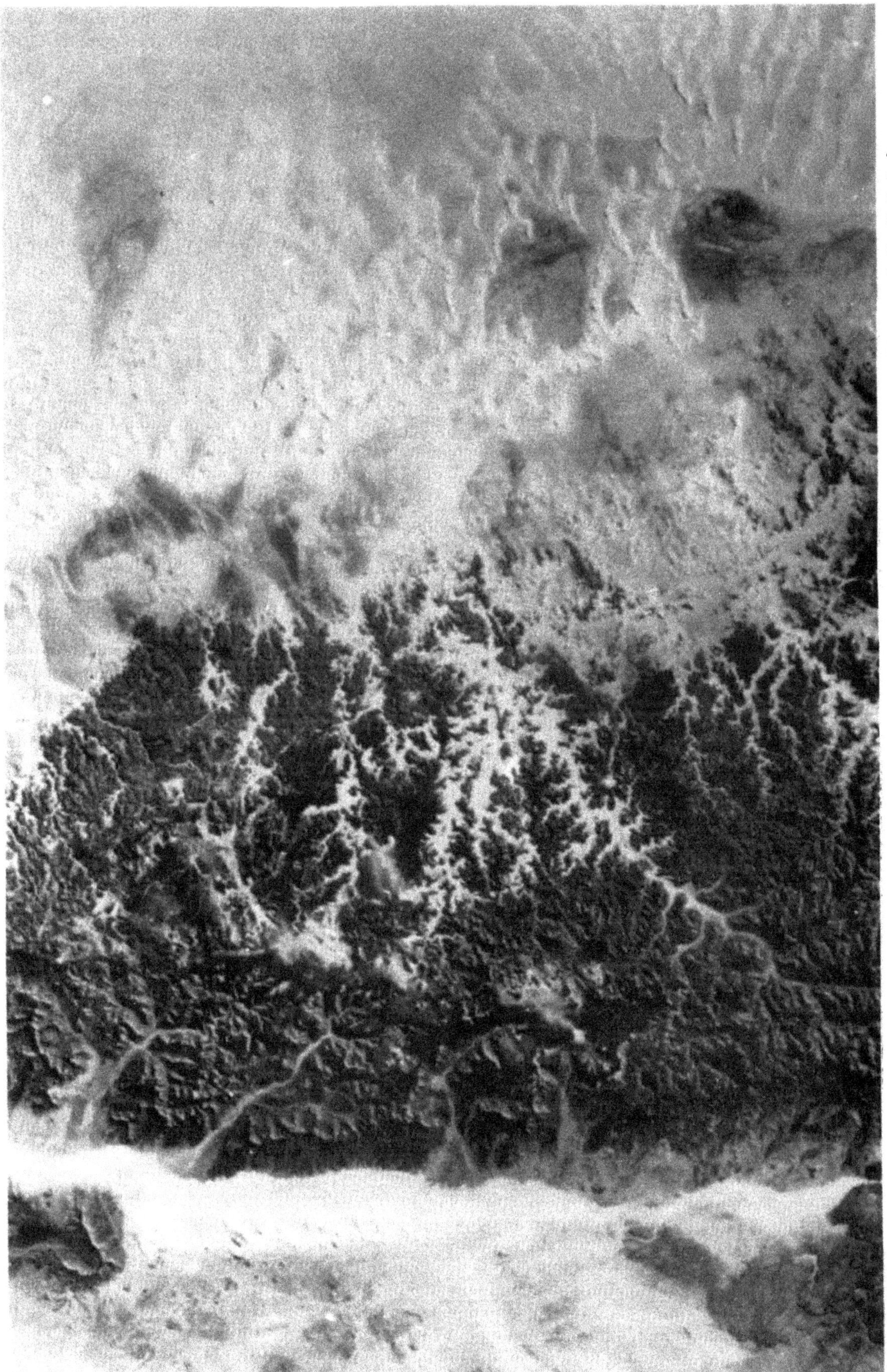

Fig. 11.2 - Tadrart Acacus. Landsat image taken in December 17th 1978, Band 7 (0.7-1μ).

of an ecosystem able to survive in an otherwise adverse environment (Marcolongo 1980).

Direct observation and the analysis of the images, resulted in three thematic maps to a 1:250,000 scale, that make up a synthetic basis for research.

The sequence of their subjects (geomorphology, geolithology and groundwater resources) following the steps of the deductive process, led to emphasize the most influential physical parameters in the history of human occupation.

Geomorphological Map

The Tadrart Acacus area is divided into three large physiographic units, namely a central mountainous part, a transit belt and some coterminous flat belts. Within each unit, due emphasis is laid on those morphological aspects that exerted a direct control over the formation of peculiar habitat conditions and that still to date dominate the environment.

In particular, the map identifies in the central massif structure some relatively broad valley bottoms, filled with alluvial deposits resulting from both sheet and gully erosion which, subsequently, were affected to varying degrees by a thin cover of eluvial or eolian material. These valley bottoms, where the "thalweg" pattern appears either distorted or uncertain (see, for instance, the intermediate section of the Wadi Teshuinat) represent intermontane troughs storing meteoric or run-off water with medium retention times. Furthermore, the map shows the colluvium and talus belts bordering the Tadrart Acacus western slope, coinciding with a structural steep slope linked to 'horst' and 'graben' tectonic patterns. This belt allows an easy and immediate infiltration by surface water that, however, re-emerges at the foot without a long-lasting retention time and, therefore, with no valid protection against evaporation in particularly arid time periods. Moreover, the map shows the alluvial fans at the outlet of the valleys, at times deeply cut or suspended along the same slope. Even these features are significant for a palaeoenvironmental reconstruction based on non-climatic elements.

The satellite images allowed extended relict strips of an ancient peneplain to be identified. These were subsequently isolated by rising phenomena quite likely dating from the Carboniferous Period and presumably reactivated in the Alpine age, that differentially affected the south-western part as compared with the others.

Finally, an extremely detailed description is provided for the mostly fossil drainage pattern that, formed in the Mesozoic and developing in the Cenozoic and Quaternary Periods, affected the various lithotypes, at times, in a selective manner. In the southern section of the area, it is a sub-dendritic-type drainage that changes into a sub-parallel pattern in the central-northern part. In any event, it seems to be significantly controlled by disjunctive tectonics, with sudden changes in direction due to the intersection of two fracture and fault systems having a NNE-SSW and a E-W orientation, respectively. A lesser influence seems ascribable to a N-S system that, at a local level, is the product of shearing stresses connected with the main NNE-SSW system (see also the direction of the Wadi Tanezzuft as well as the direction of the fossil valley at present occupied by the

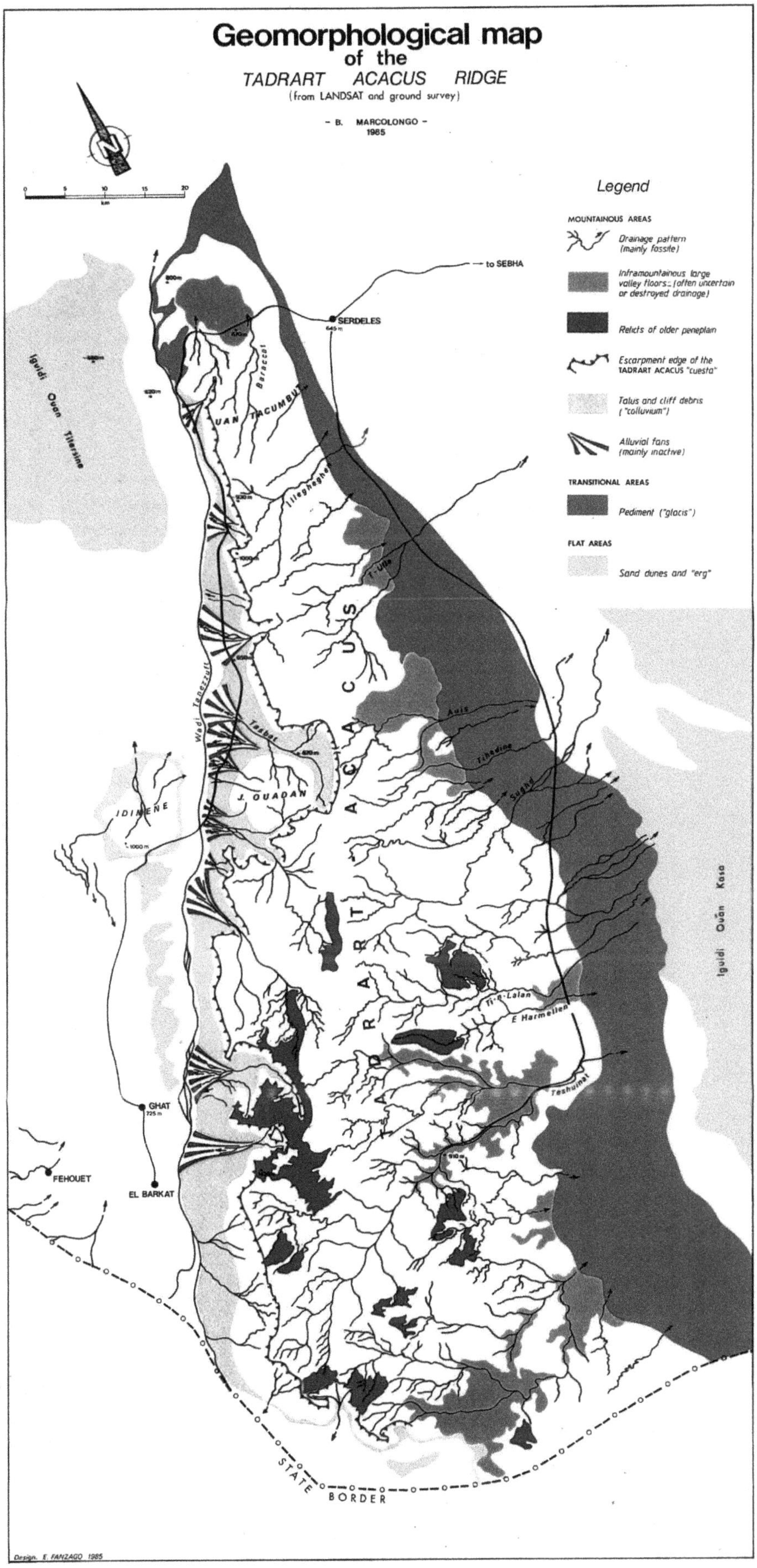

Please note that a full-size version of this map is available to download from www.barpublishing.com/additional-downloads.html
The original foldout has been reduced in size to match the A4 format of this book, the image is therefore not as clear as the original foldout. Please refer to the original foldout via the download for the original content.
Please also be aware that the image might be cut off or not complete in the printed book.

Iguidi Ouan Kasa and Erg Ti-n-Merzouga dune lines).

The evolution of the drainage pattern highlights a phase in the upper section of various catchment basins that cuts, at times markedly, into peneplain and concave morphologies connected with it, as well as a subsequent phase witnessed by the broader and more eroded medium and lower sections. Owing to the conformation of the Tadrart Acacus and to its "pinch out" structural motif, with outcrops tapering towards the North, the southernmost basins (Teshuinat, E'Harmellen, Sughd) are much more ramified and developed than the central and northern ones (Auis 'Tulla, Illegheghen, Baraccat).

This simple division suggests that the southern section of the Acacus was more hospitable to man than its outher physiographical sub-areas.

The transition belt is only present along the eastern massif slope, with the formation of an erosive pediment ("glacis") covered by a thin veneer of eluvium, locally mixed with alluvium. It features conditions favouring stagnation and infiltration of meteoric and runn-off water, formation of suspended phreatic aquifers close to the surface and persistence of a wide-spread moisture in the unsaturated soil zone. Even in this case, owing to a north-tapering structure, this pediment is wider and more significant in the central and southern section, thus contributing to the creation of a type of habitat that might sustain human life during arid climatic conditions.

Geolithological Map

The lithotypes that may be found in the upraised western side of the wide 'Murzuk' sedimentation basin belong to the Paleozoic Era and witness a whole series of marine transgressions, initially separated and subsequently concluded by phases of emersion, with continental environments characterized by various climates (see Bonadonna 1974).

A very coarse conglomerate at the bottom is followed by coarse, rather thick (600-900 m), cross-bedded sandstone typical of subaerial environments that belong to the Cambrian-Ordovician Period. A marine transgression in the Silurian period caused the deposition of graptolite deposits, now schists (variegated, finely layered clays) followed by fine sandstones (Acacus sandstone) that are well stratified with Arlania traces and are a few hundred meters thick. These sandstones, that at time present very thin interposed layers of dark gray clay, point to a rather shallow marine environment in a retreating phase.

The regression continues during the Devonian period, witnessed by three extensive layers of coarse sandstone with, once again, a cross-bedded stratification (the whole about 300 m thick), separated one from the other by levels of fine argillaceous sandstone with Arlania and Gigantostraci. The local series is closed by a Carboniferous unit with clastic and epineritic terrigenous facies, locally carbonatic, characteristic of a shallow sea.

More or less loose Quaternary deposits are wide-spread all over the area, in the form of both eluvial weathering cover and alluvial or eolic accumulation. It should be remembered, owing to long periods with particularly arid climates, that all the rocks have a darkish patina ("vernis du desert") and, quite often, hematic crusts having a thickness of up to a few decimetres.

From the structural point of view, the Tadrart Acacus massif forms a higher 'horst' to the South-West than to the North-West and to the East. A few systems of faults and fractures prevail and the major one, has a NNE-SSW trend; a shear system with approximately a N-S direction and a related system with an E-W direction are associated with it.

The overall architecture shows a rather linear tectonic history, where exogenous dynamic processes and phenomena prevail after differential uplifting phases.

Sub-Surface Water Resources

The geolithological and geomorphological understanding of the area allows the reconstruction of a groundwater circulation model and the locating of areas of potential storage.

Special attention was devoted to the identification of higher water-bearing aquifers, areas with higher soil moisture as well as those hydrogeological structures or conditions that exerted direct influence over the plant cover and, generally, the environment.

Rocks were reclassified in relation to their permeability that, as far as compact rocks are concerned, is always acquired by fracturing and/or weathering processes. A high permeability characterizes the talus deposits and alluvial fans of the western slope which, nevertheless, fail to be potentially significant structures owing to the little or no development of the catchment basins they are connected with. Of greater interest are the regions of intermontane eluvial and alluvial deposits, and the "glacis" belts along the eastern slope.

Among the Palaeozoic rocks, a good secondary permeability may be offered by the Cambrian-Ordovician and Devonian sandstones, while the finer, upper Gotlandian sandstones, (Acacus sandstone) have average permeability. The Carboniferous, fine clastic deposits are, instead, scarcely permeable and the "Tanezzuft" schists are practically impermeable.

The preferential groundwater flow directions, inferred from the tectonic pattern and fracture types, show a more significant, generalized downflow towards the East and a secondary and more limited downflow towards the West. In particular, relatively significant contributions may be concentrated, coming from SW and W, in the broad valley bottoms of the Teshuinat basin and, further South, in the Tanezzuft valley and Iguidi Ouan Kasa passage zone. Relatively interesting contributions may also be located in the medium and lower course of the Auis, at the 'Tulla mouth in the flat area at the foot of the massif and, quite diffusely, along the glacis belt included between Sughd and Teshuinat. To the West, the only area with a groundwater storage and retention potential is the terminal portion of the Tasbät valley and the detritic belt - extended for approximately fifteen km from its outlet towards the North - along the Acacus steep slope.

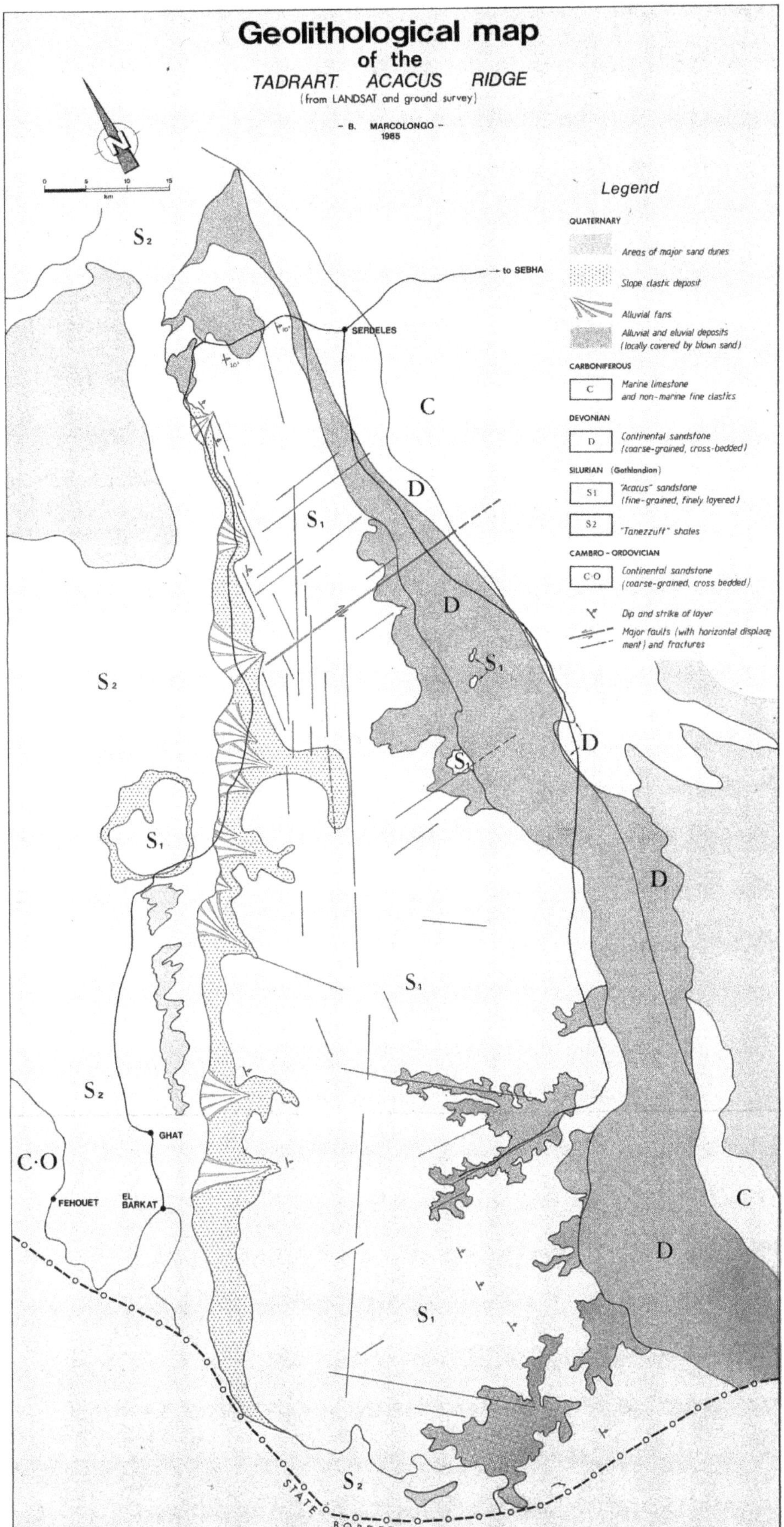

Please note that a full-size version of this map is available to download from www.barpublishing.com/additional-downloads.html
The original foldout has been reduced in size to match the A4 format of this book, the image is therefore not as clear as the original foldout. Please refer to the original foldout via the download for the original content.
Please also be aware that the image might be cut off or not complete in the printed book.

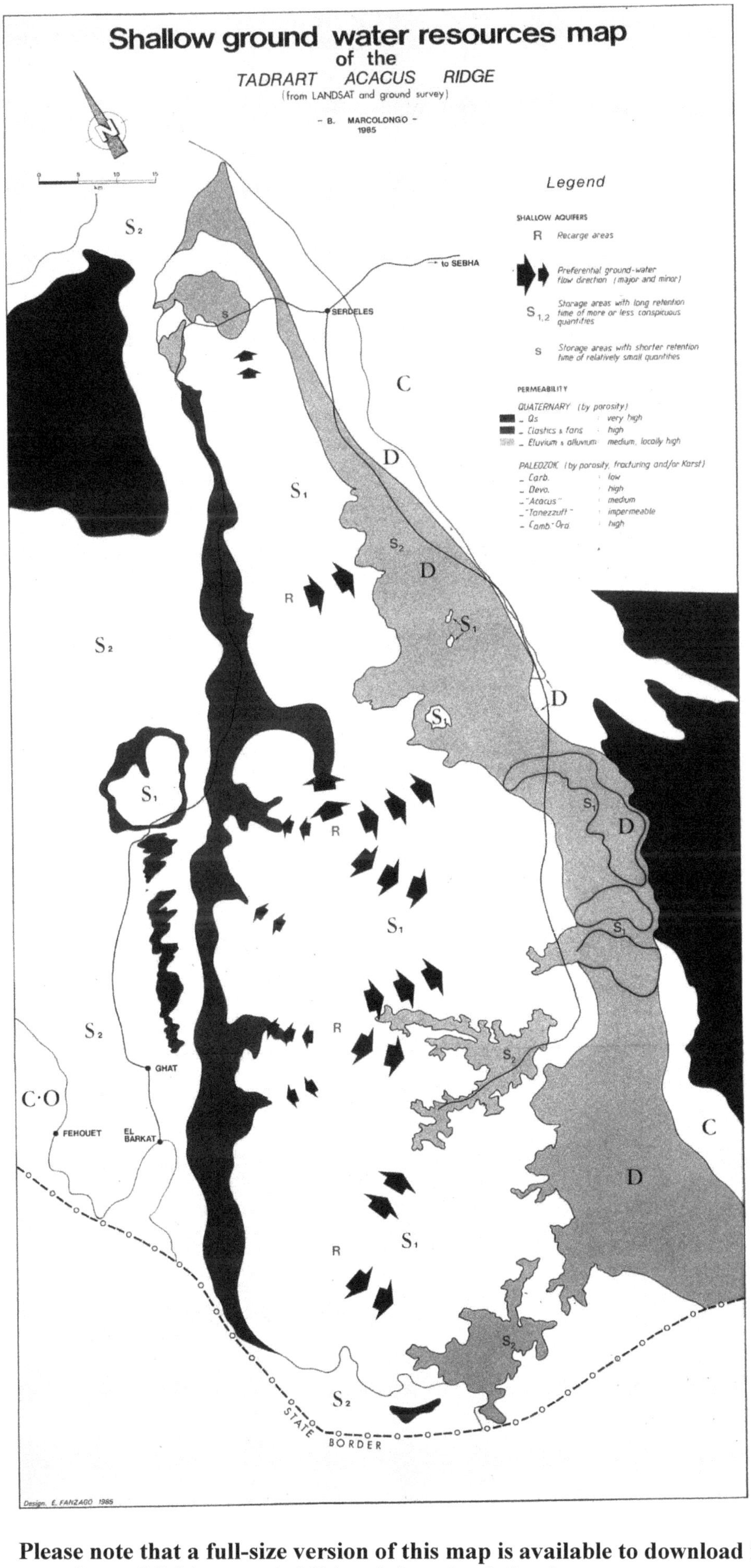

Please note that a full-size version of this map is available to download from www.barpublishing.com/additional-downloads.html
The original foldout has been reduced in size to match the A4 format of this book, the image is therefore not as clear as the original foldout. Please refer to the original foldout via the download for the original content.
Please also be aware that the image might be cut off or not complete in the printed book.

CONCLUSIONS

Palaeoenvironmental characterization allows a Tadrart Acacus zoning on the basis of a number of fundamental physical parameters.

The massif presents a marked zonal differentiation of receptivity for man, whose settlement was much easier along the eastern than along the western slope. Morphological, but also and especially hydrogeological conditions have undoubtedly favoured human occupation, particularly during the Pleistocene and Holocene times of limited aridity. The groundwater resources accumulated during the moisture-rich phases are stored over long periods of time and released slowly, thus favouring the existence of vegetation and the survival of fauna, including man. Furthermore, this determines the formation of micro-climates (valley bottoms and mountain breezes, local precipitations, greater atmospheric humidity, etc.), especially in the most protected areas within the massif (for example, the articulated Teshuinat catchment basin), and reduced the spread of arid conditions throughout the Tadrart Acacus.

Nevertheless, even on the eastern slope two sections may be differentiated and their boundary line roughly marked by the Wadi Sughd. The northern section, notwithstanding a relatively less dense and developed drainage pattern, allows easier passage to the western slope, with good connections, for example, between the Tasbät valley and the catchment basins of the Wadis Tihedine, Auis and Ti-n-Torha. Furthermore, conditions in this section favour the storage of groundwater at shallow depths in the two outer belts (both to the West and to the East). This encouraged a certain degree of human mobility, since in less favourable times, men could find means of sustenance in easily accessible nearby areas. In this section a physical stimulus to extend the occupation-related area of influence, since resources were scattered rather than concentrated, is present.

South of the Wadi Sughd, the presence of broad intermontane valley plains, linked to a much more developed and disaggregated drainage pattern, gives rise to the concentration of more extensive and long-lasting water resources, but also to a greater relative difficulty of movement.

Therefore, during the various climatic oscillations with alternate phases of moisture and aridity that characterise late Pleistocene and Holocene (Rognon 1980; Maley 1981), there has been a succession of contractions and expansions in the area of influence of individual sites, most likely however without a real break in the occupation of the territory, because of the precious water reserves scattered within or along the edges of the Tadrart Acacus.

REFERENCES

BARTOLOMEI G. 1966 - Alcuni risultati delle richerche sul Quaternario del massiccio dell'Acacus-Tadrart (Fezzan meridionale), Atti Xa Riun. Scient. Ist. Ital. Preistoria e Protostoria, Verona: 15-22.

BONADONNA F. 1974 - Schema geomorfologico dell'area dell'Auis (Acacus, Libia), in Barich B.E., La serie stratigrafica del l'Uadi Ti-n-Torha (Acacus, Libia), Origini, VIII: 163-167.

CONANT L.C., GOUDARZI G.H. 1967 - Stratigraphic and tectonic framework of Libya, The Am. Ass. of Petroleum Geologists Bull., 51: 719-730.

MALEY J. 1981 - Etudes palynologiques dans le bassin du Tchad et paléoclimatologie de l'Afrique nord-tropicale de 30.000 ans à l'époque actuelle, Travaux et Documents de l'O.R.S.T.O.M., 129 (O.R.S.T.O.M.: Paris), 586p.

MARCOLONGO B. 1980 - Photointerpretation applied to hydrogeological problems: soil moisture content, soil permeability, hydrogeological structures, in Remote Sensing Application in Agriculture and Hydrology, ed. Fraysse C., Balkema: Rotterdam.

MARCOLONGO B. 1982 - Photographic and thermic infra-red applications in hydrogeology and archaeology, 1st Course on Remote Sensing S.I.TE, Milano: M1-30.

ROGNON P. 1980 - Pluvial and arid phases in the Sahara: the role of non-climatic factors, Palaeoecology of Africa, 12: 45-62.

CHAPTER 12

THE ARCHAEOZOOLOGICAL SEQUENCE OF THE ACACUS

By Achilles GAUTIER*

1. INTRODUCTION

The present study deals with the animal remains found during the excavations at Uan Muhuggiag and Ti-n-Torha Two Caves. The vertebrate fauna from Ti-n-Torha Two Caves is very limited and has already been described in the revised study of the faunal assemblages from Ti-n-Torha North and Ti-n-Torha East (Gautier, Van Neer 1977-1982; Gautier 1984a).

The Uan Muhuggiag vertebrate fauna consists of two sets of samples. The first set was collected during the earlier excavations at the site and these samples were studied by Dr. Angelo Pasa, who died unfortunately before he could finish the analysis. Preliminary data on the fauna and other aspects can be found in Pasa and Pasa-Durante (1962) and Mori (1965). The collection is housed in the Museo Civico di Storia Naturale di Verona, together with some manuscript notes by the late Dr. Pasa. During a brief visit to Verona in August 1981, we made a rapid re-analysis of this material and obtained permission to export some specimens for a more detailed analysis. We would like to thank both Dr. F. Mori (Rome) and Dr. L. Sorbini (Verona) for giving us access to this material. The second Uan Muhuggiag collection results from the excavations in 1982 by Dr. B.E. Barich. Preliminary sorting and analysis of this material was done in Rome during two short visits there in March and December 1982; material for a more detailed analysis was sent to our laboratory.

In the following we will refer to the three collections as the Two Caves, Pasa and Barich assemblages or material. The Two Caves material is now housed in the collections of the Laboratorium voor Paleontologie under number P2576; most of the identifiable Barich material is at the same place under number P3566, while the rest of the Barich material will be stored in the collections of the Museo delle Origini in Rome. The Pasa collection will remain in the Verona Museum. It has been described succinctly in the catalogue of Quaternary fossils of this museum (Sorbini et Alii 1974: 45, number 205) on the basis of the preliminary manuscript notes by Dr. Pasa. In the course of our short description we will emend the list printed in this catalogue.

The sampling technique used to recover the Pasa collection is not known to us, but the aspect of the material is very comparable with that of the later samples; moreover the faunal spectrum also compares well with that of the material excavated by Dr. Barich. Therefore it would seem that the sampling technique was not fundamentally different from that applied during

---

* Laboratorium voor Paleontologie, Rijksuniversiteit Gent, Krijgslaan 281/S8, B-9000 Get, Belgium.

Table 12.1 : The fauna of Ti-n-Torha Two Caves (fragment counts).

Animal group \ Provenance	Totals	Surface	level I	II	IIa	III
Zootecus insularis	2	-	2	-	-	-
unidentified small landsnail	1	-	-	1	-	-
catfish (*Clarias* sp.)	3	1	1	-	-	1
Tilapia (*Tilapia* sp.)	8	-	-	3	5	-
not identified fish	2	-	-	1	1	-
birds(1)	6	-	2	-	-	4
toad or frog	3	-	1	1	1	-
small lizard	3	-	2	1	-	-
small snake	1	1	-	-	-	-
not identified bird(s)	6	-	2	-	-	4
desert hedgehog (*Paraechinus aethiopicus*)	1	-	1	-	-	-
shrew (*Crocidura* sp.)	1	-	-	-	-	1
small bat	1	-	-	-	-	1
common hare (*Lepus capensis*)	5	-	3	2	-	-
dwarf gerbil (*Gerbillus nanus*)	34	1	-	2	11	20
lesser jerboa (*Jaculus jaculus*)	9	-	-	-	-	9
mzab gundi (*Massoutiera mzabi*)	1	-	-	1	-	-
not identified small rodents	26	-	5	4	3	14
North African porcupine (*Hystrix cristata*)	1	-	1	-	-	-
golden jackal (*Canis aureus*)	1	-	1	-	-	-
dassie (*Procavia capensis*)	3	-	1	1	-	1
dorcas gazelle (*Gazella dorcas*)	4	1	2	1	-	-
Barbary sheep (*Ammotragus lervia*)	36	3	15	11	3	4
Totals	158	7	39	29	24	59
not identified	926	4	103	293	213	313

(1) Four specimens were identified by Mrs. MATTHIESEN (Gainesville; *in litt.*) One pertains to a swift (*Apus* sp.), two others to otids, one of which is a scops owl (*Otus scops*). The exact provenance of these specimens is unknown, but they are derived from layers I or III.

the second excavation when the excavators used dry sieving on screens with a 3 mm mesh. A comparable technique was used at the Two Caves site. In our opinion recovery rates are acceptable since in this kind of sites most of the microvertebrate remains which may have been lost (cf. Meadow 1980) will pertain to intrusives, the status of which may be difficult to assess.

2. DESCRIPTION OF THE MATERIAL

The material from Uan Muhuggiag is very well preserved, showing normally colours ranging from pale orange-yellow to orange-yellow and desiccation cracks as described by Behrensmeyer for her weathering stage 1 (Behrensmeyer 1978). Mummification has favoured preservation of soft tissues remains and a few bones still adhere together in articulating position because some surrounding tissue is present. The horny part of small livestock hooves and hair are also preserved, as well as a cattle horn fragment (fig. 12.1,1). The preservation of the Two Caves material is slightly worse, but still comparable to that at Uan Muhuggiag.

The identifications were made with the aid of the comparative material in our laboratory and supplemented by circumstantial evidence based on such African wildlife catalogues and descriptions as those by Meester and Setzer (1971-78) or Haltenorth and Diller (1979). Most of the material did not pose particular problems; some specimens of minor significance were not identified to the specific level. Therefore no detailed descriptions will be given, except in the case of the domestic animals.

Two Caves fauna

As said before, the fauna of Two Caves site at Ti-n-Torha is very poor; it consists mainly of small bone fragments. Its composition is summarized in Table 12.1 based on Table 1 from Gautier and Van Neer (1977-1982). This table differs somewhat from the one given in the Ti-n-Torha paper. Dama gazelle is not present at Two Caves, but in our previous paper we attributed a fairly large cannonbone fragment (level II) erroneously to it. This specimen is now included in dorcas gazelle.

The fauna compares well with that of Ti-n-Torha East in its spectrum and in the clear dominance of Barbary sheep. Some animals do not occur at Ti-n-Torha East: a small bat represented by a mandible, various remains of smaller mammals, including a shrew, gerbil and lesser jerboa, but these are no doubt all intrusive elements. As previously noted, porcupine is represented only by a fragment of a quill.

The Pasa and Barich assemblages

The qualitative and quantitative composition of each assemblage is given separately in Tables 12.2 and 12.3.

Lower vertebrates

Lower vertebrates occur only in the Pasa collection: one catfish skull fragment in layer X and perhaps a second one in layer V11; one turtle carapace fragment in layer VII and two long bone shaft fragments of birds of

Table 12.2 : Fauna of Uan Muhuggiag in the Pasa-collection (fragment counts).

Animal group \ Provenance	Totals	without provenance	A1	A2	A3	A4	I	II	III+B2	IV+B3	V	VI	VII	VIII	IX	X
catfish (Clarias sp.)	1	-	-	-	-	-	-	-	-	-	-	-	?	-	-	1
turtle	1	-	-	-	-	-	-	-	-	-	-	-	-	-	-	-
bird	2	-	-	-	-	-	-	-	-	-	-	-	-	1	1	-
baboon (Papio cynocephalus)	1	-	-	-	-	-	1	-	-	-	-	-	-	-	-	-
hare (Lepus capensis	2	1	-	-	-	-	-	-	1	-	-	-	-	-	-	-
dwarf gerbil (Gerbillus nanus)	2	-	-	-	1	-	1	-	-	-	-	-	-	-	-	-
field rat (Arvicanthis niloticus)	1	-	-	-	-	-	1	-	-	-	-	-	-	-	-	-
unidentified small rodent(s)	2	-	1	1	-	-	-	-	-	-	-	-	-	-	-	-
North African porcupine (Hystrix cristata)	3	-	-	-	-	3	-	-	-	-	-	-	-	-	-	-
golden jackal (Canis aureus)	2	-	-	-	-	-	-	1	?	-	-	-	1	-	-	-
dassie (Procavia capensis)	1	-	-	-	-	-	-	-	-	-	1	-	-	-	-	-
wild ass (Equus africanus)	3	-	1	2	-	-	-	-	-	-	-	-	-	-	-	-
Barbary sheep (Ammotragus lervia)	20	1	6	9	-	-	1	-	-	1	-	-	-	2?	-	-
domestic cattle (Bos primigenius f. taurus)	97	6	4	7	-	3	9	10	15	20	6	2	6	8	-	1
sheep/goat (Ovis ammon f. aries/Capra aegagrus f. hircus)	151	9	29	35	7	3	5	11	18	16	5	1	1	5	5	1
Totals	289	17	41	54	8	9	18	22	34	37	12	3	9	16	6	3
Not identified	±3500	75	675	850	60	+	450	225	800	165	50	40	50	30	15	-

?: mentioned by PASA in his manuscript, but not seen by the author

Table 12.3 : The fauna of Uan Muhuggiag in the Barich-collection (fragment counts).

Animal group \ Provenance	SECTOR A							SECTOR B							TOTALS
	1	1a	2	2a	2b	2c	2d	1	1a	2	2a	2b	2c	2d	
hare (Lepus capensis)	-	-	-	2	-	-	-	-	-	-	-	-	-	-	2
wild cat (Felis libyca)	-	-	-	-	-	-	-	-	2	-	-	-	-	-	2
dassie (Procavia capensis)	-	-	-	-	-	-	-	-	-	-	1	-	-	-	1
warthog (Phacochoerus aethiopicus)	-	-	-	-	-	-	-	-	-	-	1	-	-	-	1
wild ass (Equus africanus)	-	-	-	-	-	-	-	1	-	-	-	-	-	-	2
dorcas gazelle (Gazella dorcas)	2	-	-	-	-	-	-	-	-	-	-	-	-	-	2
Barbary sheep (Ammotragus lervia)	9	-	-	-	-	-	-	-	1?	-	-	-	-	1?	11
domestic cattle (Bos primigenius f. taurus)	2	1	-	3	2	-	-	1	5	3	12	9	2	2	42
sheep/goat (Ovis ammon f. aries/Capra aegragus f. hircus)	84	5	10	10	5	-	1	14	19	16*	12	23	2	2	203
total identified	94	6	10	15	7	-	1	16	27	19	26	32	4	5	262
not identified (approximate)	600	20	35	125	80	20	3	175	150	150	300	300	45	30	±2000

* including one skeleton of a juvenile sheep.

medium size in layers IX and VIII. Fragments of ostrich eggshell are also present in most of the layers in the Pasa collection, but are never frequent. In the Barich collection, we did not encounter such fragments; the excavator however informed us that numerous ostrich eggshell fragments, both worked and not worked, occur in the archaeological materials (Dr. Barich, pers. comm.; see this Volume, Chapter 7). The neck lace found on the infant mummy of Uan Muhuggiag also was made of ostrich eggshell beads (Mori, Ascenzi 1959; Arkell *et Alii* 1961; Ascenzi *et Alii* 1983).

The lower vertebrate finds in the Pasa collections are listed in the Verona museum catalogue as *Emys* sp. (= turtle), Varanidae indt (= catfish) and *Struthio camelus*. Catfish was found in Ti-n-Torha East and in the Two Caves collection (see Table 12.1). Birds are numerous at Ti-n-Torha East and occur also sporadically in the Two Caves fauna.

Wild mammals

Lagomorph remains comprise an ulna (layer III) and a mandible fragment of unknown provenance in the Pasa collection and a tibial shaft fragment (Sector A, layer 2a) in the Barich collection. These finds indicate a fairly small lagomorph and can be ascribed to the common hare, which is known to be of comparatively small size in Africa. In the Verona museum catalogue the lagomorph material from the Pasa collection is listed as *Lepus* sp. Hare was also found in the Two Caves fauna (Table 12.1) and at Ti-n-Torha East, where it is fairly frequent in the upper association.

A single, much worn lower second molar measuring 12.9 mm represents a baboon in layer I (Pasa collection). Dr. Pasa had not yet identified it and therefore it is not listed in the Verona museum catalogue. The fauna of Ti-n-Torha East contains four fragments ascribed to baboon. Mori (1965) illustrates a rock painting with four baboons from Wadi Teshuinat.

Small rodents were found only in the extensive Pasa collection. Present are a small gerbillid identifiable as *Gerbillus nanus* (layers 1, A3) and a murid, *Arvicanthis niloticus* (layer 1). Dr. Pasa identified these finds tentatively as *G. gerbillus* and *Rattus rattus*; these manuscript identifications are repeated in the Verona museum catalogue. As already indicated *G. nanus* is present in the Two Caves assemblage (Table 12.1) while *A. niloticus* occurs at Ti-n-Torha East. Some postcranial remains of small rodents (in A1 and A2, Pasa collection) were not identified, since we had no access to adequate comparative material. The only larger rodent in the Uan Muhuggiag fauna is the North African porcupine represented by a quill fragment. As noted, a comparable find is known from Ti-n-Torha Two Caves (Table 12.1); osseous remains occur at Ti-n-Torha East. In the Verona museum catalogue the species is listed as "Histrix sp." A few carnivore remains pertain to jackal and wild cat. Jackal is represented in the Pasa collection by a mandible fragment (layer VIII) and a proximal radius (layer II; TR.D. prox: 11.5mm). These are the specimens listed in the Verona museum catalogue as Carnivora indt. Two other bones, a canine and distal moiety of a humerus, represent a felid in layer 1a of Sector B dug by Dr. Barich. As to size they fall within the range of larger house cats and can no doubt be attributed to small African wild cat. It is very likely that both specimens represent the same individual. African wild cat was also found at Ti-n-Torha East.

Dassie or rock hyrax (*Procavia capensis*) is represented by a scapula (layer V) in the Pasa collection and by a edentulous maxilla (Sector B, layer 2a) in the Barich collection. As Dr. Pasa has not yet identified this material, the species is not listed in the Verona museum catalogue. It is also known from Two Caves (Table 12.1) and Ti-n-Torha East.

The equid material contains fragments of an incomplete mandible (A1 and A2, Pasa collection) with milkdentition except for I1 which is erupting. A second animal is represented in Sector B, layer 1 (Barich collection) by a fragment of a fully grown first phalanx. The specimens do not permit adequate measurements, but suggest fairly small animals, no doubt referable to the African wild ass. In the Verona Museum catalogue, this material is referred to as _Equus_ sp. and _Equus caballus_. Wild ass has already been recorded from Ti-n-Torha East.

Warthog is represented in the Barich collection (Sector B, 2a) by a fragment of a small proximal radius with undeniable suid morphology. The size of the specimen falls within the range of fairly small domestic pig, but the bone may have shrunk since it has been burned. The Ti-n-Torha East fauna contains more typical remains of the species.

Smaller wild herbivores are represented by gazelle and Barbary sheep. Some gazelle finds occur in Sector A, layer 1 (Barich collection); they consist of two proximal fragments of an anterior cannonbone which may be derived from the same animal. Most probably this material can be assigned to dorcas gazelle, which is a typical element of Saharan prehistoric faunas (cf. Gautier 1980). It also occurs at Ti-n-Torha East and in the Two Caves assemblage (Table 12.1). The few finds attributed to Barbary sheep will be discussed together with those of small livestock.

Domestic cattle (fig. 12.1,_1_,_2_)

Domestic cattle is represented in both collections by various cranial and postcranial remains. The most famous, in the Pasa collection, comes from Layer VIII (dated 5952±120 B.P.) and consists of a skull fragment with incomplete horncore (fig. 12.1._2_). Only a few specimens, all from the Pasa collection, could be measured.

Horncore, A-P.D.	±55	
Upper M3, L.	±40.5	
Lower M3, L.	±34	
radius, TR.D.prox.	±71	±72
astragalus, H.	±65	
Ph. 1, TR.D.dist.	±25	±26.5
Ph. 3, L. sole	±67.5 (burned)	

Most of the measurements can be compared with the extensive osteometric data on the Manching cattle (Boessneck _et Alii_ 1971). They suggest animals of fairly small stature (110-120 cm).

As to the figured bucranium fragment, it has a fairly well developed horncore, but Dr. Pasa classified it as _Bos brachyceros_ (Mori 1965: 225),

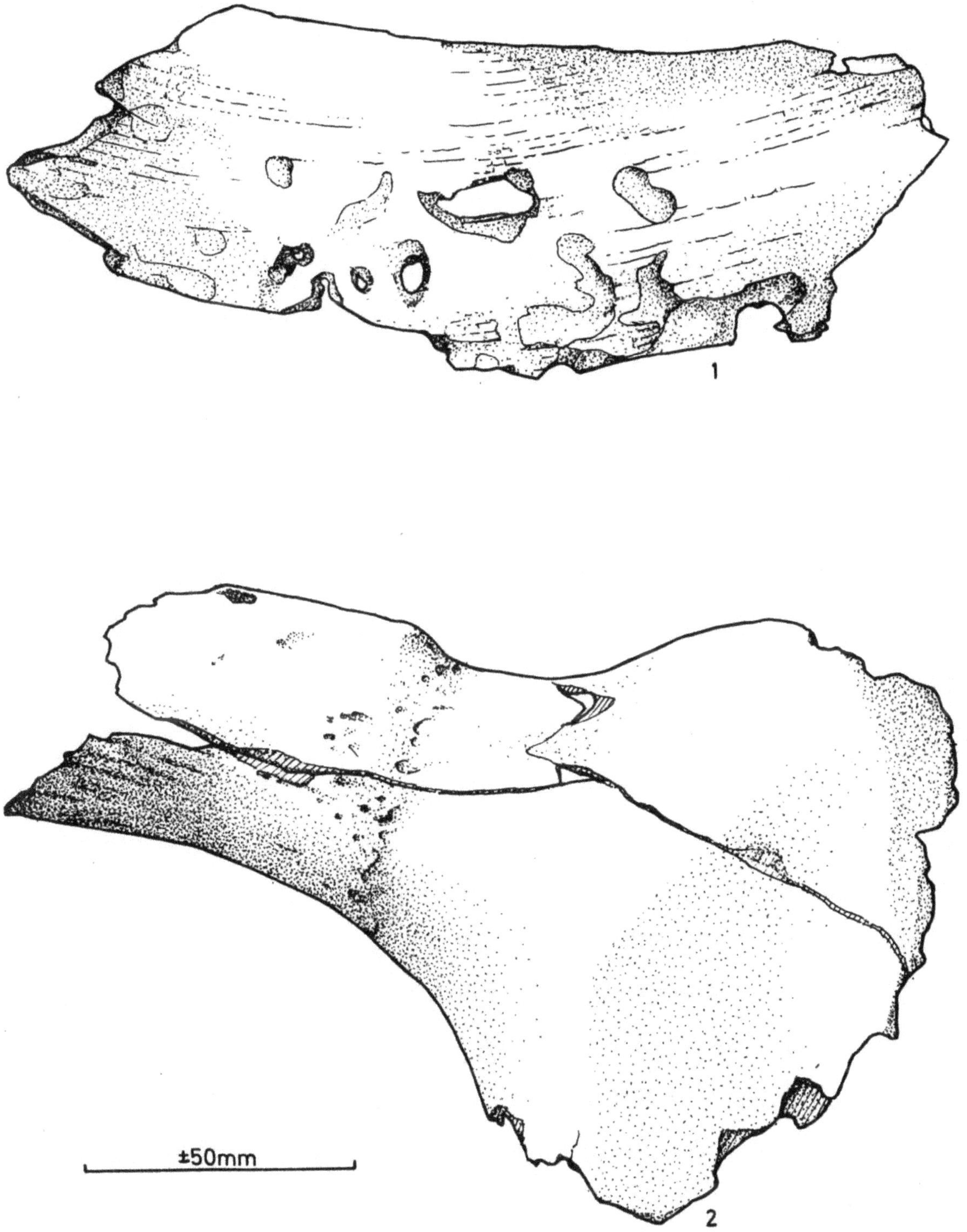

Fig. 12.1 - Uan Muhuggiag, Wadi Teshuinat. 1: horn fragment of domesticated cattle with possible traces of gnawing by insects (layer III, Pasa collection); 2: skull fragment with base of horncore of domesticated cattle, originally identified as "Bos brachyceros" (layer VIII, Pasa collection).

since he could match it with its homologs in Bronze Age cattle of the Mediterranean in Europe. In fact, it matches also fairly well its homologs in Galloroman cattle from Belgian sites in our collection. Does the foregoing mean that we are indeed dealing with a representative of a short horn breed as some have assumed? Before an answer can be given, a few concepts should perhaps be clarified.

Bos brachyceros was established by Rütimeyer (1862) as a prehistoric cattle form with relatively small horncores (length along the outer curve: 145-210 mm); other, longhorned primitive breeds identified by the same investigator would have horncores of at least 230 cm(!). We are not aware of any definition of the term shorthorn sensu lato, but Epstein (1971) describes various African breeds with hornlengths between 10 and 30 cm; the corresponding maximum horncore length would be about 25 cm. African longhorns described by the same author have minimum horn lengths of 25 cm, but typical longhorns may exceed 100 cm. Muzzolini (1983: 478) reports that in "brachycéros (récents ou actuels)" (i.e. shorthorns s.l.) the horns almost never grow larger than 20 cm. Armitage and Clutton-Brock (1976) developed a system for the classification and sexing of cattle horncores from archaeological sites with the following categories: small horned (less than 96 mm), short horned (96-150 mm), medium horned (150-200 mm) and long horned (more than 200 mm). Shorthorn breeds s.s. represent the results of breeding experiments initiated in England in the later 18th century (Mason 1969). We have no access to metrical data on the horns of these animals, but we guess that they will fall mainly in the small and short horned categories sensu Armitage and Clutton-Brock. From the foregoing we can deduce that horncore size in longhorn cattle is very variable with a poorly defined zone of overlap (in the range of 20-25 cm) with short horns sensu lato. The length of the horncore in the figured bucranium fragment cannot be estimated but could easily be more than 20 cm; if so, we are dealing with a borderline case. Moreover the circular cross section of the horncore base, the absence of pearls on this base, and the rather flat forehead may indicate that the bucranium is derived from a cow. On the other hand a horn fragment (fig. 12.1.1), in the Pasa collection, measuring 170 by 65.2 mm, is definitely derived from a large horn and hence represents a real "longhorn". The foregoing suggests that the Uan Muhuggiag cattle may have been characterized by marked sexual dimorphism of the horns.

Few osteometric data on prehistoric cattle of the Central Sahara have been published. Some data concern cattle remains from Neolithic sites (Ténérian) in the Adrar Bous (Clark et Alii 1973), where a more or less complete skeleton was excavated, dated 5760±500 B.P. on collagen. Carter and Clark (1976) estimated the height at the withers of this animal at 104 cm. We recalculated the size of the animal using the total length given for the humerus and the multiplication factor found in von den Driesch and Boessneck (1974). The results are slightly different: 104 to 107 cm, but it should be noted that the length used is only an estimate, because the distal part of the humerus is not complete. A few measurements published by Smith (1980b) concern other cattle remains from Adrar Bous (kitchenmidden remains; 4910±500 B.P.) and from Arlit (5150 B.P., western side of the Air Mountains; Smith, ibidem, 1980a). The distal widths of the humeri from Adrar Bous suggest animals with heights between about 100 and 125 cm. On the basis of published measures and some other measurements provided by Dr. Smith (pers. comm.) the Arlit cattle could be somewhat smaller (100-110 cm).

The characteristics of the horncores of prehistoric cattle in the central Sahara can be deduced only from the representations in the pastoralist rock art. The sole find which might have yielded "hard", osteological evidence is the already mentioned Adrar Bous skeleton. Carter and Clark (1976) state that the skeleton is of small size and could well represent a cattle type intermediate between a wild form such as *Bos ibericus* and the modern humpless shorthorn. Unfortunately the horncores of the specimen are very poorly preserved and nothing definite can be said about the horns. The small size of the animal however favours an identification as a cow with perhaps relatively poorly developed horncores, as in the case of Uan Muhuggiag. Shorthorn animals are very rare in the older pastoralist rock art and are no doubt extremes in a very variable longhorn stock; "real" shorthorns would make their appearance only in the first millennium B.C. (Muzzolini 1983: 471).

Dr. Pasa labelled the cattle finds in his collection as *Bos* sp. and *Bos taurus*; these names are repeated in the Verona museum catalogue.

Domestic sheep and goats, and Barbary sheep

Small livestock remains and those attributed to Barbary sheep are discussed together. The skeleton of Barbary sheep is closely comparable to that of sheep and goat, but no detailed studies are available on its diagnostic features with respect to sheep and goat. Sexual dimorphism in size appears to be very pronounced in Barbary sheep: males reach shoulderheights between 90 and 100 cm, females between 75 and 90 cm. Most primitive small livestock breeds vary in size between 50 and 75 cm (von der Driesch, Boessneck 1974). However various African ovine breeds may attain sizes exceeding 75 cm as shown by Ryder (1983: 555), especially among the savanna type thin-tailed hair sheep found from Ethiopia and Eritrea to Senegal.

Most of the ovicaprine material in our samples can be assigned fairly easily to the domestic livestock group on the basis of its reduced size. We are left, however, with a small selection of remains of larger size some of which represent undoubted Barbary sheep, whie others may either be large domestic sheep or goat, or small female Barbary sheep. A few measurements follow; the symbols behind the measurements are s(ubadult), G(oat), S(heep) and B(arbary sheep). A question mark indicates specimens of large size tentatively identified as large sheep or goat. All the measurements on the first phalanx refer to one specimen, hence the vertical line next to the column.

ml-m3, L.	34.1(?)				
P2-M3, L.	±75				
humerus, TR.D.dist.	29.0(G)	30.1(G)	±32	±35	
radius, TR.D.dist.	26.5(s)	±28(s)	30.1(sG)	±31(s)	40.0(sB)
tibia, TR.D. dist.	21.0(s)	26.2(s)			
astragalus, H.	32.0	36.0(?)	36.1(?)	39.5(B)	
Ph. 1, L.	37.4 (S)				
TR.D.prox.	13.0				
TR.D.dist.	12.2				

It is possible that some of the larger remains identified as small livestock should be added to the Barbary sheep sample. However the Barbary sheep specimens from Ti-n-Torha East and Two Caves at our disposal do not overlap in size with any of our larger ovicaprine specimens assigned to domestic livestock. We therefore are fairly certain that misidentified Barbary sheep remains constitute but a very small part of our ovicaprine sample.

Most ovicaprine cranial material is poorly preserved but several diagnostic specimens referable to goat are present. Layer VIII (Pasa collection) yielded a fragment of a twisted caprine horncore with a sharp anterior keel. Fragments of comparable horncores with a blunt keel occur in layers VIII (one specimen) and A1 (idem) of the Pasa collection and in layer 1 of Sector A (two specimens, Barich collection); one specimen in the Pasa collection is of unknown provenance. Layer V (Pasa collection) furthermore produced two ovicaprine skull fragments with fairly small horncore bases. The keeled horncore fragment probably represents a billy goat; the other specimens being smaller or without a keel may represent nanny goats. The predominance of caprine cranial fragments does not necessarily imply that goat is dominant in the small livestock sample. It is known that ovine horncores are more easily destroyed than those of goats; the same applies to skull remains of hornless animals, which constitute very often the majority of the ewes in many sheep breeds.

Our complete skeleton found in layer 2 of Sector B (Barich collection) appears to represent a juvenile sheep with milkdentition. It has very short horncores pointing outwards; their basal cross-section is suboval.

Since the remains are very fragmentary, the postcranial material does generally not allow the distinction between sheep and goat according to the diagnostic criteria established by Boessneck and collaborators (1968). The results obtained on the few diagnostic elements in the Barich collection are listed below.

	sheep	goat	total
layers A1-1a, B1	0	3	3
layers A2-2d, B1a-2d	5	2	7
totals	5	5	10

The foregoing results suggest that in the lower sequence sheep were decidedly more frequent than goats (2.5:1); in the upper sequence goats became dominant.

Pasa and Pasa-Durante (1962) write that the caprids of Uan Muhuggiag fall within the variability of "Capra mediterranea" and of shortlegged Sudanese dwarf goats. The meaning of this statement is not clear to us; moreover the very fragmentary remains in both collections do certainly not allow detailed comparisons. We judged it necessary to deny emphatically the statement by Pasa and Pasa-Durante, because it has been quoted in the literature recently (Muzzolini, Kolmer 1982).

Mammalian hair

In the Pasa collection several samples of mammalian hair were found. Most of these come from the upper sequence (A2 and A3). One sample appears to be derived from layer IV, another does not have a label. Their aspect however suggests that they may also belong to the later deposits. Some of the samples were submitted for study to Mrs. S.Y. Vons-Comis whose report is presented as an appendix to this Chapter. As can be seen there, the material does unfortunately not permit a precise identification. However cattle and small livestock are the most frequently found animals; therefore we are almost certain that the hair samples are derived from either cattle or small livestock.

Modified bones

In the Barich collection we spotted several bone fragments with typical gnawing marks left by small rodents (Sector A, layer 2a; Sector B, layer 1a). No doubt some of the small intrusive rodents mentioned from the Ti-n-Torha sites and those at Uan Muhuggiag may have left their marks on the bones. Two long bone flakes (Sector A, layer 1), with rounded edges and a surface marked with shallow dissolution hollows, may have passed through the digestive tract of a carnivore. Other and better evidence for carnivore activity is lacking.

Traces left by humans include evidence of cutting and exposure to fire. Cutting traces are not very common but beautifully illustrated by a distal humerus (Sector B, layer 2b) with numerous cuts mainly on the lateral, medial and anterior sides. An appreciable amount of bone fragments are discoloured (brown, black) by fire; a few specimens have been completely burned and are white.

Coprolites

The Pasa collection contains an appreciable number of dried coprolites. They can be divided into a carnivore and herbivore group. The carnivore group contains five specimens. They consist of a beige finely grained fluidal matrix with short hair fragments and some bone fragments, mainly toothremains of medium sized bovids. They are rather variable in size and form. Spheroid forms of different diameter (e.g. $\pm$20 mm, $\pm$ 35 mm), ovoids with a drawn out point (diameter: 17 mm) as well as remnants of what appears sausage-like coprolites occur. All of them have been collected from layers A2 (two specimens) and A4 (three specimens). These coprolites were produced by carnivores including large amounts of bone in their diet. They may have been left on the site by scavenging jackals but, given the general context, it seems more likely to us that they are indirect evidence for dogs feeding on the offal. The size of the coprolites would indicate that these were large animals.

The herbivore coprolites comprise two types both made of a orange reddish matrix of comminuted plant remains of which the enveloping smooth surface is reddish brown. Some are small subangular ovoids (max. L. = $\pm$12 mm); they are found separate except for one fragment of a cluster. The others are irregularly flattened spheroids (maximum diameter: $\pm$ 40 mm; height: $\pm$15 mm) occurring separately or irregularly stacked on each other. We found them in layers B1 (1 large type; 3 small type) and B2 (some fragmentary stacked specimens of the large type), in layer A2 (1 cluster of the small type). A comparison with recent coprolites of small livestock indicates that the larger type was produced by sheep, the smaller one by goats (see also Bang et Alii 1973: 169). Pollen analysis by Dr. Schulz on both types also indicate two herbivores with different diets (see this Volume, Chapter 13).

Insect remains

Insect remains collected during the first excavations at the site were described by Pierre (1963a). They consist mainly of tenebrionid bugs, but scarabids, dipterans (one imago and a pupa), a termite are also present. Pierre draws the attention to the scarabid bug Heliocopis gigas OL. (layer III, Pasa collection) now confined to Egypt, northern Sudan, Ethiopia and Chad, where the appearance of this bug is connected with the monsoon rains. Moreover, it is generally associated as a commensal form with large herbivores, especially (zebu) cattle. For the rest, the author hesitates to give a palaeoecological evaluation, although he insists on the presence of other southern tenebrionids (Rhytinota praelonga praelonga; Phyllognathus orion); these may now occur only in mountainous regions of the Sahara where they benefit from greater precipitation levels and water resources (cf. Pierre 1963b). Pasa and Pasa-Durante (1962: 253) record furthermore a "nest of dipterans" plastered on the shelterwall in layer A5, but they do not describe this structure. Kelner-Pillault (1983) recently recorded insect nests made or utilized by wasps, bees and spiderwasps (Pompilidae, Pseudogenia) from the early Holocene deposits in the Taoudenni Basin in Mali. The figured cattle horn fragment from the Pasa collection (fig. 12.1.1) carries ill defined traces which may have been excavated by insects.

3. TAPHONOMY

Site formation

The formation of the Uan Muhuggiag site is due to the combined action of eolian sedimentation, accumulation of desintegrating sandstone of the shelter wall and of large rock fall and anthropogenic accumulation of vegetational matter, charcoal, ash and artefactual remains. On the basis of the Barich assemblage the general "archaeozoological productivity" of the site can be estimated at approximately 150 bone fragments per cubic metre (1). This estimate has been compared with those for central Sudanese sites (Gautier 1985). The effect of the various factors influencing bone productivity of a site are difficult to gauge, but no doubt the rather low productivity of Uan Muhuggiag is mainly due to the fact that sedimentation, being anthropogenic and aeolian, was generally slow.

Taphonomic groups

As usual the material can be divided into several taphonomic groups, on the basis of such criteria as specific identity, possible economic significance and preservation. The most important group comprises naturally the animals used by people, in our case: fish, turtle, bird(s),

---

1) Total number of bone fragments divided by the total excavated volume: (2600:17) fragments per m^3.

various game animals and domestic livestock. A second much less important group contains the various intrusives: small land snails, amphibians and small reptiles, shrew, chiroptera, gerbil, lesser jerboa and field rat. Most of these are probably pene-contemporaneous intrusives as their remains are preserved very much like those attributed to the major taphonomic group. The rodent traces pertain to the same group.

The skeleton of a juvenile sheep poses a problem. It shows no evidence of butchering or cutting and no traces left by scavenging carnivores. Moreover the archaeological context had no particular characteristics. Such a find is often encountered in Neolithic and later sites where the dead bodies or skinned carcasses are left on the site; this is frequently the case with dogs, cats, horses, asses etc. It is not clear why our lamb would not have been eaten if people were around. We can speculate that it was perhaps killed by a disease which was thought to make it unfit for consumption, but would that keep the dogs away which the site occupants very probably had? Or was it a stray animal that found its way back to the abandoned site and died there? In arid conditions the dead bodies of emaciated or small animals dehydrate rapidly and mummify; therefore they do not attract scavenging carnivores, except when nothing better is available. The find reminds us of the already mentioned cattle skeleton from Adrar Bous, which also does not show evidence of butchering or scavenging (Carter, Clark 1976).

4. PALAEOECOLOGY AND PALAEOECONOMY

Present day situation

The Tadrart Acacus lies within the zone of the continental, fully arid desert climate of the central Sahara and within the central Sahara vegetational zone as defined in Maley (1983: 375, fig. 1). The average annual precipitation would remain below 20 mm (Pasa, Pasa-Durante 1962: 254). The present day vegetation appears to be restricted to the valleys and consists of a very sparse, discontinous cover of grass (*Aristida*) stunted acacias and some shrubs. Wild animals which can still be found in the area include the horned viper, some lizards, a few birds, hare, gundi, gerbil, hyrax, golden jackal, striped hyena, dorcas gazelle and Barbary sheep, but game population densities are extremely low; not doubt ruthless hunting has contributed to this situation.

Two Caves

As already pointed out in a previous paper, the Two Caves fauna can be compared with that collected at Ti-n-Torha East. The sample is much smaller and hence the faunal spectrum is less diversified. However both sites lie in the same valley and have comparable industries and dates. This suggests that the palaeoecological and palaeoeconomical adaptations deduced from the Ti-n-Torha East fauna are probably also valid for the Two Caves assemblage. A summary of the composition and evaluation of the Ti-n-Torha East fauna follows (cf. Gautier 1984a).

The fauna collected at Ti-n-Torha East comprises a few intrusive landsnails, a Mediterranean marine gastropod (*Monodonta auriculata*), some fish remains, an intrusive toad or frog, some smaller intrusive reptiles, a varied bird fauna (at least 17 species), ostrich eggshell, an intrusive

Table 12.4 : The ratio large to small livestock at Uan Muhuggiag (fragment counts).

	n	%	n	%	n	%	n	%	n	%
Pasa-collection	A1 to A4		I to V		V to X		I to X		totals	
domestic cattle	14	15.9	60	52.2	17	56.7	77	53.1	91	39.1
sheep/goat	74	84.1	55	47.8	13	43.3	68	46.9	142	60.9
totals	88	100.0	115	100.0	30	100.0	145	100.0	233	100.0
Barich-collection	A1,1a;B1		A2,2a;B1,2,2a		A/B2b to 2d		A/B1 to 2d		totals	
domestic cattle	4	3.7	23	25.6	15	31.3	38	27.5	42	17.1
sheep/goat	103	96.3	67	74.4	33	68.7	100	72.5	203	82.9
totals	107	100.0	90	100.0	48	100.0	138	100.0	245	100.0

* for the correlation see text

shrew, desert hedgehog (Paraechinus aethiopicus), baboon (Papio cynocephalus), hare (Lepus capensis), porcupine (Hystrix cristata), small intrusive rodents including grass rat (Arvicanthis niloticus) and mzab gundi (Massoutiera mzabi), various carnivores including golden jackal (Canis aureus), sand fox (Vulpes rueppelii), Egyptian mongoose (Herpestes ichneumon), another small viverrid, lion (Felis leo) and wild cat (Felis libyca), hyrax or dassie (Procavia capensis), wild ass (Equus africanus), warthog (Phacochoerus aethiopicus), dorcas gazelle (Gazella dorcas), dama gazelle (G. dama) and Barbary sheep (Ammotragus lervia). A few remains of a large unidentified bovid have been ascribed tentatively to African buffalo (Syncerus caffer) and hartebeest (Alcelaphus buselaphus). We previously suggested that these poorly preserved specimens might represent early domestic cattle comparable to the hypothetical domestic cattle from the Western Desert (Gautier 1980, 1984); it now seems much less likely that the Ti-n-Torha East people were experimenting with incipient pastoralism.

Most of the animals encountered can still live in the area, but if present their population densities are very low because of the ecological degradation due to climatic as well as human factors. The anthropogenic fauna composed of fish, birds and mammals points to an intensive utilization of local resources. The extremely poor fish fauna contains only catfish (Clarias sp.) and Tilapia, two species very resistant to adverse conditions (low oxygen, high salinity). Moreover the Tilapia appear to be stunted. The foregoing indicates that the fish may have been taken from permanent isolated water fed by springs as found in Ti-n-Torha Valley or in pools seasonally fed by Wadi Auis. The high number of predators and scavengers among the birds and mammals, suggests that the site lies in an area where these creatures tended to concentrate within an otherwise quite harsh environment. No precise reconstruction of climatic conditions can be made on the basis of the fauna, but the climate may have approached that of the northern Sudano-sahelian belt which now has a rainfall of some 200 mm. The presence of migratory, or putative migratory, birds (mallard, harrier, black kite, houbara bustard(?)) and the predominance of autochthonous game (Barbary sheep, hare) suggest occupation of the site during the dry season in late autumn, winter or early spring.

A more permanent occupation of the upper levels using stone structures may coincide with the onset of drier conditions forcing people to exploit more intensively Ti-n-Torha East at about 8500 B.P., for Maley (1977) places on the onset of a drier phase in the Tibesti at about 8550 B.P. The more intensive exploitation of Ti-n-Torha seem to be reflected in the upper faunal sequence (levels CI and CII) by a decrease of larger game (Barbary sheep, gazelle, wild ass) most likely due to increased predation and disturbance of these animals by man.

In our previous papers we did not point out that stone structures are also present at Two Caves from level II on, although these are less elaborate than those found at Ti-n-Torha East. Moreover the radiocarbon dates and the artifactual data indicate that the Two Caves assemblage would correspond to the pre-structure levels at Ti-n-Torha East (pers. comm. Dr. Barich). Nevertheless, the percentages of the major game groups (Barbary sheep; gazelle etc., small mammals (cf. Gautier, Van Neer 1977-1982: 104, Table 3) in levels I and II compare favourable with those obtained in Ti-n-Torha East, upper levels:

Barbary sheep	32	71.1%
gazelle	4	8.9%
small mammals	9	20.0%
total	45	

The sample however is restricted and perhaps not representative.

Uan Muhuggiag

If the faunal evidence from the Pasa and the Barich collections can be combined, the spectrum obtained is fairly diversified containing catfish, turtle, one avian species and 14 non-intrusive mammals (or 15 if we include the indirect evidence for the presence of dog). Of these mammals 12 are wild and also known from the older sites at Ti-n-Torha. However their frequency is very low and the bulk of the collection consists of domestic mammals, more precisely large and small livestock: 88.6% in the Pasa collection and 92.4% in the Barich collection. Among the game Barbary sheep is dominant (7.1% in the Pasa collection; 4.2% in the Barich collection). It is clear that a major change has occurred; animal husbandry has apparently become a major activity. In a previous publication (Gautier, Van Neer 1977-1982) we over-emphasized the role of hunting at Uan Muhuggiag. The quantitative analysis performed here bears out that hunting was probably practised only in a haphazard way, when the occasion presented itself.

The foregoing paragraph contains a general characterization of the Uan Muhuggiag fauna. It appears however that some quantitative changes through time are detectable. Indeed Pasa (cf. Mori 1965: 225-226) already noted a quantitative shift in the fauna. Table 12.4 summarizes the shift in the composition of large and small livestock in both collections, on the basis of the following approximate correlation.

Upper sequence

levels A1 to A4 = levels 1 and 1a (Sector A)/level 1 (Sector B)

Lower sequence

levels I to V = levels 2 and 2a (Sector A)/levels 1a, 2 and 2a (Sector B)

levels VI to X = levels 2b to 2d (both Sectors).

Both collections showed comparable slight shifts in the importance of large versus small livestock and we also observed a consistent difference in their relative importance: in the Pasa collection large cattle are much more important than ovicaprids in each subdivision used. This consistent difference points probably to a systematic difference in sampling technique or in the treatment of the samples. We already suggested that both collections were made in comparable ways but the identification rates are different (approximate 7.5% in the Pasa collection, 11.5% in the Barich collection). This could be mainly due to our rapid re-analysis of the Pasa collection being less careful with as a result that more specimens ended up in the "not-identified"-pile. These specimens may have been predominantly derived from small livestock, since this is the most frequent group and remains of small animals are more easily dismissed as non-diagnostic. (We calculated percentages of small livestock on the basis that about 4% of the unidentified bones were in fact small livestock and found that the obtained figures approach those for the Barich assemblage). The observed discrepancy can hence be explained as due to changes in the analytical procedure. It is difficult to translate the ratios in Table 12.4 into animals on the hoof and herds, since we have little knowledge of the socio-economic significance of cattle for our prehistoric pastoralists. One can imagine without too much difficulty that, without having a "cattle-complex", these people more easily killed their small livestock. The taphocoenoses found may as a result overemphasize small livestock because of a much higher turnover. In any case it seems clear that the earlier Uan Muhuggiag pastoralists had decidedly more cattle than those that occupied the site after the roof collapse. We already noted that in the upper sequence goats became dominant and the general increase of ovicaprids and of goats in particular reflects no doubt an ecological deterioration also evident in the palynological record (Pasa, Pasa-Durante 1962; Mori 1965; Schulz, this Volume, Chapter 13). The slight increase of small livestock in the lower sequence, visible in both collections, does not seem to be statistically significant.

The lower Uan Muhuggiag assemblage is culturally comparable with that of the Ti-n-Torha North. There, an upper layer (I) with dung from domestic species and a layer II with many plant remains were excavated; this sequence may be slightly younger than the lower sequence at Uan Muhuggiag. Unfortunately faunal remains are scarce in it and comprise only cattle (29 fragments) and two intrusive bones attributable to mzab gundi (Massoutiera mzabi). This poor assemblage emphasizes the importance of cattle much more than the lower assemblage of Uan Muhuggiag. The significance of this discrepancy is not clear. It may reflect real differences in composition of the livestock herds of the pastoralist groups occupying Ti-n-Torha and Uan Muhuggiag; in our opinion it is more likely to be due to chance.

On the basis of palynological and geological data Pasa and Pasa-Durante (1962: diagram in fine) suggested that Uan Muhuggiag would have been occupied during a period with a maximum of 150 mm precipitation in level X, later oscillating between about 50 and 120 mm; thus, according to the same authors, the carrying capacity of the valley would have been below or just within the lower limit for permanent herds. In our previous evaluation we estimated that a natural corral at Ti-n-Torha could easily have been used like those described by Lhote (1969), but that its carrying capacity was so limited that it could support only small herds (20-25 heads of cattle) for short periods (3 months). At that time we did not yet know that the cattle may have been mixed with an appreciable number of small livestock admittedly weighing much less than cattle, but nevertheless reducing the number of cattle (2). Moreover, we should perhaps consider that occupation of the pastoralist sites was much shorter than we presumed, perhaps only a few weeks until grazing became unsufficient. As said before, Saharan rock art emphasizes cattle very much and may offer a rather biased view of its

---

2) A herd of 20 heads of cattle can be equated roughly with a mixed herd with an approximate ratio of cattle to small livestock of 13:26.

Table 12.5 : The fauna of Amekni (fragment counts)(1).

Animal group	Quantity
unidentified freshwater bivalves(2)	R
catfish (*Clarias* sp.)	F
perch (*Lates* sp.)	R
Barbary partridge? (*Alectoris barbara*)	2
francolin (*Francolinus* sp.)	1
unknown bird	1
Nile? monitor (*Varanus niloticus*)	R
Gaboon viper? (*Bitis gabonica*)	F
spitting? cobra (*Naja nigricolis*)	F
gerbil (*Gerbillus* sp.)	1
grass rat (*Arvicanthis niloticus*)	2
tatera (*Tatera* sp.)	2
unidentified small rodents	F
marsh cane rat (*Thryonomys swinderianus*)	R
North African porcupine (*Hystrix cristata*)	2
genet (*Genetta genetta*)	4
Egyptian mongoose (*Herpestes ichneumon*)	1
dassie (*Procavia capensis*)	R
warthog (*Phacochoerus aethiopicus*)	R
dorcas gazelle (*Gazella dorcas*)	26
bohor reedbuck (*Redunca redunca*)	21
hartebeest (*Alcelaphus buselaphus*)	1
giant buffalo (*Pelorovis antiquus*)(3)	±10

(1) based on BOUCHUD (1968), CAMPS (1968), GASC (1968).

(2) *Caelatura* sp. and perhaps *Mutela nilotica*.

(3) = *Bubalus antiquus* sive *Homoioceras antiquus* auctores.

R = rare; F = frequent.

numerical importance. Cattle may indeed have had a much higher socio-economic status than sheep and goats. The faunal samples themselves however do not allow this to be tested by age distributions.

Transhumance

Neither the Pasa or Barich assemblages provide evidence for seasonality. We intimated in our previous study that the predominance of Barbary sheep in the wild fauna suggested a dry season occupation as in the case of the older hunter-gatherer assemblages; however the game fauna is restricted and we cannot uphold this opinion any more. It remains nevertheless very likely that our pastoralists did not occupy Uan Muhuggiag permanently if they wanted to avoid overgrazing. Most likely their transhumance movements were to the East because the Acacus makes westward movements difficult. Gabriel (1977: 177) described hearths attributed to Neolithic pastoralists from various regions in Libya; the most western occurrences are near Sebha, one of these with a date of 4155±165 B.P. The nearest stone fireplaces are found in Algeria, in the Serir between the Tassili-n-Ajjer and the Hoggar Massif, South of Djanet with dates from 5730±120 B.P. to 4830±145 B.P. Another excavated fireplace (German: Brandgrube) from the same region yielded a date of 4715±295 B.P. and the remains of a "bovid larger than domestic cattle" (ibidem: 40, table 5). At Ti-n-Torha East, the Mediterranean gastropod Monodonta auriculata was found, indicating contact, direct or more likely indirect, with groups living north of the Acacus.

5. COMPARISONS WITH OTHER REGIONS

Few faunas from adjacent regions in the central Saharan zone are available for comparison, although research has been carried out in many territories. It seems almost as if those working there, even within the latest decades, undervalued the importance of precise, quantified faunal studies, but perhaps we underestimate the logistic and other problems of Saharan fieldwork. The best data are still those from Amekni. As known, Amekni is situated in the Hoggar, about 40 km WNW of Tamanrasset. Table 12.5 summarizes its faunal assemblage on the basis of our appraisal of the descriptions given by Bouchud (1968), Camps (1968) and Gasc (1968). The photographs of incisor fragments of Thryonomys (Camps, ibidem) suggest that we are dealing with the marsh cane rat (T. swinderianus) as also suggested by the present distribution of this species. The identification of the gazelle is discussed in the next paragraph. The identity of the large bovid remains was verified with the aid of the work by Mr. J. Peters (Ghent, Peters in preparation) on the diagnostic features of (wild) cattle versus those of the African buffalo (Syncerus caffer); certain morphological details and size confirm the identification as Pelorovis antiquus.

Four radiocarbon dates are known for the site of which the extremes are 8670±150 B.P. (base) and 5500±250 B.P. (upper deposits) (3); moreover the fauna is rather poor and the finds have been conflated. The foregoing remarks makes clear that a detailed comparison of the Acacus faunas and that of Amekni is impossible. Anyhow, the assemblages show marked differences. First the Amekni people seem to have had much easier access to riverine

---

3) The B.C. date for the youngest sample is 3550 B.C.; a printing error in Close (1980:163) makes it into 3350 B.C.

resources, as exemplified by the high frequency of catfish, the presence of _Lates_ (needing well oxygenated water) (cf. Gautier, Van Meer 1977-1982: 119-121), and that of large freshwater bivalves indicating a perennial larger water body. Second the mammalian fauna is dominated by gazelle, bohor reedbuck and giant buffalo, Barbary sheep, the dominant species of Ti-n-Torha being virtually absent.

Bohor reedbuck, a grazing antelope of open to less open biotopes with high grass, lives never more than 5 to 8 km away from drinking water. Today it occurs in most of more humid Sudano-Sahelian belt and the interior of East Africa (Haltenorth, Diller 1979). Kingdon (1982) classifies it as a typical inhabitant of grasslands other than those near rivers (valley grasslands) and dry to subdesert "grasslands". Camps (1968: 202; 1974: 225), refers to the species as "cobe" and states that "les cobes (_Redunca redunca_) ont une prédilection pour les forêts galeries ou les savanes à fonds marécageux". Such a statement would rather refer to other reduncines such as "le cobe du Buffon" (_Kobus kob_) or, even better, "le cobe defassa" (_Kobus ellipsiprymnus_). These reduncines are larger than the one identified at Amekni and much more linked to water. Anyhow, the Amekni reduncine certainly suggests more humid conditions than today around the site, but not gallery forests or large expanses of swampy terrain.

Giant buffalo is also an indicator of good grazing grounds, but until now, no real effort has been made to evaluate the requirements of this large bovid. The most frequent antelope at Amekni, however, is gazelle, which Bouchud (1968) identifies tentatively as dorcas gazelle. His measurements do indeed suggest a small gazelle such as the dorcas gazelle, the redfronted gazelle being generally somewhat bigger. Dorcas gazelle is a typical inhabitant of the dry Sahelian savanna and any suitable locality of the Saharan desert, preferring biotopes with stony soil. The redfronted gazelle is less well adapted to very dry conditions (Haltenorth, Diller 1979). Amekni is situated within granite country with stony soils, granitic inselbergs and boulder fields; such a habitat suits dorcas gazelle very well.

All in all, the Amekni fauna suggests a climate more humid than that in the Acacus. On the basis of the present distribution of bohor reedbuck and African buffalo (as a substitute for giant buffalo), the annual precipitation during the occupation can be estimated at some 250-300 mm. Bouchud (_ibidem_) reports that today the region around Tamanrasset has an average rainfall of less than 200 mm; a more precise recent estimate gives about 46 mm (Messerli _et Alii_ 1980: 94, Table 5.2), which is more than the amount given for the Acacus by Pasa and Pasa-Durante (1962). No doubt, in the past comparable pluviometric differences existed and account in part for the observed faunal differences.

If the fauna is representative of the whole occupation period and if this period lasted indeed until the middle of the 6th millennium B.P., it would seem that at Amekni hunting persisted as an important subsistence activity; in the Acacus people had already fully adopted pastoralism in the 6th millennium. As to seasonality at Amekni, the available data are too incomplete to allow a detailed evaluation. In large rivers, catfish can be most easily harvested in the alluvial plain during the floodseason, but we have no idea of the characteristics of the Wadi Amekni during the occupation of the site. The birds are apparently all resident species and the mammals could all be autochthonous or are too poorly represented to evoke migration patterns.

Another site in the Hoggar, at Méniet, situated NNW of Tamanrasset yielded also a fauna, identified by Arambourg (Hugot 1963). Unfortunately no quantitative data are given. The following species seem to be present: freshwater fish, turtle, Nile? monitor (*Varanus niloticus*), Nile crocodyle (*Crocodylus niloticus*), an unknown murid of the size of rat (*Rattus sp.*), marsh cane rat (*Thryonomys swinderianus*), striped hyaena (*Hyaena hyaena*), bohor reedbuck (*Redunca redunca*), oryx (*Oryx gazella*), giant buffalo (*Pelorovis antiquus*), dorcas gazelle (*Gazella dorcas*), a canid either golden jackal (*Canis aureus*) or dog and a medium sized *Bos* identified as *Bos ibericus*?, which in our view could be domestic cattle. Camps (1968) lists also *Capra* sp., which could be small livestock. The radiocarbon date for Méniet is 5400±150 B.P. and the middle could be contemporaneous with the upper deposits of Amekni; adoption of pastoralism with both large and small livestock appears to have begun but the absence of quantitative data precludes any precise estimation of how far it has progressed. As it stands, the combined data of Amekni and Méniet indicate a late persistence of hunters in the Hoggar and may reflect the slightly better ecological conditions there. Indeed, if the adoption pastoralism was linked with some form of ecological stress due to climatic change, it could be appreciably diachronic.

Archaeozoological data are also available from the Taoudenni Basin (Petit-Maire, Riser 1983), Adrar Bous (Clark *et Alii* 1973), Tibesti, Serir Tibesti, Serir Calanscio (cf. Gabriel 1977), Gilf Kebir, Nabta Playa and the Bir Kiseiba region (Gautier 1980, 1984). A comparison with our finds is beyond the scope of this paper and will be included in a more general evaluation of the origin of pastoralism in North Africa (Gautier in press), but it is already clear that the results of this evaluation are disappointing: most of the data are very imprecise and come from widely separated regions. However, they do not contradict the hypothesis that the adoption of pastoralism may have been markedly diachronic due to the ecological differences between the various regions involved. The oldest dates for domestic cattle would be those from the Western Desert. Both at Nabta and Bir Kiseiba, large bovid remains were found in a faunal spectrum dominated by dorcas gazelle and hare, suggesting very poor ecological conditions. We identified these remains tentatively as domestic cattle. Most scholars can probably accept, without too much reluctance, these identifications for material from the later period (Middle or Late Neolithic: 7000 B.P. and later), but our putative domestic cattle would go back to the late 10th millennium! We suggested that the cattle originated from the Nile Valley where it had been one of the major game species for several thousand years (cf. Gautier 1968) and that it was brought to the desert in response to the very poor living conditions. Sheep and goats, coming from outside Africa, was adopted later and at Nabta cattle is accompanied by sheep and goat from Middle Neolithic times on. In the light of the foregoing hypothesis, our very tentative identification of ill defined large bovids as early domestic cattle at Ti-n-Torha East loses much of its possible significance. It seems now much more reasonable to accept that the earlier occupants of Ti-n-Torha were pure hunters-and-gatherers and that, if early cattle domestication experiments were being made, these were restricted to the Western Desert.

SUMMARY AND CONCLUSIONS

The Acacus has provided us until now with faunal evidence for three subsistence patterns. The first is documented at Ti-n-Torha East and Two

were fishing, fowling and hunting, the most important animal resource apparently being Barbary sheep, a typical caprid of mountainous regions. Occupation was probably seasonal during the dry season (in winter?) and rendered attractive by the presence of a 'guelta' fed by a source. In the later occupation period at Ti-n-Torha people built stone structures suggesting more permanent or regular occupation, which may have had a negative effect on the local fauna and which itself may have been triggered by worsening ecological conditions.

During the second period (6th millennium?: dates from 7438±220 B.P. to 5260±130 B.P.) Ti-n-Torha North was occupied by pastoralists, who are much better represented at Uan Muhuggiag, where we find, in the lower sequence a diversified faunal assemblage dominated by small and large livestock with a ratio of about 3:1. Among the small livestock sheep predominates (ratio sheep/goat 2.5:1).

A third period is represented by the upper sequence (late 5th millennium B.P. or 4th millennium) with again a pastoralist faunal spectrum, this time dominated by small livestock (more than 90%) apparently mainly attributable to goat. This shift suggests a much drier environment.

No direct evidence for seasonal occupation is available for the pastoralist phases, but it can be assumed that the pastoralists moved into the valleys of the Acacus during the dry season. Faunal data from other regions (Amekni, Méniet; Taoudenni Basin; Adrar Bous; Tibesti, Serir Tibesti; Western Desert) are often very incomplete, imprecise or hypothetical. They suggest appreciable regional differences in wildlife during the early Holocene and do not contradict the hypothesis that the adoption of pastoralism may have been markedly diachronic, as this change of life style may represent a response to worsening climatic conditions. In view of the dearth of precise data, it does not yet seem warranted to present reconstructions of the origin of large and small livestock in the different regions. Moreover this process may have been complex, involving local "domestication experiments" (cattle) as well as the acceptance of domesticates (cattle, sheep and goats) brought from outside Africa.

REFERENCES

ARKELL A.J., CORNWALL J.W. e MORI F. 1961 - Analisi degli anelli componenti la collana della mummia infantile di Uan Muhuggiag, Riv. di Antropologia, 48: 161.

ARMITAGE P.L., CLUTTON-BROCK J. 1976 - A System for Classification and description of the Horncores of Cattle from archaeological sites, Journ. Archaeol. Sci., 3: 329-348.

ASCENZI A., COCKBURN A. and KLEISS E. 1983 - Miscellaneous mummies, in Mummies, Disease and Ancient Cultures, eds. Cockburn A., Cockburn E., Abridged Edition, Cambridge University Press: Cambridge, pp. 224-228.

BANG P., DAHLSTROM P. and VAN WIJNGAARDEN A. 1973 - Elseviers Diersporengids, Elsevier: Amsterdam/Brussel.

BARICH B.E. 1974 - La serie stratigrafica dell'Uadi Ti-n-Torha (Acacus, Libia), Origini, VIII: 7-184.

BEHRENSMEYER A.K. 1978 - Taphonomic and ecologic information from bone weathering, Paleobiol., 4: 150-162.

BOESSNECK J., MÜLLER H.-H. und TEICHERT M. 1964 - Osteologische Unterscheidungsmerkmale zwischen Schaf (Ovis aries Linné) und Ziege (Capra hircus Linné), Kühn-Archiv., 78: 1-129.

BOESSNECK J., von den DRIESCH A., MEYER-LEMPPENAU U. und WECHSLER-VON OHLEN E. 1971 - Die Tierknochenfunde aus dem Oppidum von Manching, Steiner Verlag: Wiesbaden.

BOUCHUD J. 1968 - Etude des mammifères et des oiseaux d'Amekni, in Camps G., Amenki. Néolithique Ancien du Hoggar, Mémoire du C.R.A.P.E., 10 (A.M.G.: Paris): 173-177.

CAMPS G. 1968 - Amekni. Néolithique Ancien du Hoggar, Mémoire du C.R.A.P.E., 10 (A.M.G.: Paris), 230 p.

CAMPS G. 1974 - Les Civilisations préhistoriques de l'Afrique du Nord et du Sahara, Doin: Paris.

CARTER P.L., CLARK J.D. 1976 - Adrar Bous and African cattle, in Proceedings of Panafrican Congress, VIIth Session 1971, Addis Ababa: 487-493.

CLARK J.D., WILLIAMS M.A.J. and SMITH A.B. 1973 - The Geomorphology and Archaeology of Adrar Bous, Central Sahara: a Preliminary Report, Quaternaria: 245-297.

CLOSE A.E. 1980 - Current Research and Recent Radiocarbon Dates from Northern Africa, Journ. Afric. Hist., 21: 145-167.

EPSTEIN H. 1971 - The Origin of the Domesticated Animals of Africa, I - II, Africana Publ. Corp.: New York/London/Munich.

GABRIEL B. 1977 - Zum ökologischen Wandel in Neolithikum der östlichen Zentral - Sahara, Berliner Geograph. Abhand., 27 (Institut für Physische Geographie der Universität: Berlin), 111 p.

GASC C. 1968 - Etude des reptiles de la faune d'Amekni, in Camps G., Amekni. Néolithique Ancien du Hoggar, Mémoire du C.R.A.P.E., 10 (A.M.G. : Paris): 171-172.

GAUTIER A. 1968 - Mammmalian remains of the Northern Sudan, in The Prehistory of Nubia, Vol.I, ed. Wendorf F., Fort Burgwin Research Centre and Southern Methodist University Dallas, pp. 80-90.

GAUTIER A. 1980 - Contributions to the Archaeozoology of Egypt, in Wendorf F., Schild R., Prehistory of the Eastern Sahara, Academic press: New York, pp. 317-344.

GAUTIER A. 1983 - Animal life along the prehistoric Nile : the evidence from Saggai 1 and Geili (Sudan), Origini, XII: 50-115.

GAUTIER A. 1984 - Archaeozoology of the Bir Kiseiba region, Eastern Sahara, in Cattle-Keepers of the Eastern Sahara: the Neolithic of Bir Kiseiba, ed. Close A., assemblers Wendorf F., Schild R., Department of Anthropology, Institute for the Study of Earth and Man, Southern Methodist University: Dallas, pp. 49-72.

GAUTIER A. 1984 a - New data concerning the prehistoric fauna and domestic cattle from Ti-n-Torha (Acacus, Libya), Palaeoecology of Africa, 16:305-309.

GAUTIER A. 1985 - La faune de l'occupation néolithique d'el Kadada (secteurs 12-22-32) au Soudan Central, Archéologie du Nil Moyen, 1: 9-111.

GAUTIER A. in press - Prehistoric men and cattle in North Africa: a dearth of data and a surfeit of models, in Prehistory of Arid North Africa - Essays in honor of Fred Wendorf, ed. Close A.E., Southern Methodist University: Dallas.

GAUTIER A., VAN NEER W. 1977 - 1982 - Prehistoric fauna from Ti-n-Torha (Tadrart Acacus, Libya), Origini, XI:87-127.

HALTENORTH TH., DILLER H. 1979 - Elseviers Gids van de Afrikaanse Zoogdieren, Elsevier: Amsterdam/Brussel (Dutch edition revised by C. Smeenk).

HUGOT H.J. 1963 - Recherches préhistoriques dans l'Ahaggar nord-occidental 1950 - 1957, Mémoire du C.R.A.P.E.., 1 (A.M.G.: Paris), 206 p.

KELNER-PILLAULT S. 1983 - Insectes, in Sahara ou Sahel?, eds. Petit-Maire N., Riser J., Laboratoire de Géologie du Quaternaire du C.N.R.S.: Marseille, pp. 153-155.

KINGDON J. 1982 - East African mammals IIIc, Bovids, Academic Press: New York.

LHOTE H. 1969 - Recherches sur les voies de migrations et de la zone d'expansion des populations pastorales préhistoriques du Sahara, Actes du Premier Colloque International d'Archéologie Africaine, Fort-Lamy: 169-285.

MALEY J. 1977 - Paleoclimates of Central Sahara during the early Holocene, Nature, 269 (5629): 573-577.

MALEY J. 1983 - Histoire de la végétation et du climat de l'Afrique nord tropicale au Quaternaire récent, Bothalia, 14, 3-4: 377-389.

MASON I.L. 1969 -A World Dictionary of Livestock Breeds, Types and Varieties, Commonwealth Agric. Bur.: Farmham Royal, Bucks (2nd ed.).

MEADOW R. 1980 - Animal bones : problems for the archaeologist together with some possible solutions, Paléorient., 6: 65-77.

MEESTER J., SETZER W. eds. 1971-78 - The mammals of Africa. An identification manual, Smithsonian Institution Press: Washington.

MESSERLI B., WINIGER M. and ROGNON P. 1980 - The Saharan and East African uplands during the Quaternary, in The Sahara and the Nile, eds. Williams M.A.J., Faure H., Balkema: Rotterdam, pp. 87-132.

MORI F. 1965 - Tadrart Acacus. Arte rupestre e culture del Sahara prehistorico, Einaudi: Torino.

MORI F., ASCENZI A. 1959 - La mummia infantile di Uan Muhuhhiag. Osservazioni antropologiche, Riv. di Antropologia, 46: 125-148.

MUZZOLINI A. 1983 - L'art rupestre du Sahara Central: classification et chronologie. Le boeuf dans la préhistoire africaine (Thèse 3ème cycle) Université de Provence (Aix-Marseille I).

MUZZOLINI A., KOLMER H. 1982 - Les peintures des "Têtes Rondes" et les peintures de l'ère "pastorale" dans l'Acacus (Libye). Chronologie relative et chronologie absolue, Ars prehistorica, 1: 99-118.

PASA A., PASA-DURANTE M.V. 1962 - Analisi paleoclimatiche nel deposito di Uan Muhuggiag, nel Massiccio dell'Acacus (Fezzan Meridionale), Mem. Mus. Civ. di Storia Nat. di Verona, 10: 251-255.

PETIT-MAIRE N., RISER J. eds. 1983 - Sahara ou Sahel? Quaternaire récent du bassin de Taoudenni (Mali), Laboratoire de Géologie du Quaternaire, C.R.N.S.: Marseille.

PIERRE F. 1963a - Note sur une faune d'insectes subfossiles recueillie au Fezzan par le professeur Angelo Pasa, Mem. Mus. Civ. di Storia Nat. di Verona, 11: 197-204.

PIERRE F. 1963b - Sur quelques ténébrionidés recueillis au Fezzan par le professeur A. Pasa, Mem. Mus. Civ. di Storia Nat. di Verona, 11: 205-209.

RÜTIMEYER L. 1862 - Die Fauna der Pfahlbauten in der Schweiz, Neue Denkschr. Schweiz. Gesell. Naturwiss., 19 (Zürich), pp. 1 -248.

RYDER M.L. 1983 - Sheep and Man, Duckworth: London.

SMITH A.B. 1980a - The Neolithic tradition in the Sahara, in The Sahara and the Nile, eds. Williams M.A.J., Faure H., Balkema: Rotterdam, pp. 451-465.

SMITH A.B. 1980b - Domesticated cattle in the Sahara and their introduction into West Africa, in The Sahara and the Nile, eds. Williams M.A.J., Faure H., Balkema: Rotterdam, pp. 489-501.

SORBINI L., PASA-DURANTE M.V., BONATO G., MONTOLLI A. e TOMMASINI BADIA P. 1974 - Le collezioni paleontologiche quaternarie del Museo Civico di Storia Naturale di Verona, Museo Civ. di Storia Nat. di Verona: Verona.

von der DRIESCH A., BOESSNECK J. 1974 - kritische Anmerkungen zur Widerristhöhenberechnung aus Längenmassen vor-und Frügeschichtlicher tierknochen, Säugetierk.Mitt., 22(4): 325-348.

APPENDIX

HAIR FINDS FROM UAN MUHUGGIAG, TADRART ACACUS

By Sandra Yolanda VONS-COMIS*

The earlier excavations at Uan Muhuggiag yielded various organic remains, which were deposited in the Museo Civico di Storia Naturale di Verona. They included a few tufts of hair, some with fragments of skin attached. Most of the hair samples come from layers A2 and A3 in the Upper Sequence (see Gautier, this Chapter, Table 12.2). One sample does not have a label, another would be derived from layer B3 (= layer IV). These samples do not differ from the first set. As suggested by Gautier, they are probably also derived from the Upper Sequence, which is dated 4730±310 B.P.

Through the agency of Dr. A. Gautier (Ghent, Belgium) the hair samples were sent to the Albert Egges-van Giffen Instituut voor Prae- en Protohistorie (IPP, Amsterdam) where the author is conducting research into wool and hair remains from excavated contexts using various methods of analysis, including microscopic examination. Some of the characteristics of hair remains can be registered by the unaided eye. These are, present colour, fibre length, curliness, gloss and the extent of damage. For the intact fibres, the profile, the medulla type, the distribution of pigment and the scale pattern are examined by a magnification of 300X; the diameter of the hair is measured at the same time. The description of scale pattern and the medulla type relies on the terminology used by Wildman (1954). The identification, where possible, was based on Appleyard (1978) and the author's comparative collection.

In general, the tufts of hair are well preserved, save that the gloss has disappeared in the course of time. In size, the samples vary from relatively large plucks of hair (V3873) to only a couple of hair strands (V3876).

1. DESCRIPTION/CATALOGUE

Provenance unknown. Several tufts of white hair, fibres only slightly waved, 2 cm long, medium to coarse hairs with wide continuous medulla. No pigment. The fibres are undamaged and the scale pattern regular waved with near margins, is clearly visible. Fibre diameter: 30-50 μm.

V3874 (level A2). Several tufts of light brown hair, slightly waved, fibre length 1.5-2.0 cm, medium to coarse hair, with wide continuous medulla. Sparse pigment distribution. Scale pattern, regular waved with near margins. Fibre diameter: 30-60 μm.

V3873 (level A2). a. A tuft of white hair with skin remains. Fibre length 2.5:3.0 cm, medium to coarse hair without waves. Wide continuous medulla. No pigment. Scale pattern, irregular mosaic waved. Fibre

* Rijksdienst voor het Oudheidkunding Bodemonderzoek (ROB), Kleine Haag 2, 3811 HE Amersfoort, Netherlands.

diameter: 80-135 μ m.

b. Registered under the same number is a tuft of brown hair, also with skin remains. Fibre length, 2.5 cm. Medium to coarse hairs, slightly waved, with wide continuous medulla. Dense pigment distribution. The fibres are too severely damaged for the scale pattern to be identifiable. Fibre diameter varies between 40-130 μ m.

V3876 (level A2). A few white hairs without waves and probably with some skin remains. The present fibre length is only 0.7 cm, but because the tips have been broken off, the original length can no longer be ascertained. The fibres are coarse with wide continuous medulla. No pigment. Scale pattern, interrupted waved. Fibre diameter: 50-160 μ m.

V3905 (level A3). A few white hairs with scanty skin remains (?). Length of stiff fibres is 4 cm. Coarse fibres with wide continuous medulla. No pigment. Although the fibres are slightly damaged, the scale pattern is still visible in certain places. The scales are irregular, almost mosaic waved with very close margins. Fibre diameter: 80-100 μ m.

V2882 (level B3?). Tuft of dark brown hair, slightly waved. Fibre length 2.5 cm. Medium to coarse hairs with medium to wide continuous medulla, sometimes interrupted. Pigment is absent in some fibres, in others there is an even pigment distribution. The fibres are so damaged that the scale pattern is nowhere intact. Fibre diameter: 40-70 μ m.

2. IDENTIFICATION

The identification of the fibres is severely hampered by the absence of any indication as to which part of the coat the hairs came from. Hairs from, for example, the tail differ considerably from those on the belly or back. The variation within a single animal species may also be very great. Furthermore, there is a general lack of information on ancient animal coats. Although the coats of most wild animals are unlikely to have changed much through archaeological time, it is a fact that the fleeces of sheep and possibly also the hairs of goats and cattle have been considerably modified by human interference (Ryder 1969).

Some hair tufts retain fragments of the original dried skin, and in these cases at least it is certain that the hair comes from a dead animal and does not represent a pluck of hair from an animal in moult. It is known that wild sheep and goats shed fine wool fibres particularly in early summer, while the coarser fibres are only shed in the autumn. Such tufts would result in an inaccurate picture of the original fleece structure.

Unfortunately, the skin remains are all extremely small and only the V3873a fragment (area 2 cm^2) was at all suitable for identification. The follicle structure, which is characteristic for each species, points not to cattle but rather to sheep or goat. Since the fibres are too thick for wool sheep, it would appear that we are concerned here with the remains of hair- sheep or goat (my thanks to Mrs. C. Van Driel who was kind enough to identify this skin remnant). The fine fibres (diameter 15-30 μ m) are also absent from the other tufts of hair with adhering skin.

None of the fibres are likely to belong to either hare or members of the cat family because both of these possess a ladder type of medulla. This type is totally absent from the Uan Muhuggiag finds.

In general it may be said that the find consists of relatively short, medium to coarse hairs belonging to large animals. It is, however, possible to divide the fibres with intact scale pattern into three groups: 1) medium to coarse hairs, regular waved scale pattern, fibre diameter 30-60 μ m (V3874 and un-numbered find); 2) coarse hairs, irregular mosaic waved scale pattern, fibre diameter 80-135 μ m (V3873a and V3905); 3) medium to coarse hairs with interrupted waved scale pattern, fibre diameter 50-160 μ m (V3876).

It seems probable that the hairs of the second group are a coarser version of the hairs of the first group. The hairs available from V3876 were so few in number and so incomplete that no further conclusions may be drawn.

The characteristics of first two groups are so general that no specific identification is justified, beyond the statement that the hairs may belong to hair sheep, goat and cattle. Hairs can only be identified as to animal species with certainty if larger areas of hair are available, preferrably together with an identifiable skeleton. Until such time, it is advisable to describe fibres according to a standard system, to enable comparisons to be made with other archaeological finds.**

**Translated by Mrs. C. van Driel.

REFERENCES

APPLEYARD H.M. 1978 - Guide to the Identification of Animal Fibres, Wira: Leeds (2nd ed.).

RYDER M.L. 1969 - Changes in the fleece of sheep following domestication (with a note on the coat of cattle), in The domestication and exploitation of plants and animals, eds. Ucko P.J., Dimbleby G.W., Duckworth: London, pp. 495-521.

WILDMAN A.B. 1954 - The Microscopy of Animal Textile Fibres, Wool Industries Research Association: Leeds.

CHAPTER 13

HOLOCENE VEGETATION IN THE TADRART ACACUS: THE POLLEN RECORD OF TWO EARLY CERAMIC SITES

By Erhard SCHULZ*

1. INTRODUCTION

The excavations in the northern and central Acacus (Barich 1974, 1978) resulted in a rich collection of both faunal and floral remains (Gautier 1984; Gautier, van Neer 1977 - 82; Pasa, Pasa-Durante 1962).

Analytical pollen investigations have been made in sediments from the Uan Muhuggiag and Ti-n-Torha sites, in the central and northern Tadrart Acacus respectively. Both sites are rock shelters with sequences of archaeological layers containing botanical macro- and micro-remains.

As the deposits cover approximately the period between 9200 and 5000 B.P., they allow their palaeoenvironmental information to be compared with that of contemporary deposits in the middle terrace of the central Sahara (Jäkel 1977; Maley 1981; Morel 1983; Schulz 1980).

2. PRESENT ENVIRONMENT

The Tadrart Acacus massif belongs to the geological Murzuk basin and is built up of sand and siltstones dating from the Devonian-Silurian period. From a geomorphological point of view, it is characterized by deeply cut wadis following sandstone fissures and tower-like sandstone weathering phenomena. From a climatic point of view, the Tadrart Acacus is sited in the region of the polar front and monsoon interaction. Precipitations are very scarce (25.3 mm Walter, Lieth 1960) with a maximum rainfall in the summer and a minimum in the winter; the most striking phenomenon is the rainfall irregularity. Moreover, the Tadrart Acacus lies in the rainfall shadow of the Tassili-n-Ajjer. The temperature range is great; maximum daily temperatures during the summer exceed 40°C while frost is not rare in the winter time.

Perennial plant life occurs only in the form of restricted vegetation along wadis and in depressions where groundwater is available. There is no diffuse vegetation in deserts, except for the short-lived achab floras, consisting of annual herbs and grasses growing after irregular rainfalls.

As part of the central Saharan mountains, the Tadrart Acacus has a typical wadi vegetation that, depending on the nature of sediments, may be of two different types. Wadis filled with coarse sand or gravel bear an Acacia-Maerua-Panicum vegetation, whereas those with fine sand or silt have

---

* Geographisches Institut der Universität Würzburg, Am Hubland, D 8700, Würzburg, Germany.

a Tamarix-Stipagrostis vegetation. Both types show the characteristic aspect of groups or lines of trees with an undergrowth of tufted grasses as well as some shrubs and herbs. Waterholes are often surrounded by Typha-Phragmites-Cyperaceae belts.

In the larger wadis, like the Tanezzuft or Isein, larger populations of Artemisia and Pulicaria are often mixed with Acacia-Panicum vegetation (Corti 1942). In some wadis, Nerium represents another Mediterranean/montane pre-Saharan element besides Artemisia. In dune areas, Calligonum shrubs are very abundant.

No altitudinal zoning with spectacular remains of former vegetation - such as Olea, Cupressus or Myrtus - as it is described in the Hoggar Mountains and Tassili-n-Ajjer (Leredde 1957; Quezel 1965) and no diffuse Artemisia-Ephedra shrubs as in the higher parts of the Hoggar Mountains were ever reported in the Acacus Mountains (Corti 1942; Le Houerou, Martel 1965).

3. PALAEOVEGETATION

Samples for pollen investigations were collected by B.E. Barich in the sites of Uan Muhuggiag and Ti-n-Torha North and East. They consist of fine and coarse sand or silt with a varying content of organic detritus. The stratigraphy is described by Barich (this Volume) (see also fig. 13.1).

Two coprolites from goat and sheep, collected at the Uan Muhuggiag site and transmitted by A. Gautier, were also analysed.

All the samples were prepared with a combination of classical treatment of HCl, HF, KOH, including a high density of fluid ($ZnCl_2$) treatment. The sediment residues were stained with fuchsine and stored in glycerol. Spectra were counted to the point of consistency. The diagrams (figs. 13.2,3) show the types of pollen arranged by trees, shrubs, herbs.

The sequence does not represent natural deposit and most of the layers are influenced by human activities. Therefore, the presentation in the diagrams should not suggest a continuous sequence. The samples were collected in different layers of the shelters, so they must be regarded as separate spectra arranged only after datation or stratigraphic position.

There are similar traits in the pollen content of both shelter sediments. Both diagrams show a bulk of pollen from Gramineae, Cyperaceae, Typha and Compositae. Pollen of trees and shrubs have only minor significance, with the exception of Maerua and Artemisia. The portion of pollen transported over long distances is extremely limited (Juniperus, Pinus, Ostrya, Quercus) and most of it comes from the North. The long distance transport of tropical elements is negligible.

The Uan Muhuggiag Shelter (fig. 13.2)

Sediments from former excavations of the site were analysed by Pasa and Pasa-Durante (1962).

The sediments used in this investigation result from two excavations in the shelter. The sandy deposits A1 and A2 were taken in the lower sequence of Sector A and may have an approximate age of 6000 B.P. (see Belluomini,

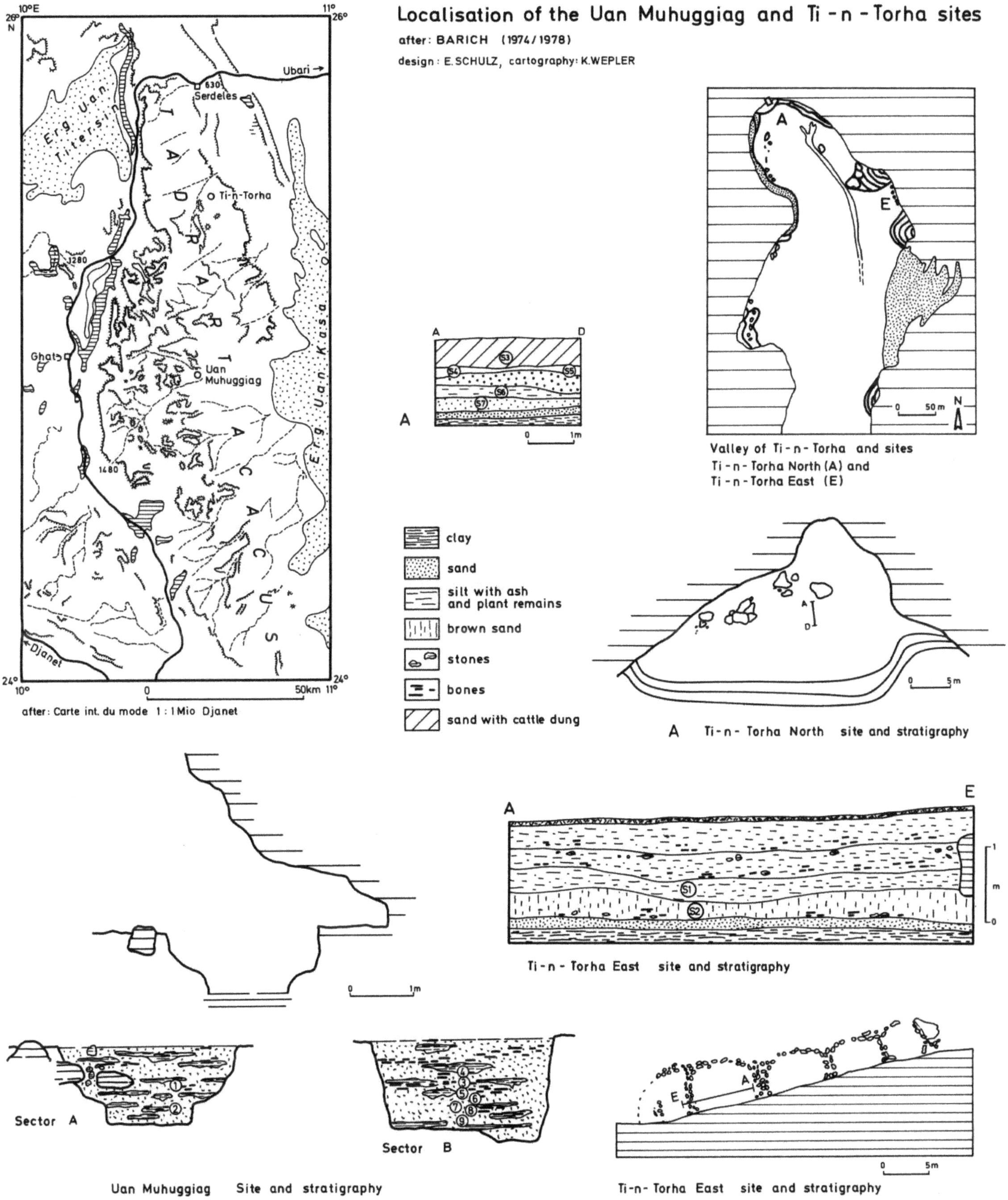

Fig. 1

Localisation of the Uan Muhuggiag and Ti-n-Torha sites

after: BARICH (1974/1978)

design: E. SCHULZ, cartography: K.WEPLER

Please note that a full-size version of this figure is available to download from www.barpublishing.com/additional-downloads.html. The original foldout has been reduced in size to match the A4 format of this book, the image is therefore not as clear as the original foldout. Please refer to the original foldout via the download for the original content.

Chapter 14). Samples 3-9 were collected in Sector B. From their stratigraphic position, A1 and A2 may be compared with samples B 5-7, since the samples 3-4 are younger than 6000 and the samples 8-9 are most likely older. (For more detailed information, see Barich in this Volume). The two coprolites of goat (B1) and sheep (B2), submitted by A. Gautier had a stratigraphic position comparable to B5 (Gautier pers. comm and in this Volume).

The samples had a sufficient pollen content for analysis , with the exception of sample B9 which was very poor.

Gramineae, together with Compositae, point to a predominance of open vegetation. Among the grasses, the panicoid pollen type (dark part of the column) is the most prominent, while within the Compositae tubuliflorae, a significant value should be ascribed to the Pulicaria type.

Maerua and Tamarix are the only trees with somewhat high values.

Shrubs like Artemisia, Cassia and Fagonia show higher percentages in the middle and upper layers, while in the lower ones their presence is certainly limited by the local predominance of Typha.

Chenopodiaceae/Amaranthaceae have only very low values in the sediments but their presence in the vegetation is proved by the high values in the B1 coprolite spectrum. These differences may be explained by the different flowering period of grasses and chenopods.

The lower horizon of Sector B and sample A2 point to a significant presence of Typha which, together with Cyperaceae, represents the vegetation surrounding a former nearby waterhole. In the upper layers of Sector B, the influence of the aforementioned waterhole was less significant, although it might have been still present.

Compared with the present plant cover - the Acacia-Panicum or Tamarix-Stipagrostis type wadi vegetation - these pollen spectra indicate no major difference; the modern pollen spectra from the southernmost Acacus is similar (Schulz 1980, 1980a).

During the sediment deposition, the prevailing vegetation was open except in the wadis which may have had a configuration comparable to the present one. Cassia, Fagonia and the recorded herbs still to date characterize the wadi vegetation of the central Sahara. Artemisia, in combination with Ephedra and Olea, shows the existence of a eMediterranean/montane pre-Saharan element in the wadis as well as the surrounding mountains.

The higher values of the Compositae in the middle and upper levels are probably due to local over-representation.

The pollen spectra of the two coprolites fit very well with the sediments. The resulting high values of certain shrubs, especially Chenopodiaceae/Amaranthaceae and Cornulaca in the first coprolite and, conversely, the high Gramineae values observed in the second, are caused by a different type of feeding. Goats (B1), as typical browsers, prefer shrubs and some herbs whereas sheep (B2) prefer grasses. Furthermore, the B2 spectrum shows that, in fact, both (pollen) types of Gramineae were present in the former vegetation. Apart from these facts, the pollen spectra of the two coprolites indicate no major difference from the vegetation type represented in the middle layers of the sector.

The Ti-n-Torha Shelters (fig. 13.3)

These samples derive from two excavations in the Wadi Ti-n-Torha, in the northern Tadrart Acacus. Samples s1 and s2 were collected in the lower levels of the eastern site close to the base, while samples s3-s7 were collected from the northern site. The following radiocarbon dates came from the levels where the samples were collected (Barich *et Alii* 1984):

s1: 8640±70 B.P. (R-1035α) s3: 5600±50 B.P. (R-1028)

s2: 9080±70 B.P. (R-1036α) s4: 5970±50 B.P. (R-1031α)

s7: 7070±60 B.P. (R-1032)

These pollen spectra are generally similar to those obtained from the Uan Muhuggiag shelter. Elements of an open-type vegetation prevail: *Gramineae* with a large portion of the panicoid-type (black part of the column) *Compositae* and *Cyperaceae*. As for the *Cyperaceae*, most of them belong to the *Cyperus conglomeratus* type (black part of the column). No local influence of waterholes may be detected (as was possible in the Uan Muhuggiag samples), and only *Adianthum* points to the presence of wetter places in the rocks. The high values of the *Compositae luguliflorae* certainly result from ground disturbance perhaps by insects (Bottema 1975).

There is a very low representation of trees and shrubs. Only *Acacia*, *Cassia*, *Fagonia* and *Chenopodiaceae/Amaranethaceae* may be noted, with higher *Acacia* values. Boraginaceae, like *Trichodesma* and *Moltkiopsis*, are typical elements of the Saharan vegetation.

Again, these spectra may be interpreted as a typical wadi vegetation of the *Acacia-Panicum* type. The special situation of the Ti-n-Torha East Shelter is shown by the high values of the *Compositae liguliflorae*. Compared to the spectra of the Uan Muhuggiag samples, there is a greater share of open vegetation elements - such as *Gramineae* and *Cyperus conglomeratus* - and a lower amount of trees and shrubs. This may be caused by the more marginal (northern) situation of these sites, near the dunes of Erg Ouan Kasa and Erg Titersit.

In view of the special conditions of each sample and the situation of the shelters - especially the fact that sediments are not deposited naturally - no major change in the vegetation type of that region may be recorded for the 9000-5600 B.P. period. Nevertheless, changes relative to the density of vegetation, or its diffusion into the forelands, and even some changes in the floral composition may have occurred. The pollen spectra point to the plant persistence in that area of the Saharan vegetation.

A comparison of these results with those provided by Pasa and Pasa-Durante (1962) shows that it is quite unlikely that a continuous reconstruction of the climatic evolution, in terms of precipitations and temperature, can be made on the basis of the flora; this is due to the irregularities in the deposition of the different samples.

Fig. 2 Pollen diagram of the Uan Muhuggiag site

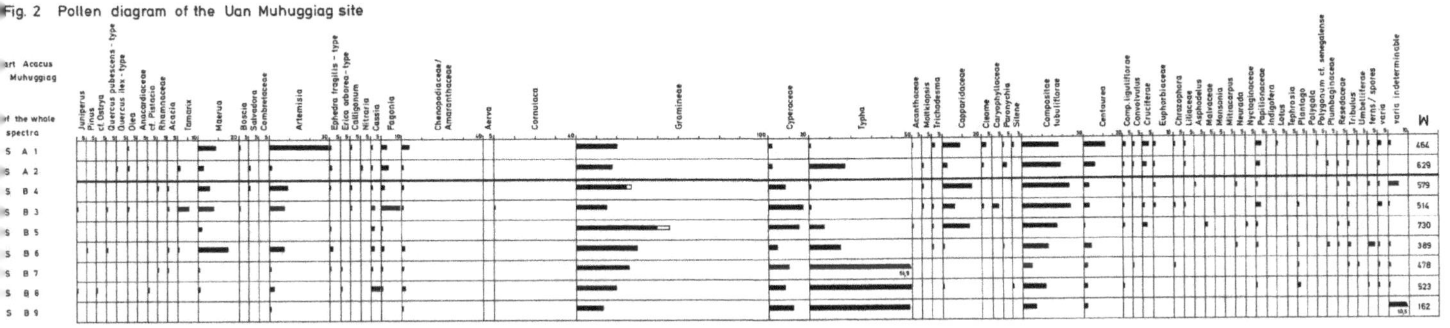

Please note that a full-size version of this figure is available to download from www.barpublishing.com/additional-downloads.html
The original foldout has been reduced in size to match the A4 format of this book, the image is therefore not as clear as the original foldout. Please refer to the original foldout via the download for the original content.
Please also be aware that the image might be cut off or not complete in the printed book.

4. COMPARISON WITH MACRO-REMAINS

The following macro-remains were reported (Pasa, Pasa-Durante 1962) from the Uan Muhuggiag shelter: Acacia albida, Balanites aegyptiaca, Artemisia campestris, Zilla spinosa, Aristida sp., Colocynthis vulgaris.

Aumassip (1985) reports some macro-remains from the Ti-n-Hanakaten excavation, a comparable site in the Tassili-n-Ajjer: Acacia albida, A. seyal, A. raddiana, A. flava (ehrenbergiana), Tamarix sp., Olea sp., Ephedra sp., Ziziphus lotus.

In the lower section of that site, Aumassip (1985) noted a predominance of Acacia albida and A. flava (ehrenbergiana), replaced in the upper part of the sector by A. seyal, A. raddiana and Ziziphus. With the exception of A. albida, that Leredde (1957) reports to be very rare in the Tassili-n-Ajjer, all the elements are still common in the regional flora.

Even the Gao-tree (A. albida), the northernmost occurrence of which is reported in the Tassili-n-Ajjer (Leredde 1957), is common in the large wadis of the Air Mountains where A. albida occurs in combination with A. nilotica, A. raddiana, A. ehrenbergiana, Maerua crassifolia, Balanites aegyptiaca and others. Although A. albida grows in the wadis where more groundwater is available, it still belongs to the Saharan-type restricted vegetation.

No significance discrimination of different Acacia types could be made in the pollen record.

The presence of these macro-remains in both the Tadrart Acacus and the Tassili-n-Ajjer at a similar time suggests a vegetation model comparable to the present-day plant cover of the Air Mountains.

CONCLUSIONS

The available pollen counts evidence allows the reconstruction of the type of Holocene vegetation present in the Tadrart Acacus. The plant cover belonged to the Acacia-Panicum or Tamarix-Stipagrostis type wadi vegetation system and different trees and shrubs, like Balanites, Maerua, Salvadora, Cassia, Fagonia and others were also present. Extensive populations of Artemisia, together with other Compositae were certainly present in large wadis. Olea and Ephedra may have occurred in rocky areas or smaller wadis. The grass and herb cover was much more widespread, although subject to significant seasonal changes.

The presence of macro-remains of Acacia albida and A. seyal in the Tassili-n-Ajjer and Tadrart Acacus leads to conclude that even the Sahelian element was then important. These findings suggest that the earlier Holocene vegetation resembled the recent vegetation of the Air Mountains, some 800 km to the Southwest. However, the presence of these Sahelian elements in the Tassili-n-Ajjer - as rare as they are - shows that the difference between the former and the present vegetation is not as dramatic as one may think. In the interpretation of these elements, one should also take into account the present over-exploitation, by overgrazing and the severe drought of the past decade.

The pollen analyses of the middle terrace sediments of the Air Mountains gave similar results (Morel, Schulz 1982).

The upper part of the mountains was certainly covered by a shrub and herb vegetation, dominated by Artemisia, Olea and Ephedra, but the pollen counts and scarce macro-remains are not sufficient to establish whether there was a permanent diffuse (semi-desert) vegetation as one finds in the upper parts of the Hoggar Mountains.

The surrounding plains and Ergs had certainly a more extensive grass and herb cover with Calligonum shrubs. It included both the perennial plants and the shorter-lived achab floras which occurred more often and to a larger extent.

More frequent run-offs in the wadis may have resulted in more groundwater in the fine sediments and some longer lasting waterholes surrounded by belts of Phragmites, Typha and Cyperaceae.

This type of plant cover still belonged to the wadi vegetation system of the Acacia-Panicum or the Tamarix-Stipagrostis type, depending on the sediment nature. It was richer in its flora and certainly extended into the surrounding plains; it is even likely that a Mediterranean/montane pre-Saharan vegetation, characterized by Artemisia, Ephedra and Olea, occurred in the upper part of the Mountains.

There was a regional difference between the northern and the central part of the Acacus Mountains. Vegetation in the North was more open and dry than in the central part.

At the time of sedimentation, this type of vegetation was persistent, even if there were changes in its density.

Neither the pollen counts nor the few known macro-remains indicate a change towards a Mediterranean or a Sahelian woodland.

Obviously, this type of vegetation provided sufficient pasture for both a longer stay in the massif of herdsmen and their animals and a more extensive use of the surrounding region.

Fig. 3 Pollen diagram of the Ti-n-Torha site

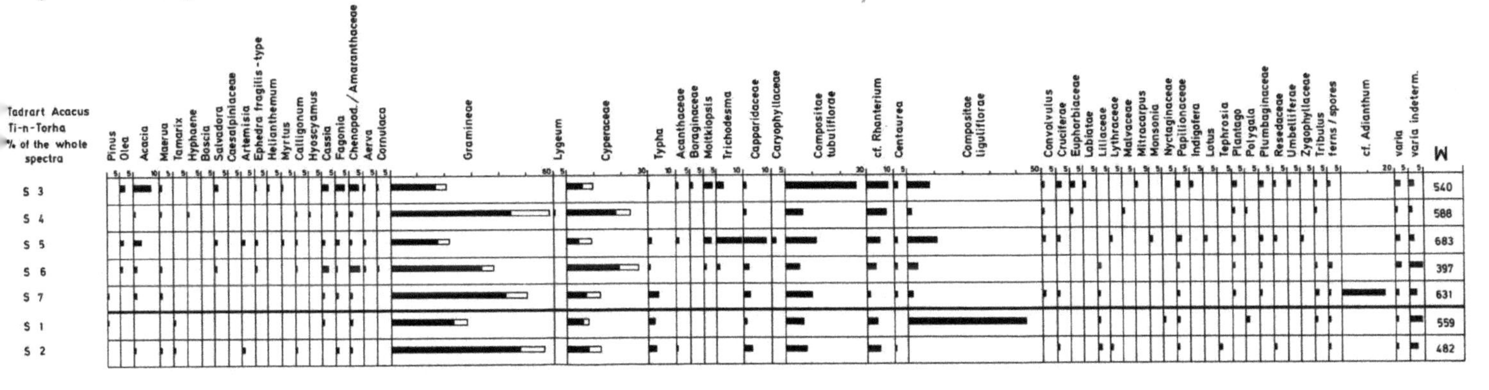

Please note that a full-size version of this figure is available to download from www.barpublishing.com/additional-downloads.html
The original foldout has been reduced in size to match the A4 format of this book, the image is therefore not as clear as the original foldout. Please refer to the original foldout via the download for the original content.
Please also be aware that the image might be cut off or not complete in the printed book.

REFERENCES

AUMASSIP G. 1985 - Le site Ti-n-Hanakaten et la néolithisation sur les marges orientales du Sahara central, Cahiers O.R.S.T.O.M. (in press).

BARICH B.E. 1974 - La serie stratigrafica dell'Uadi Ti-n-Torha (Acacus, Libia). Per una interpretazione delle facies a ceramica saharo-sudanesi, Origini, VIII:7-184.

BARICH B.E. 1978 - Neue Ausgrabungen im Acacus-Gebirge, in Sahara: 10.000 Jahre zwischen Weide und Wüste, Museen der Stadt: Köln, pp. 222-245.

BARICH B.E. et Alii 1984 - Ecological and cultural relevance of the recent new radiocarbon dates from Libyan Sahara, in Origin and Early Development of Food-producing Cultures in North-Eastern Africa, eds. Krzyzaniak L., Kobusiewicz M., Polish Academy of Sciences, Poznan Archaeological Museum: Poznan, pp. 411-417.

BOTTEMA S. 1975 - The interpretation of pollen spectra from prehistoric settlements (with special attention to liguliflorae), Palaeohistoria, XVII: 17-35.

CORTI R. 1942 - Flora e vegetazione del Fezzan e della regione di Ghat, Reale Società Geografica Italiana: Firenze.

GAUTIER A. 1984 - New data concerning the prehistoric fauna and domestic cattle from Ti-n-Torha (Acacus, Libya), Palaeoecology of Africa, 16: 305-309.

GAUTIER A., VAN NEER W. 1977-1982 - Prehistoric fauna from Ti-n-Torha (Tadrart Acacus, Libya), Origini, XI: 87-127.

JAKEL D. 1977 - Abfluss und fluviale Formungsvorgänge im Tibesti-Gebirge als Indikatoren zur Rekonstruktion einer Klimageschichte der Zentral Sahara in Spätpleistozän und Holozän, Paper presented at the Xth INQUA Congress, Birmingham.

LE HOUEROU H.N., MARTEL A. 1965 - De Ghadames à Ghat, Trav.Inst.Rech.Sah., 24: 191-204.

LEREDDE C. 1957 - Etude écologique et phytogéographique du Tassili-n-Ajjer, Trav.Inst.Rech.Sah., Sér.du Tassili, II, 455 p.

MALEY J. 1981 - Etudes palynologiques dans le bassin du Tchad et paléoclimatologie de l'Afrique nord-tropicale de 30.000 ans à l'époque actuelle, Travaux et Documents de l'O.R.S.T.O.M., 129 (O.R.S.T.O.M.: Paris), 586 p.

MOREL A. 1983 - Erosion et sédimentation dans le massif de l'Air (Sahara meridional): essai d'interpretation paléoclimatique de la moyenne terrasse, Cahiers O.R.S.T.O.M., Sér.Geol., XIII, 2: 111-118.

MOREL A., SCHULZ E. 1982 - Recherches palynologiques et géomorphologiques dans le massif de l'Air (Niger): note préliminaire, Revue de Geographie Alpine, LXIX, 4: 583-592.

PASA A., PASA-DURANTE M.V. 1962 - Analisi paleoclimatiche nel deposito di Uan Muhuggiag nel Massiccio dell'Acacus (Fezzan meridionale), Mem.Mus.Civ.di St.Nat.Verona, X: 251-255.

QUEZEL P. 1965 - La végétation du Sahara, Gustav Fisher: Stuttgart.

SCHULZ E. 1980 - Zur Vegetation der östlichen zentralen Sahara und zu ihrer Entwicklung im Holozän, Wurzburger Geographische Arbeiten, 51, 194 p.

SCHULZ E. 1980a - An investigation of current geologic processes and an interpretation of Saharan palaeoenvironment, in The Geology of Libya, III, eds. Salem M.J., Busrewil M.T., Academic Press: London, pp. 789-796.

WALTER H., LIETH H. 1960 - Klimadiagramm, Weltatlas, Gustav Fisher: Jena.

CHAPTER 14

RADIOCARBON DATES FROM THE TADRART ACACUS MASSIF

By Giorgio BELLUOMINI*
Luigia MANFRA**

This chapter presents synthetically (Table 14.1) the complete series of radiocarbon dates collected by the Libyan-Italian Joint Mission during its research in the Tadrart Acacus.

Scintillation counters (Belluomini _et Alii_ 1978) were used to date samples from two settlements located 300 m from each other in the Wadi Ti-n-Torha, Auis area (East Shelter and North Shelter) (Alessio _et Alii_ 1978) and from another in the same area called Two Caves (Barich, Belluomini _et Alii_ 1984).

Date estimations were made on the following samples: 3 samples of charcoal and 3 samples of manure from North Shelter; 9 samples of carboniferous material from East Shelter; and 9 samples of carboniferous material from Two Caves. From Uan Muhuggiag, 3 recent analyses on carboniferous material conducted in our new dating laboratory in Udine, as well as age estimations carried out in Pisa and Cambridge (U.S.A.) several years ago are reported.

Discussion

Table 14.1 shows that human settlements have been present in the Acacus area from at least the late 10th to the 4th millennium B.P.

A very careful examination of age values allows the identification of two frequency peaks around 8,500 years B.P. (Two Caves and East Shelter) and 6,000 years B.P. (North Shelter and Uan Muhuggiag), respectively.

Generally speaking the moister climate of the Sahara during the Early Holocene clearly created very favourable conditions for human settlements in all the Saharan environments (Barich, Belluomini _et Alii_ 1984). In the Acacus, the archaeological record has revealed a continuity in the human inhabitation except for short periods of climatic crisis. Settlements must have diminished in the area slightly after 7,500 B.P. and then been established again with different characteristics. The manure samples analysed point to a society of shepherds instead of the fishermen-hunters of

* Centro di Studio per la Geocronologia e Geochimica delle Formazioni Recenti, C.N.R., Dipartimento di Scienze della Terra, University of Rome "La Sapienza", Rome, Italy.

** Dipartimento di Scienze della Terra, University of Rome "La Sapienza", Rome, Italy.

Table 14.1 - Radiocarbon dates from the Tadrart Acacus Massif.

SAMPLE	SITE	5.568 HALF-LIFE		MATERIAL
		B.P.	B.C.	
R-1028	Uadi Ti-n-Torha (North Shelter)	5600±50	3650	Pen
R-1029	"	5260±130	3310	"
R-1030	"	5360±120	3410	"
R-1031 a	"	5970±50	4020	Charcoal
R-1164 a	"	6540±50	4590	"
R-1032	"	7070±60	5120	"
R-1033 a	Uadi Ti-n-Torha (East Shelter)	8460±60	6510	Charcoal
R-1034 a	"	7990±70	6040	"
R-1035 a	"	8640±70	6690	"
R-1036 a	"	9080±70	7130	"
R-1158 a	"	7330±80	5380	"
R-1160 a	"	8540±140	6590	"
R-1161 a	"	8460±50	6510	"
R-1162 a	"	8440±50	6490	"
R-1163 a	"	8670±60	6720	"
R-1405	Uadi Ti-n-Torha (Two Caves)	8840±60	6890	Charcoal
R-1407	"	8520±60	6570	"
R-1408	"	8450±60	6500	"
R-1404	"	8630±50	6680	"
R-1404 bis	"	8400±90	6450	"
R-1403	"	8230±50	4280	"
R-1406	"	8620±50	6670	"
R-1409	"	8650±105	6700	"
R-1402	"	9350±110	7400	"
Geo	Uan Muhuggiag	4730±310	2780	Charcoal
Pi	"	5405±180	3455	Leather
Pi	"	5952±120	4002	Charcoal
Pi	"	7438±220	5488	"
UD-224	"	3770±200	1820	"
UD-225	"	6035±100	4085	"
UD-226	"	5350±200	3400	"

R: Carbon-14 Laboratory, Rome, Italy.
Geo: Geochron. Laboratories, Cambridge, Mass., USA.
Pi: Pisa, Italy.
UD: Udine, Italy.

the 9th millenium.

The youngest samples of the chronological series, belonging to approximately the second half of the 6th and first half of the 4th millennium B.P. (UD-226, UD-224), were recovered from the Uan Muhuggiag Shelter, near the centre of, and in a more elevated area, of the Acacus Massif. At that time, this region was likely to offer a more favorable climate than Wadi Auis, which was geographically more peripheral and where desertification had surely already started. In any event the drying process must have been discontinuous and some density discontinuities identified within the site suggest that there were some intervals with more favorable climatic conditions amid the increasing aridity, but this fact did not reverse the general desertification trend.

The two samples R-1402 and R-1036, collected in the Two Caves and East Shelter respectively, can be ascribed to the 10th millennium B.P. These are some of the oldest Holocene dates of all the Saharan-Sudanese regions, and corroborate the assumptions of the age of Saharan Neolithic cultures based on findings of other researchers (cf. Camps 1974a, 1974b).

REFERENCES

ALESSIO M., ALLEGRI L., BELLA F., IMPROTA S., BELLUOMINI G., CALDERONI G., CORTESI C., MANFRA L., PETRONE V. and FRUSCALZO A. 1978 - University of Rome Carbon-14 date XV, Radiocarbon, V, 20,1:68-78.

BARICH B.E., BELLUOMINI G., BONADONNA F., ALESSIO M. and MANFRA L. 1984 - Ecological and cultural relevance of the recent new radiocarbon dates from Libyan Sahara, in Origin and Early development of food producing Cultures in North-Eastern Africa, eds. Krzyzaniak L., Kobusiewicz M., Polish Academy of Sciences, Poznan Archaeological Museum: Poznan, pp. 411-417.

BELLUOMINI G , DELFINO A., MANFRA L. and PETRONE V. 1978 - Benzene synthesis for radiocarbon dating and study of the catalyst used for acetylen trimerization, Int.Journ.of Appl. Radiations Isotopes, 39:453-459.

CAMPS G. 1974 a - Les Civilisations préhistoriques de l'Afrique du Nord et du Sahara, Doin: Paris.

CAMPS G. 1974 b - Tableau chronologique de la préhistoire récente du Nord de l'Afrique (Deuxième synthèse des datations absolues obtenues par le Carbon 14), Bull. Soc. Préh. franç. (Etud. et Trav.), 71: 261-278.

15

CONCLUSIONS

THE TADRART ACACUS CONTRIBUTION TO THE STUDY OF CULTURAL CHANGE IN THE SAHARA

By Barbara E. BARICH

Until recently, the pastoral economies of the 6th-4th millennia B.P. represented the principal event so far recognized in the development of human settlement in the Sahara. Even though a few regions were characterized by economies which consisted primarily of small-game hunting and gathering, they were all included under the common "Saharan Neolithic" label. This resulted in an apparent uniformity in data which was in reality very varied and that has been given individuality only after long periods of comprehensive studies.

The investigations in the Tadrart Acacus Massif have contributed to the clarification of these regional differentiations, by bringing to light, through archaeological and paleoenvironmental data, two socio-economic spheres in relation to two different ecosystems that were, respectively, marginal to and inside the massif. They represent two distinct ways of life, far removed in time and space that, following a detailed comparison of their cultural subsystems, appear to be diachronic elements in a unitary process that took place principally but not exclusively _in situ_. The studies relative to the habitat and the whole of the archaeological data have pointed quite clearly to the existence of conditions favouring long-term settlement and continuous technological and economic progress.

The vegetation study has not detected major differences in the pollen spectra that were obtained from the two major areas being investigated. In both cases one may note the predominance of an open-type vegetation: the _Acacia-Panicum_ or _Tamarix-Stipagrostis_ wadi vegetation system which remained essentially unchanged throughout the stratigraphic series at Ti-n-Torha East, Ti-n-Torha North and Uan Muhuggiag. The only differences relate to the distribution of the vegetation - that is, more open in the north-eastern section (and thus favouring a more extensive use of the territory) and the marked decrease in the _Gramineae_ group moving from the Torha East (Level RInf-RSup) to the Torha North (Level I) spectra.

The durability of the plant-cover resulted from groundwater resources that made it possible for conditions favouring human occupation to be maintained even in times of marked aridity. The study of the non-climatic physical parameters has clearly pointed to the existence, especially along the eastern slope, of underground water resources, accumulated during moisture-rich phases, which were stored over long periods of time and slowly released. Furthermore, in view of the fact that conditions favouring the preservation of these groundwater resources only existed in the outer belts - both to the East and to the West -, it may be assumed that this stimulated a certain degree of mobility in the human occupation patterns, in order to reach those favoured areas in times of stress, besides stimulating an extension of the settlement area in the exploitation of scattered rather than concentrated resources.

It is evident that these latter assumptions match those resulting from the pollen study that have clearly pointed to a more markedly open vegetation at Ti-n-Torha favouring a more extensive human land-use. Conversely, in the more sheltered areas, such as the Teshuinat basin, the influence exerted by the hydrogeological situation of the eastern slope provided micro-climates with a greater atmospheric humidity and local precipitations. This led to a greater concentration of water resources, with the consequent development of a thicker vegetation (Typha) that endured even when the resources of the marginal belts began to diminish.

The hydrogeological and vegetation conditions allow us to consider the possibility that the three economic models identified by the archaeozoological study might have developed uninterruptedly. Evidence of this continuity is also provided by the absence of actual breaks in the regional stratigraphic sequence (except when a few sites were abandoned for local reasons), and by the technological pattern in which the lithic sequence of the Wadi Ti-n-Torha witnessed the preservation of a stable and successful settlement pattern, from which A.E. Close can refer to "a developing process of adaptation" (see Chapter 3).

In the light of these findings, there is no reason to characterize food production in this area as abruptly innovative or even revolutionary in character. Indeed, it would seem more useful to direct the inquiry towards a reconstruction of the mechanism leading to change.

The attainment of new economic patterns by the groups in desert areas may be viewed as aspects of adaptive strategy, since the phenomena of cultural change occurred in those areas at the very time of the environmental changes. It was indeed a new awareness of the processes of interaction between the sphere of culture and the sphere of the environment that allowed different aspects to come together - as stages in a development process - within the mixture that for so long represented the Saharan Neolithic.

So long as the above aspects were accepted as established 'blocks', and their "Neolithic" characters set against "pre-Neolithic" ones, there could only be a fracture of the process of development that, as previous pointed out, has proved to be a unitary process. The starting point for a change in perspective - the result of the effort of a number of research teams - was represented by the acknowledgment of a context of pre-adaptation to the productive economy and the fact that investigations focused on the fluid amount that represented "the beginnings" of change.

The latter conditions, related to the whole eastern, central and central-eastern Sahara (Western Egypt, North Sudan, South Libya, Niger), with dates going back even beyond 9000 B.P. (Table 15.1). The "Early Post-Paleolithic" definition, formerly suggested by A.B. Smith, would seem particularly suited to the artifactual facies documented in these regions (1). We are indeed dealing with contexts that - although featuring

---

1) The definition is used with special reference to a few sites in Mali (Asselar) and Niger (Outeidat, Tamaya Mellet) (Smith 1980: 452).

ceramics - preserve a Paleolithic life-style except for the eventual and sporadic presence of plant or animal species at a level of incipient domestication. Compared to this, the advent of pastoral economies appears as a later achievement that may date back no further than the 6th millennium B.P.: from more archaic aspects, as at Uan Muhuggiag in Libya, to more evolved aspects as at Adrar Bous in Niger.

Table 15.1 - Chronology of the principal sites in Niger and Libya.

NIGER	Temet	9550±100 B.P.
	Tamaya Mellet	9350±170
	Tagalagal	9370±130
LIBYA	Tin-n-Torha Two Caves	9350±110
	Ti-n-Torha East	9080±70
LIBYA	Ti-n-Torha North	5970±50
	Uan Muhuggiag	6035±100
NIGER	Agorass-n-Tast, Adrar Bous	5760±500
	" " "	4440±120

In the light of the above, a unifying definition of the different pottery-making groups of the Saharan area has no longer any validity and indeed might deceive. Conversely, all the archaeological contexts for several millennia make up, one by one, specific aspects of a unitary if diversified process of development.

When we, however, attempt to probe further into the two levels of development discussed above, especially in respect of their relationships, it becomes evident that the bases of change were already present in the Late Paleolithic system. Consequently, the Saharan environment may be viewed as an area of fundamental importance for exploring the beginnings of food production, just as the Western Asian areas have long been. The natural mechanisms involved in the transition from the Pleistocene to the Holocene period are the same in both these regions and, as a matter of fact, they may be explained by the same general principles recognized by ecology and the historical geography. Quite evidently, they differ in the specific situations and developments, for example, the Nile Valley - with its peculiar climatic events may be related only indirectly to those of the Sahara, but the flourishing Late Paleolithic and Epipaleolithic industries of Upper Egypt and Nubia at the close of the Pleistocene age were only slightly in advance of the Saharan developments and exhibited aspects of pre-adaptation to food production of the type that subsequently developed in the Sahara.

The finds of the last few years have highlighted a few areas, that are of great moment for our investigations, and that may be viewed as three adjoining though separate clusters; the Air-Ténéré Acacus; the Egyptian Sahara; and the Northern Sudan (fig. 15.1). This clustering is shown by the respective archaeological contexts and differentiated chronologies. As a matter of fact, one must maintain very carefully a chronology, with dates approximately ranging from 10000 to 7000 B.P., to avoid mixing levels. Even though it is unquestionable that contacts among the different groups took place, it is extremely difficult to identify them in detail, since each series of finds has its peculiar characterizations, within a context that is generically common.

The Tadrart Acacus provides one of the most comprehensive sequences owing to the frequency, state of preservation and depth of the archaeological deposits that allow us to trace processes of development through time. To this can be added a close association with the rock art evidence, that may give an ethnological contribution to reconstruction. It also allows the art itself to acquire chronological certainties as well as stylistic interpretation keys. From an archaeological point of view, the two ecosystems in the Acacus, the north-eastern and central, correspond to two differentiated archaeological facies: those of the Wadi Ti-n-Torha and the Wadi Teshuinat (in which is the Uan Muhuggiag shelter).

The Ti-n-Torha facies, which was found to have a preparatory role to the forms of artifact production that were to follow within a relatively short time (see Chapter 5), represents the earliest horizon of gatherers and fishers that had preferentially settled on the eastern slope of the massif owing to its special advantages for human habitation (a drainage pattern favouring settlement, areas of shallow groundwater, and crossings towards the western slope).

This level of exploitation lasted from the 10th to the 7th millennium B.P., as witnessed by the three major stratigraphies of Two Caves, Torha East and Torha North (V-IV) among which evident links may be detected. Further South, always on the eastern slope, this horizon may be identifiable in the sites of Fozzigiaren, Tihedine and Sughd which have only been surveyed.

A feature shared by all these settlements was a strategic location with respect to the wadis that had concentrations of both water, vegetation and animals. The site of Ti-n-Torha East, already defined as a "proto-village", provides outstanding evidence that semi-sedentary habits were being acquired. It most likely began as a seasonal settlement (during the dry season that, in this area of monsoon rains, is winter), while it acquired a more permanent character during the dry interval (8500-8200 B.P.) ascertained in the sedimentary deposits of the Middle Terrace (see Maley 1981: 214-229) (2).

---

2) This author refers specifically to the stratigraphy of Ti-n-Torha and points to a number of specific correlations wih the climatic sequence of the Tibesti.

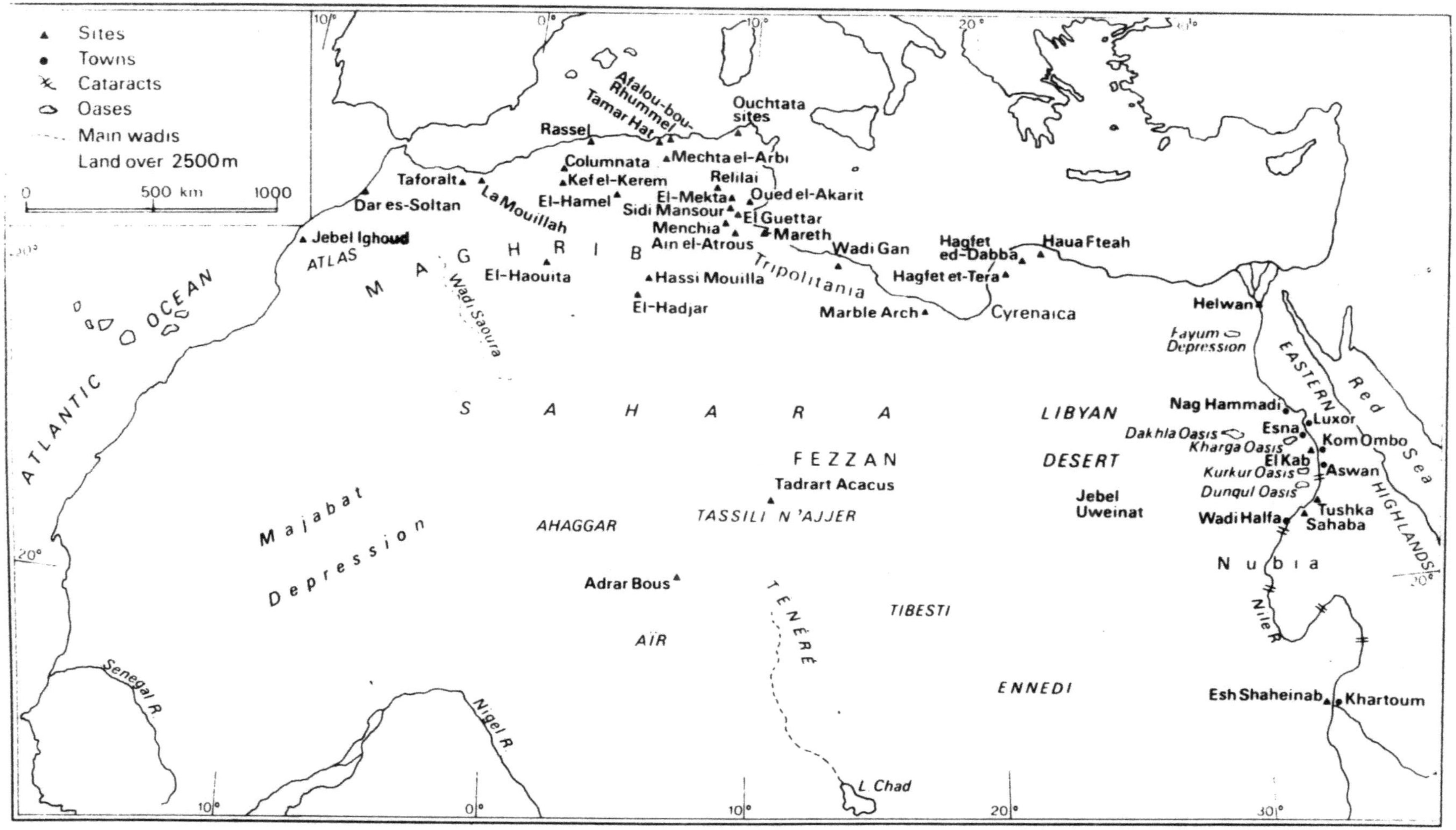

Fig. 15.1 - The principal Late Paleolithic and Epipaleolithic sites in Northern Africa and Sahara.

In fact, if the custom of fishing favoured the acquisition of residential habits by human groups, until the summer monsoon rains lasted, the latter would have been compelled to undertake seasonal wanderings. On the other hand, the transformation that took place at Ti-n-Torha East (Level CII: R-1033 α, 8460±60 B.P.; R-1034 , 7990±70 B.P.) with the construction of rudimentary huts by the use of stone blocks as boundary walls points to the discontinuing of the wandering and the beginning of an extensive exploitation of that part of the environment within easy reach. The evident increase - further to these changes - in the number of Barbary sheep (Ammotragus lervia) and, to a lesser extent, of gazelle, may be an indication of the evolution of forms of control involving flock animals (cf. Close 1980: 151). Furthermore, the presence of millstones in Level CI-CII, witnesses the practice of gathering activities probably involving the herbaceous species that grew around the settlement (3). This new propensity may be related to similar elements characterizing the Nabta and Bir Kiseiba areas in the Eygptian Sahara. Just as in our case, those sites gave rise to conditions favouring the domestication of animal species, Bos in particular which was present in environments still Late-Paleolithic in character. Indeed, as it will be detailed later on, in both cases this new propensity may have been stimulated in times of stress on the resources provided by the environment.

The investigations carried out in the Ti-n-Torha area have been complemented and enriched by those in the Wadi Teshuinat area. If at the end of the Ti-n-Torha cycle, a change may be inferred in the original subsistence pattern, owing to the shift of interest towards selective domestication activities with plants and animals, the sequence of Uan Muhuggiag documents a further development.

Chapter 10 placed emphasis on the possibility of connecting the Uan Muhuggiag sequence with the final phase of the Ti-n-Torha horizon. This is proved by the recovery in the lowest levels at Uan Muhuggiag, of lithic and ceramic types which were, like some climatic and environmental ones, different from what followed. Even leaving these aside, the whole technology of the other levels at Uan Muhuggiag is seen to be a derivation of the Ti-n-Torha facies though with some modifications. These involved, in the ceramics, a progressive standardization of the decorative patterns due to two-toothed implements being used in pivoting fashion (see Chapter 9) and, in the flaked stone tools, the progressive disappearance of backing and bifacial retouch and a restricted range of the tool-classes. This, together with the change in economic activities, suggests a local tendency towards domestication. At the end of the process, the changes that took place are easily detected through the differences in land-use and settlement patterns.

At first sight the dwellings of the pastoralists may be likened to the shelters of hunter-gatherer-fishers, but a more thorough examination showed their temporary nature and a utilization that differs from that of the home-bases of groups practising a broad-spectrum exploitation of the local environment. Hunting was still practiced, as is shown both by the presence

---

3) It was shown (Chapter 13, this Volume) that there was a predominance of open vegetation, consisting for the moist part of herbaceous vegetation - Gramineae, Cyperaceae, Typha, Compositae - and to a much less extent, trees and shrubs.

of bifacial points in the basal section of the stratigraphy and in the scenes depicting hunters on the back wall of the Uan Muhuggiag Shelter which may be linked with the 'Round-Head' styled scene depicting a boat (see Chapter 6). There is no doubt, however, that cattle, sheep and goats were the most important economic resource of the group.

The association of large and small domestic livestock from Levels 2d-2c of Uan Muhuggiag (prior to 6035±100 B.P.) led A. Gautier to revise his identification of a "large bovid" in the Ti-n-Torha East fauna as domesticated. It is well known that sheep/goat were imported species which appear at Nabta in Middle Neolithic contexts (4). Therefore, he would see the Uan Muhuggiag cattle as the result of introduction that would make the previous determination of them as indigenous meaningless.

In my opinion, however, even though the presence of an archaic type of _Bos_ in the Ti-n-Torha horizon should be viewed as a result of introduction, it may prove to be an indication of contacts established _ab antiquo_ by groups of hunters and may indeed explain the acquisition of small livestock, together with a more evolved form of cattle, within a relatively short time.

This difficulty shows that, however rich and exhaustive the data might be, it is unlikely that they allow a detailed reconstruction of the prehistoric system. What the data is called to do instead is to confirm hypotheses relative to more general explanations concerning the working of the system, whether they were formulated on the basis of the observational data or independently, having recourse to working principles related to different fields.

Recent investigations have emphasized the critical state of studies concerning hunter/gatherer societies (Bettinger 1980), stressing the need for generalizations that extend to the human environment adaptive configurations of biogeographical and economic nature (Kirch 1980; Lightfoot 1983). The value of them is inherent in the "universal" character of the explanation they provide. As far as the environmental models are concerned, they generally view the development of food production as the adjustment of the groups with a purely extractive economy to the presence of special environmental constraints. Furthermore, they stress in particular the role played by the fluctuation of resources as a factor in the increase - or diversification - of the exploitable base (5). This type of explanation may contribute to reconstructing the final stage of the process when, towards the close of the Pleistocene, groups came to realize the

---

4) As for the penetration of sheep/goat in Africa - where it is known that there are no wild prototypes - reference is made (Gautier 1980: 336) to the very early date in which these species made their appearance in the Western Asia sites (Zawi Chemi Shanidar, Iraq, from about 11000 B.P.). This leaves sufficient time for them to spread from their places of origin but makes it difficult to trace the routes followed.

5) Reference is made in particular to the explanatory model of B. Hayden (1981). This model provides a specific evaluation of the usefulness of the systemic model as an instrument supplementing the documentation data, with specific reference to field experiences in the Sahara (Barich 1985).

nutrient potential of the small sized species that abounded in their habitat (6). This determined a series of adaptation- related changes, such as: a) the elaboration of special implements suitable for exploiting the new resources; b) the increase in these resources due to their greater reliability; c) the earliest form of sedentariness resulting from a generalized exploitation of the environment that was locally accessible to the human group.

The passage to the actual domestication is viewed by this model as a state of further growth - or technological innovation - following the attempts to increase the availability of resources by increasing their diversification. This did not involve areas extensively provided with resources but other areas where it was still possible, somehow, to see to their use or "manipulation" (Hayden 1981: 527-529). It is well-known that there is no reason for a group to stimulate productivity in times - or territories - naturally well endowed to support hunting-gathering societies (Harlan 1967; Binford 1968).

By comparing our factually analysed situation with general-type explanations we are led to infer that the level characterized by our early ceramic-users corresponds to the final Late-Paleolithic stage of utmost diversification of the subsistence base when there was an increase in the resource reliability through the utilization of small-sized species, a greater technological complexity and early sedentary living. From a structural point of view, an intermediate arrangement between the "simple movement societies" and the "sedentary complex societies" of the type suggested by K. Lightfoot, would appear the most suitable to lead to the development of a preliminary storage of goods and their circulation through exchange with nearby populations, in order to offset the effects of shortages (Lightfoot 1983: 198-199).

In Chapter 10, in referring to an hypothesis made by Gautier (1984: 71), it was stated that it is likely that the earliest domestic stock might have been exchanged and circulated among communities where the pre-requisites for production already existed. This is all the more likely if one considers that, on the basis of a comparison of finds, there is evidence for the exchange of technological elements or, perhaps, only ideas. At the same time, however, due emphasis was laid on the difficulty of reconstructing one-way movements.

With reference to the three site-clusters previously mentioned, there is no doubt that the Nabta area, in the Egyptian Sahara, presents a few analogies with the situation characterizing the Tadrart. Firstly, in the general cultural sequence of the region, the Terminal Paleolithic period allowed the recognition of bovids in an incipient domestication phase (cf. Gautier 1980: Sites E-77-3 and E-77-7) and was followed by a ceramic-using facies. The recognition of aligned huts, corresponding to family units, within the earliest settlement associated with ceramics (E-75-6, upper level), dated to around 8100 B.P. At the same site, the high frequency of gazelle leads their control, if not a full domestication, to be suggested. This type of convergence in the two areas is related to a similar economic

6) R-selected species, according to the definition provided by E.R. Pianka (1974). Ecologists use the concept of the r- (signifying "rate of population increase") and the k- (signifying "carrying capacity of the environment") selection as a means to describe reproductive strategies.

model, and is supported by similarities in the lithic assemblages, where points and backed bladelets are the most used types. Pottery has its own peculiar characteristics: it becomes more frequent in the Middle Neolithic (for example site E-75-8, lower layer, dating around 7100 B.P.) and presents features that may be likened to the Early Khartoum ceramics.

If all this makes the exchanges between the Egyptian and Libyan Sahara spheres quite probable, the same degree of affinity should also be acknowledged with Niger (Roset 1982, 1983). The semi-residential hunting-gathering level is represented in this region by the sites of Greboun, Temet and Tagalagal, where a generalized exploitation of habitat, with special emphasis being placed on fishing and gathering practices is found. The tool assemblages included elements reported at Ti-n-Torha and were developing towards a bifacial technique of the type reported in the Acacus. What is also very striking is the presence of ceramics with decorative patterns closely approximating to the motifs in the Acacus, and dating from a similarly early period. Furthermore, the Agorass-n-Tast witnessed its earliest domestication at the beginning of the 6th millennium B.P., like the Acacus and the initial domestication was followed by a phase of developed pastoralism (the Tenerian) that spread throughout the central Saharan belt during the 6th-5th millennia.

The finds of the Egyptian Sahara as well as those of Niger give rise to a sequence that may be likened to that suggested for the Tadrart Acacus. A similar pattern of development may also be found in the Khartoum region of Sudan - the third cluster - even though in this case there are at present fewer possibilities of specific comparisons in the documentation.

In conclusion, all these situations present the same adaptive arrangements in the presence of similar factors of interaction and this lends validity to the model of explanation of the Pleistocene/Holocene transition suggested on the basis of ecological factors. It goes without saying that the general reconstruction scheme may be adjusted to the special situations by establishing specific elements of exchange. As far as the Acacus Massif is concerned, the data allow us to view the region as a long-term settlement area, without ruling out movements and circulations of goods by task groups. It would seem that in the earliest phases (represented by Two Caves and the Levels RInf-RSup of Torha East), short-distance exchanges, were preferentially directed towards the more easily accessible eastern areas. Subsequently, at the time of the settlement in Torha East/CI and the lower horizon of Torha North, groups that were better organized and spurred by the renewed climatic difficulties may have ventured as far as the western Sahara, thus acquiring sheep and goats and more evolved cattle. In my opinion, however, the Acacus remained the home-base and was always inhabited. This allows the changes observed in the subsistence pattern - through the acquisition of more evolved species - to be reconciled with the slow, step-by-step transformation of the technological sub-system.

The Tadrart Acacus over a very broad time-span provided evidence of the transformation of societies of hunters into developed pastoral societies. This phenomenon, which is related to the major phases of rock art, for the diachronic development pattern that has been suggested provides a reasonable framework into which at least two of the artistic styles of the Tadrart may be fitted:

be fitted: the 'Round-Heads' and the 'pastoral' phases (7).

It is most likely that in the 4th and 3rd millenia B.P. the pastoral economies of the Sahara represented a strong attraction for the warriors-merchants crossing the region. The nomadic way of life makes it difficult to find evidence of their passage through the Acacus, except for a few depictions of chariots going at full speed and scenes in the 'Horse-style' with figures of a schematic type. It would seem appropriate to ascribe to these groups the megalithic structures - of the type analysed by M. Baistrocchi in Chapter 4 - for they are located in relation to the strategic points of the access wadis. These "Equidians" met, most likely, the later aspects of pastoralism in the form witnessed by the upper horizon of Uan Muhuggiag, that is a pastoralism based on small sized species that subsequently characterized more modern groups.

---

7) The 'wild fauna phase' might be related to a Terminal Paleolithic period, at the peak of the hunting system characterized by the use of the Levallois technique, witnessed by surface findings in the Wadi Tihedine.

REFERENCES*

ARKELL A.J. 1949 - Early Khartoum, Oxford University Press: London.

ARKELL A.J. 1953 - Esh-Shaheinab, Oxford University Press: London.

ARKELL A.J. 1964 - Wanyanga and an archaeological reconnaissance of the South-West Libyan Desert. The British Ennedi expedition, University Press: Oxford.

ASCENZI A., COCKBURN A. and KLEISS E. 1983 - Miscellaneous mummies, in Mummies, Disease and Ancient Cultures, eds. Cockburn A., Cockburn E., Cambridge University Press: Cambridge, pp. 224-238.

AUMASSIP G. 1973 - Néolithique sans poterie de la région de l'Oued Mya, Mémoire du C.R.A.P.E., 20 (S.N.E.D.: Alger), 235 p.

BAILLOUD G. 1969 - L'évolution des styles céramiques en Ennedi (République du Tchad), in Actes du Prémier Colloque International d'Archéologie Africaine, Fort-Lamy: 31-45.

BANKS K.M. 1984 - Climates, Cultures and Cattle: the Holocene Archaeology of the Eastern Sahara, Department of Anthropology, Institute for the Study of Earth and Man, Southern Methodist University: Dallas.

BARICH B.E. 1974 - la serie stratigrafica dell'Uadi Ti-n-Torha (Acacus, Libia). Per una interpretazione delle facies a ceramica saharo-sudanesi, Origini, VIII: 7-184.

BARICH B.E. 1978 in press - Nuove evidenze nell'area del Tadrart Acacus (Fezzan), Libya Antiqua, XV.

BARICH B.E. 1980 a - Pour une définition du Néolithique en Afrique du Nord et au Sahara, Proceedings of the 8th P.C.P.Q.S., Nairobi: 271-272.

BARICH B.E. 1980b - Aperçu sur le phénomène pastoral dans le milieu saharien, Proceedings of the 8th P.C.P.Q.S., Nairobi: 273-274.

BARICH B.E. 1984a - The Epipalaeolithic-ceramic groups of Libyan Sahara: notes for an economic model of the cultural development in the West-central Sahara, in Origin and Early Development of Food-producing Cultures in North-Eastern Africa, eds. Krzyzaniak L., Kobusiewicz M., Polish Academy of Sciences, Poznan Archaeological Museum: Poznan, pp. 399-410.

BARICH B.E. 1984b - Fieldwork in the Tadrart Acacus and the "Neolithic" of the Sahara, Current Anthropology, 25: 683-686.

BARICH B.E. 1984c - La tradizione autoctona nella definizione storica dell'area fezzanese (Atti del II Congresso Euro-Africano, Parigi 1983), L'Universo, 5, 2: 539-548.

* The authors listed here below refer exclusively to Chapters 1,2,5,6,7,10, and Conclusions by B.E. Barich. The other contributions bear their own references at the end of each chapter.

BARICH B.E. 1985 - la spiegazione sistemica: sua rilevanza per la archeologia sahariana, in Studi di Paletnologia in onore di Salvatore M. Puglisi, a cura di Liverani M., Palmieri A. e Peroni R., Università di Roma "La Sapienza": Roma, pp. 95-104.

BARICH B.E., BELLUOMINI G., BONADONNA F., ALESSIO M. AND MANFRA L. 1984 - Ecological and cultural relevance of the recent new radiocarbon dates from Libyan Sahara, in Origin and Early Development of Food-producing Cultures in North-Eastern Africa, eds. Krzyzaniak L., Kobusiewicz M., Polish Academy of Sciences, Poznan Archaeological Museum: Poznan, pp. 411-417.

BARICH B.E., MORI F. 1970 - Missione Paletnologica Italiana nel Sahara Libico. Risultati della Campagna 1969, Origini, IV: 79-144.

BARTOLOMEI G. 1966 - Alcuni risultati delle ricerche sul Quaternario del massiccio dell'Acacus-Tadrart (Fezzan meridionale), Atti Xa Riun.Scient.Ist.Ital. Preistoria e Protostoria, Verona: 15-22.

BETTINGER R.L. 1980 - Explanatory/predictive models of hunter-gatherer adaptation, in Advances in Archaeological Method and Theory, Vol. 3, ed. Schiffer M.B., Academic Press: New York, pp. 189-255.

BINFORD L.R. 1965 - Archaeological systematics and the study of culture process, American Antiquity, 31: 203-210.

BINFORD L.R. 1968 - Post-Pleistocene adaptations, in New Perspectives in Archaeology, eds. Binford S., Binford L.R., Aldine: Chicago, pp. 313-341.

BINFORD L.R. 1977 - General introduction, in For Theory Building in Archaeology, ed. Binford L.R., Academic Press: New York, pp. 1-10.

BONADONNA F. 1974 - Schema geomorfologico dell'area dell'Auis (Acacus, Libia), in Barich B.E., La serie stratigrafica dell'Uadi Ti-n-Torha (Acacus, Libia), Origini, VIII: 163-167.

CADENAT P. 1963 - Sondage au "Cubitus", station épipaléolithique des environs de Tiaret, Libyca, 11: 75-122.

CADENAT P. 1970 - Le Columnatien, industrie épipaléolithique d'Algérie, Bull.de la Soc.d'Etud.et de Recherches Préh., Les Eyzies, 20: 40-50.

CAMPS G. 1968 - Amékni, Néolithique Ancien du Hoggar, Mémoire du C.R.A.P.E., 10 (A.M.G.: Paris), 230 p.

CAMPS G. 1974a - Les Civilisations Préhistoriques de l'Afrique du Nord et du Sahara, Doin: Paris.

CAMPS G. 1974b - Tableau chronologique de la préhistoire récente du Nord de l'Afrique. (Deuxième synthèse des datations absolues obtenues par le carbone 14), Bull.Soc.Préh.Franc., (Etud.et Trav.), 71: 261-278.

CAMPS G. 1982 - Beginnings of pastoralism and cultivation in North-West Africa and Sahara: origins of the Berbers, in The Cambridge History of Africa, Vol. I, ed. Clark J.D., Cambridge University Press: Cambridge, pp. 548-623.

CAMPS-FABRER H. 1966 - Matière et Art Mobilier dans la Préhistoire nord-africaine et saharienne, *Mémoire du C.R.A.P.E.*, 5 (A.M.G.: Paris), 557 p.

CAMPS-FABRER H. 1975 - *Un Gisement Capsien de Faciès Sétifien Medjez II, El-Eulma (Algérie)*, C.N.R.S.: Paris.

CANEVA I. ed. 1983 - Pottery using gatherers and hunters at Saggai (Sudan): preconditions for food production, *Origini*, XII: 7-278.

CARTER P.L., CLARK J.D. 1976 - Adrar Bous and African cattle, in Proceedings of P.C.P.Q.S., VIIth Session 1971, Addis Ababa: 487-493.

CLARK J.D. 1976 - Epi-palaeolithic aggregates from Greboun Wadi, Air, and Adrar Bous, north-western Tenere, Republic of Niger, in Proceedings of P.C.P.Q.S., VIIth Session 1971, Addis Ababa: 67-78.

CLARKE D.L. ed. 1977 - *Spatial Archaeology*, Academic Press: New York.

CLOSE A.E. 1980 - Current research and recent radiocarbon dates from Northern Africa, *Journal of African History*, 21: 145-167.

CLOSE A.E. ed. 1980 (Wendorf F., Schild R. assemblers) - *Loaves and Fishes: the Prehistory of Wadi Kubbaniya*, Department of Anthropology, Institute for the Study of Earth and Man, Southern Methodist University: Dallas.

CLOSE A.E. ed. 1984 (Wendorf. F., Schild R. assemblers) - *Cattle-Keepers of the Eastern Sahara: the Neolithic of Bir Kiseiba*, Department of Anthropology, Institute for the Study of Earth and Man, Southern Methodist University: Dallas.

CLOSE A.E., WENDORF F. and SCHILD R. 1979 - *The Afian: a study of stylistic variation in a Neolithic industry*, Department of Anthropology, Institute for the Study of Earth and Man, Southern Methodist University: Dallas.

COURTIN J. 1969 - le néolithique du Borkou, Nord-Tchad, Actes du Prémier Colloque d'Archéologie Africaine, Fort Lamy: 147-159.

CUSTER J.F., EVELEIGH T. 1983 - A landsat-generated predictive model for prehistoric archaeological sites in Delaware's coastal plain, *Bull. of the Archaeol. Society of Delaware*, 14: 19-38.

DUNNEL R.C., DANCEY W.S. 1983 - The siteless survey: a regional scale data collection strategy, in *Advances in Archaeological Method and Theory*, Vol. 6, ed. Schiffer M.B., Academic Press: New York, pp. 267-287.

FROBENIUS L. 1937 - *Ekade Ektab, die Felsbilder Fezzans*, Harrassowitz: Leipzig.

GABRIEL B. 1977 - Zum ökologischen Wandel im Neolithikum der östliken Zentralsahara, *Berliner Geograph. Abhand.*, 27 (Institut für Physische Geographie der Universität: Berlin), 111 p.

GABRIEL B. 1981 - Die östliche Zentralsahara im Holozän - Klima, Landschaft und Kulturen (mit besonderer Berücksichtigung der neolithischen Keramik), in Préhistoire Africaine, Mélanges offerts au Doyen Lionel Balout, reunis par Roubet C., Hugot H.J. et Souville G., A.D.P.F.: Paris, pp. 195-211.

GABRIEL B. 1984 - Great plains and mountain areas as habitats for the Neolithic man in the Sahara, in Origin and Early Development of Food-producing Cultures in North-Eastern Africa, eds. Krzyzaniak L., Kobusiewicz M., Polish Academy of Sciences, Archaeological Museum Poznan: Poznan, pp. 391-398.

GAUTIER A. 1980 - Contributions to the Archaeozoology of Egypt, in Wendorf F., Schild R., Prehistory of the Eastern Sahara, Academic Press: New York, 317-344.

GAUTIER A. 1984 - Archaeozoology of the Bir Kiseiba region, Eastern Sahara, in Cattle-Keepers of the Eastern Sahara: the Neolithic of Bir Kiseiba, ed. Close A.E., assemblers Wendorf R., Schild R., Department of Anthropology, Institute for the Study of Earth and Man, Southern Methodist University: Dallas, pp. 49-72.

GAUTIER, A., VAN NEER W. 1977-1982 - Prehistoric fauna from Ti-n-Torha (Tadrart Acacus, Libya), Origini, XI: 87-127.

GOETZ C. 1967 - la Station de Bou Aichem (Oran), Libyca, 15: 15-63.

GRAZIOSI P. 1942 - Arte rupestre della Libia, Ed. Mostra d'Oltremare: Napoli.

GRAZIOSI P. 1962 - Arte rupestre del Sahara libico, Vallecchi: Firenze.

HARLAN J.R. 1967 - A wild wheat harvest in Turkey, Archaeology, 30, 3: 197-201.

HARRIS D.R. 1969 - Agricultural systems, ecosystems and the origin of agriculture, in The Domestication and Exploitation of plants and animals, eds. Ucko P.J., Dimbleby G.W., Aldine: Chicago, pp. 3-15.

HAYDEN B. 1981 - Research and development in the Stone Age: technological transitions among hunter-gatherers, Current Anthropology, 22: 519-548.

HIGGS E.S. ed. 1972 - Papers in Economic Prehistory, Cambridge University Press: Cambridge.

HUGOT H.J. 1957 - Essai sur les armatures de pointes de flèches du Sahara, Libyca, 5: 89-236.

HUGOT H.J. 1963 - Recherches préhistoriques dans l'Ahaggar nord-occidental 1950-1957, Mémoire du C.R.A.P.E., 1 (A.M.G.: Paris), 206 p.

KIRCH P.V. 1980 - The archaeological study of adaptation: theoretical and methodological issues, in Advances in Archaeological Method and Theory, Vol. 3, ed. Schiffer M.B., Academic Press: New York, pp. 101-148.

KUPER R. 1981 - Untersuchungen zur Besiedlunsgeschichte der östlichen Sahara (Vorbericht über die Expedition 1980), Beiträge zur Allgemeinen und vergleichenden Archäeologie, 3: 215-275.

LIGHTFOOT K. 1983 - Resource uncertainty and buffering strategies in an arid, marginal environment, in Ecological Models in Economic Prehistory, ed. Bronitsky G., Arizona State University : Tempe Ariz., pp. 189-218.

LUBELL D., SHEPPARD P. and JACKES M. 1984 - Continuity in the Epipaleolithic of Northe,rn Africa with emphasis on the Maghreb, in Advances in World Archaeology, Vol. 3, eds. Wendorf F., Close A.E., Academic Press : New York, pp. 143-191.

LUPACCIOLU M. 1978 in press - Le due nuove stazioni rupestri di arte preistorica di Istanen I e Uan Ashraf (In Araien) : orientamento delle ricerche sul campo e nel settore delle scienze applicate, Libya Antiqua, XV.

MAITRE J.P. 1971 - Contribution à la Préhistoire de l'Ahaggar. I (Tefedest central), Mémoire du C.R.A.P.E.., 17 (A.M.G. : Paris), 225 p.

MALEY J. 1981 - Etudes palynologiques dans le bassin du Tchad et paléoclimatologie de l'Afrique nord-tropicale de 30.000 ans à l'époque actuelle, Travaux et Documents de l'O.R.S.T.O.M., 129 (O.R.S.T.O.M. : Paris), 586 p.

MARMIER F., TRECOLLE G. 1968 - stratigraphie du gisement d'Hassi-Mouillah, Région de Ouargla (Algérie), Bull.Soc.Préh.Franc. (c.r. des séanc.mens), 65:121-127.

MC BURNEY C.B.M. 1960 - The Stone Age of Northern Africa, Penguin Books : Harmondsworth.

MC BURNEY C.B.M. 1967 - The Haua Fteah (Cyrenaica) and the Stone Age of the South East Mediterranean, Cambridge University Press : Cambridge.

MORI F. 1961 - Aspetti di cronologia sahariana alla luce dei ritrovamenti della V Missione Paletnologica nell'Acacus, Ricerca Scientifica, 31:204-215.

MORI F. 1965 - Tadrart Acacus. Arte rupestre e culture del Sahara preistorico, Einaudi : Torino.

MORI F. 1971 - Proposta per una attribuzione alla fine del Pleistocene delle incisioni della fase piu antica dell'arte rupestre sahariana, Origini, V:7-19.

MORI F., ASCENZI A. 1959 - La mummia infantile di Uan Muhuggiag. Osser vazioni antropologiche, Rivista dell'Ist. di Antropologia, XLVI :125-148.

MUNSELL SOIL COLOR CHARTS 1973 - Macbeth Division of Kollmorgen Corporation : Baltimore.

PASA A., PASA-DURANTE M.V. 1962 - Analisi paleoclimatiche nel deposito di Uan Mahuggiag, nel massiccio dell'Acacus (Fezzan meridionale), Mem.Mus.Civ. di Storia Nat. di Verona, 10:251-255.

PIANKA E.R. 1974 - Evolutionary Ecology, Harper : New York.

PIERRE F. 1963a - Note sur une faune d'insectes sub-fossiles recueillie au Fezzan par le professeur Angelo Pasa, Mem.Mus.Civ. di Storia Nat. di Verona, 11:197-204.

PIERRE F. 1963b - Sur quelques ténébrionidés recueillis au Fezzan par le professeur A. Pasa, Mem.Mus.Civ.di Storia Nat. di Verona. 11: 205-209.

PUGLISIGS S.M. 1976 - coordinamento delle ricerche paletnologiche nel Sahara e nell'Alto Nilo, in AA.VV., Civiltà Preistoriche del Sahara e dell'Alto Nilo., C.N.R.: Roma, pp. 3-12.

PUGLISI S.M. 1981 - Nuove prospettive nelle ricerche preistoriche in Libia, Culture, 7-8: 109-114.

ROCHE J. 1963 - L'Epipaléolithique marocain, 2 Voll., Bertrand: Lisbonne.

ROSET J.P. 1978 - Poteries néolithiques du Ténéré. I. La region de l'Adrar Chiriet, Cahiers O.R.S.T.O.M., sér.Sc.Hum., XV, 4: 379-406.

ROSET J.P. 1981 - Les plus vieilles ceramiques du Sahara, Archéologie, Oct.: 43-50.

ROSET J.P. 1983 - Nouvelles données sur le probléme de la Néolithisation du Sahara méridional: Air et Ténéré, au Niger, Cahiers O.R.S.T.O.M., sér. géol., XIII, 2: 119-142.

ROUBET C. 1968 - Nouvelles observations sur l'Epipaléolithique de l'Alg½rie Orientale. Le gisement de Koudiat Kifen Lahda, Libyca, 16: 55-101.

ROUBET C. 1979 -Economie pastorale préagricole en Algérie orientale: le Néolithique de Tradition Capsienne, exemple: l'Aurés, C.N.R.S.: Paris.

SMITH A.B. 1976 - A microlithic industry from Adrar Bous, Tenere Desert, Niger, Proceedings of P.C.P.Q.S., VIIth Session 1971, Addis Ababa: 181-196.

SMITH A.B. 1980 - The Neolithic tradition in the Sahara, in The Sahara and the Nile, eds. Williams M.A.J., Faure H., Balkema: Rotterdam, pp. 451-465.

SOLEILHAVOUP F. 1979 - L'étude, la degradation et la protection des peintures rupestres préhistoriques. Examples du Tassili N'Ajjer (Sahara Algerien), Caesaraugusta, 49-50: 115-153.

SORBINI L., PASA-DURANTE M.V., BONATO G., MONTOLLI A. e TOMMASINI BADIA P. 1974 - Le collezioni paleontologiche quaternarie del Museo Civico di Storia Naturale di Verona, Museo Civ.di Storia Nat. di Verona: Verona.

STEMLER A., FALK R. 1980 - A scanning electron microscope study of cereal grains from Nabta Playa, in Wendorf F., Schild R., Prehistory of the Eastern Sahara, Academic Press: New York: 393-399.

TIXIER J. 1962 - Le "Téneréen" de l'Adrar Bous III, in Mission Berliet Ténéré-Tchad, A.M.G.: Paris, pp. 333-348.

TIXIER J. 1963 - Typologie de l'Epipaléolithique du Maghreb, Mémoire du

C.R.A.P.E., 2 (A.M.G.: Paris), 209 p.

TIXIER J. 1975 - Le campment préhiostorique de Bordj Mellala (Ourgla) Algérie, Ed. du Cercle de Rech. Et. Préh.: Paris.

UCKO P.J., DIMBLEBY G.W. eds. 1969 - The Domestication and Exploitation of Plants and Animals, Aldine: Chicago.

WENDORF F. 1984 - A brief summary of the earliest Neolithic settlements in the Eastern Sahara, in La Néolithisation au Sahara, c.r. du Semin. du Lab. de Géol. du Quat., Marseille-Luminy, Cahiers O.R.S.T.O.M., sér. géol., XIV, 2: 197-198.

WENDORF F., SCHILD R. 1980 - Prehistory of the Eastern Sahara, Academic Press: New York.

www.ingramcontent.com/pod-product-compliance
Lightning Source LLC
LaVergne TN
LVHW070045021026
849160LV00027B/1616